D0852850

Springer Series on

SIGNALS AND COMMUNICATION TECHNOLOGY

## SIGNALS AND COMMUNICATION TECHNOLOGY

**Advances in Wireless Ad Hoc and Sensor Networks**
M.X. Cheng and D. Li (Eds.)
ISBN 978-0-387-68565-6

**Wireless Sensor Networks and Applications**
Y. Li, M. Thai, and W. Wu (Eds.)
ISBN 978-0-387-49591-0

**Multimodal User Interfaces: From Signals to Interaction**
D. Tzovaras (Ed.)
ISBN 978-3-540-78344-2

**Handover in DVB-H: Investigations and Analysis**
X. Yang
ISBN 978-3-540-78629-0

**Passive Eye Monitoring: Algorithms, Applications, and Experiments**
R.I. Hammoud (Ed.)
ISBN 978-3-540-75411-4

**Digital Signal Processing: An Experimental Approach**
S. Engelberg
ISBN 978-1-84800-118-3

**Digital Video and Audio Broadcasting Technology: A Practical Engineering Guide**
W. Fischer
ISBN 978-3-540-76357-4

**Foundations and Applications of Sensor Management**
A.O. Hero III, D. Castañón, D. Cochran, and K. Kastella (Eds.)
ISBN 978-0-387-27892-6

**Three-Dimensional Television: Capture, Transmission, Display**
H.M. Ozaktas and L. Onural (Eds.)
ISBN 978-3-540-72531-2

**Human Factors and Voice Interactive Systems, Second Edition**
D. Gardner-Bonneau and H. Blanchard
ISBN 978-0-387-25482-1

**Wireless Communications: 2007 CNIT Thyrrenian Symposium**
S. Pupolin
ISBN 978-0-387-73824-6

**Blind Speech Separation**
S. Makino, T.-W. Lee, and H. Sawada (Eds.)
ISBN 978-1-4020-6478-4

**Continuous-Time Systems**
Y. Shmaliy
ISBN 978-1-4020-6271-1

**Cognitive Radio, Software Defined Radio, and Adaptive Wireless Systems**
H. Arslan (Ed.)
ISBN 978-1-4020-5541-6

**Adaptive Nonlinear System Identification: The Volterra and Wiener Model Approaches**
T. Ogunfunmi
ISBN 978-0-387-26328-1

**Wireless Network Security**
Y. Xiao, X. Shen, and D.Z. Du (Eds.)
ISBN 978-0-387-28040-0

**Satellite Communications and Navigation Systems**
E. Del Re and M. Ruggieri
ISBN 0-387-47522-2

**Terrestrial Trunked Radio – TETRA: A Global Security Tool**
P. Stavroulakis
ISBN 978-3-540-71190-2

**Multirate Statistical Signal Processing**
O. Jahromi
ISBN 978-1-4020-5316-0

**Face Biometrics for Personal Identification: Multi-Sensory Multi-Modal Systems**
R.I. Hammoud, B.R. Abidi, and M.A. Abidi (Eds.)
ISBN 978-3-540-49344-0

**Positive Trigonometric Polynomials and Signal Processing Applications**
B. Dumitrescu
ISBN 978-1-4020-5124-1

**Wireless Ad Hoc and Sensor Networks**
A Cross-Layer Design Perspective
R. Jurdak
ISBN 978-0-387-39022-2

**Ad-Hoc Networking Towards Seamless Communications**
L. Gavrilovska and R. Prasad
ISBN 978-1-4020-5065-7

**Cryptographic Algorithms on Reconfigurable Hardware**
F. Rodriguez-Henriquez, N.A. Saqib, A. Díaz Pérez, and C.K. Koc
ISBN 978-0-387-33883-5

**Acoustic MIMO Signal Processing**
Y. Huang, J. Benesty, and J. Chen
ISBN 978-3-540-37630-9

**Algorithmic Information Theory: Mathematics of Digital Information Processing**
P. Seibt
ISBN 978-3-540-33218-3

**Multimedia Database Retrieval**
A Human-Centered Approach
P. Muneesawang and L. Guan
ISBN 978-0-387-25627-6

*(continued after index)*

Edited by
Maggie Xiaoyan Cheng
Deying Li

# Advances in Wireless Ad Hoc and Sensor Networks

*Editors*
Maggie Xiaoyan Cheng
Department of Computer Science
Missouri University of Science
and Technology
Rolla, MO 65409
USA

Deying Li
School of Information
Renmin University of China
100872 Beijing
China

ISSN: 1860-4862
ISBN: 978-0-387-68565-6 e-ISBN: 978-0-387-68567-0
DOI: 10.1007/978-0-387-68567-0

Library of Congress Control Number: 2007938692



Printed on acid-free paper

springer.com

# Contents

# Foreword

Wireless ad hoc networks, mobile or static, have special resource requirements and different topology features, which make them different from classic computer networks in resource management, routing, media access control, and QoS provisioning. Some issues are unique to ad hoc wireless networks and sensor networks, such as self-organization, mobility management, energy efficient design, and so on. The purpose of this book is not to provide a complete survey of the state-of-the-art research on all areas of ad hoc and sensor networks, but rather to focus on the theoretical and experimental study of a few advanced topics. We carefully selected papers around the following four topics: security and trust, broadcasting and multicasting, power control and energy efficency, and QoS provisioning.

Chapters 1–3 are about QoS routing in Mobile Ad hoc NETworks (MANET): Chapter 1 discusses QoS routing for heterogeneous mobile ad hoc networks; Chapter 2 proposes a link state QoS routing protocol for ad hoc networks using bandwidth and delay as routing metrics; Chapter 3 studies the interworking between a mobile ad hoc network and the Internet, extending the Differentiated Services (DiffServ) model to a wireless environment.

Chapters 4–6 are related to security and trust issues in ad hoc environments: Chapter 4 addresses secure communication in ad hoc networks, providing a threshold decryption scheme that allows different mobile nodes to use public keys of several different cryptosystems; Chapter 5 proposes a secure group communication protocol for ad hoc wireless networks and maintenance processes for topology changes; Chapter 6 addresses routing in ad hoc networks from the trust and security perspectives, and proposes a direct trust model that establishes and manages trust without using cryptographic mechanisms, which is suitable in ad hoc networks.

Chapters 7–9 address power control and energy efficient design: Chapter 7 proposes a power optimization scheme that improves both power consumption and throughput of multihop wireless networks; Chapter 8 presents mechanisms to save energy in sensor networks without losing sensing area, which control the network density based on the Voronoi diagram, and deterministically deploy sensors after the initial network has been used; Chapter 9 addresses the self-organization of MANET from the message optimality perspective and proposes an algorithm that is message-efficient for initial configuration and message-optimal for self-configuration under mobility.

Chapters 10–13 focus on broadcast and multicast in MANET: Chapter 10 considers energy-efficient multicast with mobility support for ad hoc networks by using two multicast trees; Chapter 11 proposes novel approaches to construct multicast trees in mobile ad hoc networks and to maintain the trees under rapid topology changes. The resulting multicast trees have the LAST property; that is, the cost of each path from the source to any terminal in the multicast tree does not exceed a given constant factor $\alpha$ from the corresponding shortest-path cost in the original graph, and the total cost of the multicast tree does not exceed a given constant factor $\beta$ from the total cost of the Minimum Spanning Tree (MST); Chapter 12 proposes three techniques to improve the counter-based broadcasting scheme in mobile ad hoc networks; Chapter 13 addresses energy-efficient broadcasting and multicasting schemes in ad hoc networks that balance the energy consumption in broadcasting and multicasting by minimizing the maximum energy consumption.

We would like to thank all the authors whose contributions made this book possible, and all the anonymous reviewers whose valuable suggestions ensure the high quality of this book. We hope this book will serve as a useful reference for studying mobile ad hoc and sensor networks.

January, 2008

*Maggie Cheng*
*Deying Li*

Chapter 1

# Backbone Quality-of-Service Routing Protocol for Heterogeneous Mobile Ad Hoc Networks

Xiaojiang Du
*Department of Computer Science*
*North Dakota State University, Fargo, ND 58105, USA*
E-mail: `xdu@web.cs.ndsu.nodak.edu`

## 1 Introduction

Mobile Ad hoc NETworks (MANETs) form a class of dynamic multihop networks consisting of a set of mobile nodes that intercommunicate on shared wireless channels. MANETs are self-organizing and self-configuring multihop wireless networks, where the network structure changes dynamically due to node mobility. Quality-of-Service (QoS) routing is important for a mobile network to interconnect wired networks with QoS support (e.g., Internet). QoS routing is also needed in a standalone mobile ad hoc network for real-time applications, such as voice, video, and so on.

QoS routing requires not only finding a route from a source to a destination, but a route that satisfies the end–to–end QoS requirement, often given in terms of bandwidth or delay. QoS routing in wired networks has been well studied. Some of the recent works are listed below. In [5], Xue proposed an efficient approximation algorithm for minimum-cost QoS multicast routing problems and an efficient heuristic algorithm for unicast routing problems in communication networks. In [25], Orda and Sprintson proposed efficient precomputation schemes for QoS routing in networks with topology aggregation by exploiting the typical hierarchical structure of large-scale

M.X. Cheng, D. Li (eds.) *Advances in Wireless Ad Hoc and Sensor Networks.*
Signals and Communication Technology, doi: 10.1007/978-0-387-68567-0_1.

networks. In [16], Cao et al. studied QoS for Voice-over-IP, and they proposed measurement-based call admission control to guarantee the QoS. In [21], Li and Mohapatra proposed QoS-aware Routing protocols for Overlay Networks (QRONs).

Quality of service is more difficult to guarantee in ad hoc networks than in most other types of networks, because the network topology changes as the nodes move and network state information is generally imprecise. This requires extensive collaboration between the nodes, both to establish the route and to secure the resources necessary to provide the QoS. In recent years, several researchers have studied QoS support in ad hoc networks [1–4,11–14,25]. QoS needs a set of service requirements to be met by the network while transporting a packet stream from source to destination. The ability to provide QoS heavily depends on how well the resources are managed at the MAC layer. Some QoS routing protocols [4,25] use generic QoS measures and are not tuned to a particular MAC layer. Some QoS routing protocols [2,3,14,15] use CDMA to eliminate the interference between different transmissions. In this chapter, we develop a QoS routing protocol: B-QoS for heterogeneous mobile ad hoc networks using pure TDMA. In [2,3], CDMA is overlaid on top of the TDMA to reduce interference; that is, multiple transmissions can share TDMA slots via CDMA. With some minor modification, our B-QoS routing protocol can also be applied to ad hoc networks with a MAC layer using CDMA/TDMA.

In MANETs, QoS issues include delay, delay jitter, bandwidth, probability of packet loss, and so on. In this chapter, we are mainly concerned about bandwidth. The goal is to establish bandwidth-guaranteed QoS routes in MANETs. Our B-QoS is an on-demand routing protocol, and builds QoS routes only as needed. A flow specifies its QoS requirement as the number of transmission time slots it needs from a source to a destination. For each flow, the B-QoS routing protocol will find both the route and the transmission time slots for each node on the route.

The rest of the chapter is organized as follows. In Section 2, we review the related work of QoS routing, best-effort backbone routing, and routing with location information. In Section 3, we discuss the algorithm that calculates the available maximum bandwidth in a given path. We describe our B-QoS routing protocol in Section 4 and give a routing example there. In Section 5, we discuss the simulation experiments performed with the B-QoS routing protocol and we compare the performance with another QoS routing protocol [1] and the best-effort AODV routing protocol [17]. In Section 6, we estimate the probability of having backbone nodes in one cell by both simulation and computation. And we conclude the chapter in Section 7.

## 2 Related Works

Most existing QoS routing protocols assume homogeneous MANETs: all nodes have the same communication capabilities and characteristics. They have the same (or similar) transmission power (range), bandwidth, and processing capability, and the same reliability and security. However, a homogeneous ad hoc network suffers from poor scalability. Recent research has demonstrated its performance bottleneck both theoretically and through simulation experiments and testbed measurement [7]. In many realistic ad hoc networks, nodes are not homogeneous. For example, in a battlefield network, there are soldiers carrying portable wireless devices, there are vehicles and tanks carrying more powerful and reliable communication devices, and there may be aircraft and satellites flying above, covering the whole battlefield. They have different communication characteristics in terms of transmission power, bandwidth, processing capability, reliability, and so on. So it would be more realistic to model these network elements as different types of nodes. Also there are many advantages that can be utilized to design better routing protocols when nodes in heterogeneous MANETs are modeled as different types.

The major difference between our B-QoS routing protocol and other QoS routing protocols [1–4,11–14,25] is: B-QoS considers heterogeneous MANETs, whereas other QoS routing protocols consider homogeneous MA-NETs. B-QoS routing takes advantage of the different communication capabilities of heterogeneous nodes in many ad hoc networks. Some physically more powerful nodes are chosen as backbone nodes for routing. The idea of using backbones in routing has appeared in several previous works. The CEDAR [13] algorithm establishes and maintains a routing infrastructure called core-in ad hoc networks. And routing is based on the core. There are several differences between CEDAR and our B-QoS routing protocol. We list some of the differences in the following. (1) CEDAR considers homogeneous nodes, whereas B-QoS considers heterogeneous nodes. The heterogeneous node model is more realistic and provides efficient routing. (2) In CEDAR, a complex algorithm is used to generate and maintain the core nodes, and the algorithm introduces large overhead, because every node needs to broadcast messages to its neighbors periodically. While in B-QoS, the election of backbone nodes is very simple, the first backbone-capable (more powerful) node that sends out a claim message becomes the backbone node. (3) In addition, CEDAR needs to broadcast a route probe packet to discover the location of a destination node. While in B-QoS, a Global Positioning System (GPS) is used to provide node location information, and an efficient algorithm is used to disseminate node location information. The idea of using backbone nodes in routing has also appeared in [8], where Butenko et al. proposed to compute a virtual backbone (a minimum connected dominating set) based on physical topology. In addition, Butenko et al. [8] consider homogeneous node models.

There are several best-effort (non-QoS) routing protocols that consider heterogeneous MANETs. One obvious difference is that B-QoS is a QoS support routing protocol, whereas these routing protocols do not consider the QoS issue. Besides, there are some other differences. We compare our B-QoS with some of these best-effort routing protocols in the following.

In [7], Xu, Hong, and Gerla proposed an MBN routing protocol with backbone nodes. Besides the above difference, the major differences between our B-QoS routing and MBN are the way to deploy backbone nodes and the routing algorithm for backbone nodes. In MBN, a multihop clustering scheme is used to form clusters in the network, and the cluster heads become the backbone nodes. However, the multihop clustering algorithm is complex. In B-QoS routing, the backbone node deployment is based on node location information. The entire routing area is divided into several small equal-size squares—cells—and one backbone node is elected in each cell. A simple algorithm is used for backbone node election. In MBN, routing among backbone nodes is based on another routing algorithm, LANMAR [19], which is not trivial. Furthermore, LANMAR uses a logical group concept to aid routing. However, the logical group is not applicable to all MANETs. In B-QoS routing, routing among backbone nodes is based on node location information and the cell structure: some cells between source and destination are chosen as routing cells, and a route is discovered among backbone nodes in the routing cells. Details are given in Section 3.

Several papers have discussed the node heterogeneity problem [14,15,30]. However, they mainly discuss how to solve the unidirectional link problem in ad hoc networks. In B-QoS routing, we consider how to take advantage of the different communication capabilities of heterogeneous nodes and provide a better QoS routing strategy. The unidirectional link problem also exists in B-QoS routing, that is, the connection from source or destination to a nearby backbone node. Usually the source (or destination) is close to the nearest backbone node, and it is only a small number of hops to the backbone node.

We solve the unidirectional link problem as follows. When there is a packet that needs to be sent, the source (or destination) node floods the packet within a small area to find a path to the nearest backbone node.

In [9], Ye et al. proposed a scheme to build a reliable routing path by controlling the positions and trajectories of some reliable nodes. Ye et al. [9] mainly consider how to build reliable best-effort routes. In our B-QoS routing, we do not assume control of the positions and trajectories of backbone (reliable) nodes, and our goal is to establish efficient and effective QoS routes.

Research has shown that geographical location information can improve routing performance in ad hoc networks. Routing with assistance from geographic location information requires each node to be equipped with a GPS device. This requirement is quite realistic today because such devices are inexpensive and can provide reasonable precision. Several routing algorithms based on location information have been proposed. The well-known location-based routing algorithms are the Location-Aided Routing (LAR) protocol [10], Distance Routing Effect Algorithm for Mobility (DREAM) [28], and Greedy Perimeter Stateless Routing (GPSR) [27], among others.

B-QoS routing utilizes node location information to simplify the routing strategy. The entire routing area is divided into several cells. The cell or grid structure has been utilized in some routing algorithms such as GRID [20], GAF [24], and so on. There are several differences between B-QoS and these algorithms. The major difference is that B-QoS considers QoS routing in heterogeneous MANETs, whereas GRID and GAF consider best-effort routing in homogeneous MANETs. The design of B-QoS is based on the following assumptions.

1. In B-QoS routing, we assume the routing area is fixed (i.e., nodes move around in a fixed territory). This is true for many MANETs, such as ad hoc networks in military battlefields, disaster relief fields, conferences, convention centers, and so on. They all have a fixed routing territory.

2. We consider MANETs whose topologies do not change very quickly. We also assume the routing area is fixed in 1; this means that we mainly consider MANETs where nodes do not move very quickly. If the topology of an ad hoc network changes too quickly, the provision of the QoS can be even impossible [4]. In [22], the authors called an ad hoc network combinatorially stable if and only if the topology changes occur sufficiently slowly to allow successful propagation of all topology updates as necessary. Combinatorial stability follows directly when the geographical distribution of the mobile nodes do not change much relative to one another during the time interval

of interest. In this chapter, we only study the type of ad hoc networks whose topologies do not change so quickly that they make the QoS routing meaningless.

3. We assume there are a reasonable number of backbone-capable nodes in the network: for example, the number of backbone-capable nodes is close to (or larger than) the number of cells in the network.

# 3 The Path Bandwidth Calculation Algorithm

In a time-slotted network (e.g., TDMA), to provide a bandwidth of B slots on a given path P, it is necessary that every node along the path find at least B slots to transmit to its downstream neighbor, and that these slots do not interfere with other transmissions. Because of these constraints, the end-to-end bandwidth on the path is not simply the minimum bandwidth on the path.

In general, to compute the available bandwidth for a path in a time-slotted network, one not only needs to know the available bandwidth on the links along the path, but also needs to determine the scheduling of the free slots. To resolve slot scheduling at the same time as available bandwidth is searched on the entire path is equivalent to solving the Satisfiability Problem (SAT) which is known to be NP-complete [23]. In [1], Zhu and Corson developed a heuristic algorithm — Forward Algorithm (FA) — to compute the available bandwidth in a path. In [3] Lin and Liu also proposed a heuristic approach to calculate the path bandwidth. In this chapter, the focus is not on developing a new bandwidth calculation algorithm. Instead, we use the existing bandwidth calculation algorithms. Most bandwidth calculation algorithms that are developed for time-slotted networks can be incorporated into B-QoS. In the current design and simulation, our B-QoS routing protocol adopts the FA algorithm in [1] to calculate the available path bandwidth and slot scheduling at each node in the path. The FA algorithm is a greedy scheme that finds the local maximal bandwidth from the source to the next hop, given the sets of slots used on the three links closest to the current node. We briefly state the FA algorithm in the following. Consider a given path $P = (n_m \rightarrow, \ldots, n_{k+3} \rightarrow n_{k+2} \rightarrow n_{k+1} \rightarrow n_k \rightarrow, \ldots, n_1 \rightarrow n_0)$, where $n_m$ is the source, and $n_0$ is the destination. Based on the input from the upstream node $n_{k+2}$, an intermediate node $n_{k+1}$ computes the slot allocations at links $n_{k+3} \rightarrow n_{k+2}$ and $n_{k+2} \rightarrow n_{k+1}$, and determines the available bandwidth from the source to itself. Then node $n_{k+1}$ passes the two slot allocations and its free transmission slots as the input to the next node, $n_k$. Node $n_k$ computes the slot allocations at links $n_{k+2} \rightarrow n_{k+1}$ and $n_{k+1} \rightarrow n_k$, and determines the available bandwidth from the source to itself. Note the slot allocation at link $n_{k+2} \rightarrow n_{k+1}$ is computed twice, by both node $n_{k+1}$ and $n_k$. Only the one computed at $n_k$ is used to determine the final slot allocation. (The one computed at $n_{k+1}$ is just used as an input to node $n_k$.) Node $n_k$ stores the slot allocation at link $n_{k+2} \rightarrow n_{k+1}$. Then node $n_k$ passes the input to the next node, and the process continues until the destination $n_0$ is reached. In the QoS routing, the destination $n_0$ will send a route reply message via the reverse path to source $n_m$, and each intermediate node with reserve the slot according to the computed slot allocation. The details of the FA algorithm can be found in [1].

## 4 QoS Routing Based on Backbone Nodes

Many real-world ad hoc networks are heterogeneous MANETs, where physically different nodes are present. Thus it would be more realistic to model nodes in such networks as different types of nodes. For simplicity, we consider there are only two types of nodes in the network. One type of node has a larger transmission range (power) and bandwidth, better processing capability, and is more reliable and robust than the other type. We refer to the more powerful nodes as Backbone-Capable nodes, in short as BC-nodes. In B-QoS routing, BC-nodes can be elected to serve as Backbone nodes (B-nodes). Other nodes are referred to as general nodes. For example, in a battlefield MANET, tanks and vehicles can be considered as BC-nodes, and soldiers can be considered as general nodes. There might be more than two types of nodes in a heterogeneous MANET. It is possible to extend B-QoS routing to consider more than two types of nodes, and this will be our future work. In this chapter, we only consider the two types of node model.

The main idea of B-QoS routing is to find a QoS route mainly based on B-nodes. There are several advantages of using B-nodes in QoS routing.

- B-nodes have larger bandwidth than general nodes. The large bandwidth of B-nodes increases the chance of satisfying the QoS requirement.
- B-nodes have larger transmission range than general nodes, which reduces the number of hops in routing, and thus reduces the routing overhead and latency.

- B-nodes have better processing capability than general nodes. Routing packets via B-nodes is more efficient than via general nodes.
- B-nodes provide better reliability and fault tolerance, because they are more reliable than general nodes.

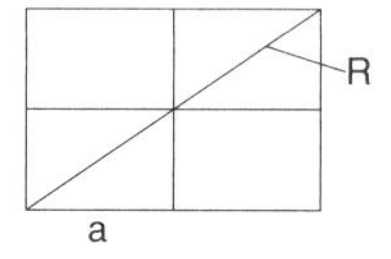


Figure 1: The relationship between a and R.

Usually the transmission range of B-nodes, R, is much larger than that of general nodes, r. For simplicity, we assume the transmission ranges of B-nodes and general nodes are fixed. The routing area is divided into several small, equal-sized squares referred to as cells. An example with nine cells is shown in Figure 2.

If the side length of a cell is set as $a = R/2\sqrt{2}$, as shown in Figure 1, where it is the worst case (longest distance) between B-nodes in two nearby cells, then a B-node can always directly communicate with B-nodes in all nearby cells, including the diagonal one. Because most of the time, two nearby B-nodes are not in the two opposite corners, a larger cell size can be used (i.e., $a > R/2\sqrt{2}$), and still usually ensure the connection of nearby B-nodes. A more detailed discussion of cell size is given in Section 5.6. All the cells form a grid structure, and the grid structure is fixed for a given cell size. One and only one B-node is elected and maintained in each cell if there are BC-nodes available in the cell. In B-QoS routing, we assume the routing area is fixed, thus for a given cell size a, the position of each cell is also fixed. Given the location (coordinates) of a node, there is a predefined mapping between the node location and the cell in which it lies. For simplicity, we assume the routing area is a two-dimensional plane. The grid is created starting from the left-top point of the routing area. The B-QoS routing protocol is presented below.

## 4.1 The Backbone QoS Routing Protocol

The basic operation of the B-QoS routing protocol is now described.

1. There is a unique ID for each cell. In Figure 2, the number is the ID for each cell. One (and only one) B-node is elected and maintained in each cell, and each B-node has a second address, which is the same as the ID of the cell where it stays. So a B-node can send a packet to a B-node in a nearby cell by using the second address, even though the identity of that B-node may change.

2. There is no routing table maintained among B-nodes and general nodes. The QoS route is discovered on demand. When a B-node moves out of a cell, it initiates a B-node election process in the cell and a BC-node will be elected as the new B-node. The B-node election algorithm is described in Section 4.2.

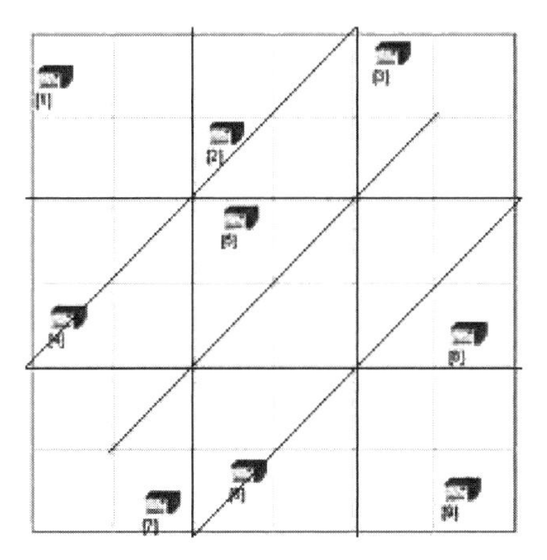

Figure 2: Routing cells.

3. Routing among B-nodes. In this step, we discuss the scheme by which B-nodes find routes (without QoS requirement) to other B-nodes. This scheme is used in step 10 for the dissemination of node location information, and it is also the base for QoS routing in step 4. B-nodes use their second addresses to communicate with each other. Assume B-node $B_s$ (in cell $C_s$) wants to send a packet to the B-node in cell $C_d$ (denote the B-node as $B_d$). Although nodes move around, the cells are fixed. $B_s$ knows $B_d$'s second address because it is the same as the ID of cell $C_d$. A straight line H is drawn between the centers of cell $C_s$ and cell $C_d$. An example is given in Figure 2, where B-node 7 wants to send a packet to the B-node in cell 3. The center line is line H.

Two border lines (outside lines in Figure 2) which are parallel to line H with distance of W from H are drawn from cell $C_s$ to $C_d$. The set of all the cells that are (fully or partially) within the two border lines is defined as routing cells. The value of W determines the width of the routing cells. The proper value of W depends on the density of BC-nodes in the network. If there are enough BC-nodes in the network (i.e., with high probability there is at least one BC-node in each cell), then W can be small. For QoS routing, W also depends on the available bandwidth of the B-nodes and the bandwidth requirement of the QoS session. If the B-nodes have enough available bandwidth, or if the QoS bandwidth requirement is low, then a QoS route can be found easily, and W can be small. Otherwise, large W should be used. After determining the routing cells, source B-node $B_s$ can start sending packets to $B_d$. If the packet is a short one (like the location request/update packet in step 10), $B_s$ will flood the packet to all B-nodes in the routing cells, and the packet will be forwarded to $B_d$. If it is a long packet, like a data packet, first a route request packet is flooded to all B-nodes in the routing cells, then the data packet is sent via the discovered route. In both cases, some B-nodes in the routing cells form a route from $B_s \rightarrow B_d$. Consider the example in Figure 2: if W is set to zero, then the routing cells are only the cells that intercept with red line H, cells 7, 5, 3. The B-nodes from the routing cells form a route: $B_7 \rightarrow B_5 \rightarrow B_3$. And if W is set as $a\sqrt{2}/2$, where a is the side length of a cell, then the routing cells are cells 7, 5, 3; 4, 2; and 8, 6. The routing cells are used to balance the chance of finding a (QoS) route and the overhead from route discovery. The width of the routing cells is based on the network state. In the current B-QoS, W is based on the number of BC-nodes in the network and the QoS requirement of a session. The information of the available bandwidth in other B-nodes may also be used to determine W, with some scheme to disseminate such information among B-nodes. However, for simplicity, we do not use such information in our current design. A proper W should provide a high probability of finding the (QoS) route while limiting the routing overhead.

4. QoS route discovery starting from B-nodes. Assume a source node S (in cell $C_s$) wants to set up a QoS route for a flow to a destination node D (in cell $C_d$). We first discuss the case where S is a B-node. And we discuss the QoS routing scheme when S is a general node in

step 9. S is the first B-node in the QoS route, and is referred to as the starting B-node. In B-QoS routing, the starting B-node S needs to know the current location of the destination node D. The scheme by which S obtains D's location is described in step 10. With D's location information, S knows the cell $C_d$ in which D stays, and S knows the B-node in cell is $B_d$ (using the second address of the B-node). First S determines the width of the routing cells (2W), based on the number of BC-nodes in the network and the QoS requirement from the flow. Then S determines the routing cells between cell $C_s$ and $C_d$ as in step 3. The routing cells may also include the circle that centers at node D with the radius being the expected moving distance, like the scheme used in the LAR routing protocol [10]. Based on assumption 2 that nodes do not move very fast, usually node D is within the transmission range of the B-node in $C_d$.

5. The starting B-node S floods Route Request (RR) packets to all the B-nodes in the routing cells. The RR packet includes the following fields: starting B-node, sequence-$n$, route, routing-cells, slot-set-list, destination-cell, RB, where RB is the required bandwidth. Each B-node maintains a sequence-$n$, and the sequence-$n$ increases for each RR flooding. Starting B-node plus sequence-$n$ uniquely determines a route request session. The route field records the path that the RR packet traversed. At each node, the slot-set-list records the free slots at the node and the slot allocations at the two upstream links. For example, consider a route ($n_m \rightarrow \ldots, n_{k+3} \rightarrow n_{k+2} \rightarrow n_{k+1} \rightarrow n_k \rightarrow \ldots,$ $n_1 \rightarrow n_0$). At node $n_{k+1}$, the slot-set-list records the free slots at $n_{k+1}$, plus the slot allocations at link $n_{k+3} \rightarrow n_{k+2}$ and $n_{k+2} \rightarrow n_{k+1}$, which are computed by the FA algorithm. When receiving a RR packet, an intermediate B-node uses the FA algorithm to compute the maximum bandwidth from the source to itself, based on its free slots and the slot-set-list from its upstream node. If the maximum bandwidth is less than the required bandwidth, then the QoS cannot be satisfied, and the RR packet is dropped. Otherwise, the B-node $n_k$ computes the slot allocations at two upstream links $n_{k+3} \rightarrow n_{k+2}$ and $n_{k+2} \rightarrow n_{k+1}$, stores the slot allocation of link $n_{k+2} \rightarrow n_{k+1}$, updates the slot-set-list, and forwards the RR packet to neighbor B-nodes that are in the routing cells (except the incoming B-node). When a B-node receives duplicate RR packets of the same route request session, it also processes the RR packet in the same way. This is to increase

the chance of finding a QoS route. It is possible that a detour path may have larger bandwidth than a direct link, such as the example in Figure 3. The direct link on the top only has two slots, whereas the detour path below has a bandwidth of three slots.

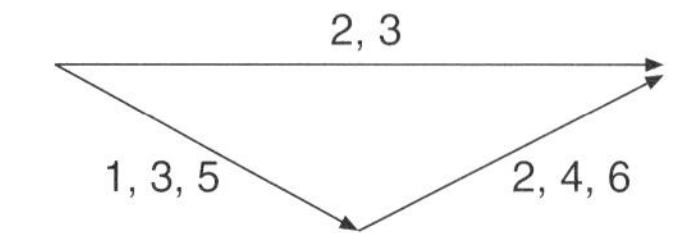


Figure 3: Detour path with larger bandwidth.

6. When the RR packet arrives at the B-node $B_d$ in cell $C_d$, $B_d$ will first send a probe packet to search the destination node D. The probe packet includes $B_d$'s location. Because $B_d$ has a large transmission range, the transmission of the probe packet can reach all nodes in the neighbor cells. If node D is still in the cell, or in a neighbor cell, D will receive the probe packet. And D will send an Ack (acknowledge) packet including its free slots to $B_d$. If D is a general node, there is a unidirectional link problem here. D may not be able to send the Ack packet to $B_d$ in one hop. Instead, based on the location of $B_d$ and itself, node D knows the direction to node $B_d$ and the distance between itself and $B_d$. Node D sends the Ack packet to $B_d$ via limited-hop, small-area directional flooding. When $B_d$ receives the Ack (with D's free slots) from D, $B_d$ computes the maximum bandwidth from source S to destination D and the slot allocations at links $B_{d-1} \rightarrow B_d$(where $B_{d-1}$ is the upstream node of $B_d$) and $B_d \rightarrow D$. Note the last bandwidth computation is done by B-node $B_d$, not by D. This approach is more efficient because of the possible unidirectional link between $B_d$ and D. If the required bandwidth cannot be satisfied, $B_d$ will send a route-failure (RF) packet to source S via the reversed route. If the required bandwidth is satisfied, $B_d$ will reserve the slots, and send a route reply (RP) packet (including the uplink slot allocation) along the reversed route back to source node S. Consider the example in step 5: the RP packet from node $n_k$ to upstream node $n_{k+1}$ includes the slot allocation of link $n_{k+1} \rightarrow n_k$, calculated in step 5. An intermediate node $n_{k+1}$ reserves the slots for this route, according to the slot allocation at link $n_{k+1} \rightarrow n_k$ and the required bandwidth. Note the reserved slots (referred to as slot assignment) at $n_{k+1}$ is a subset of the slot allocation.

For example, if the slot allocation is 1,2,3 and the required bandwidth is 2, then only two slots are reserved. The starting B-node $B_1$ needs to send its slot assignment to source node S, which may be used in route repairing when source S moves away (details are discussed in Section 4.2). When the RP packet arrives at the source node S, the QoS path is set up and the bandwidth is reserved. If node $B_d$ receives multiple RR packets from the same route discovery session, node $B_d$ will reply to two or three of them, and discard the rest. This is to set up one or two backup QoS routes in addition to the primary route. In the case when the primary route is broken, the backup route can be used. If source node S does not receive any RP packet for a route request timeout (or S receives a route-failure packet), S assumes the route discovery within the routing cells failed. S will then flood the RR packet to all B-nodes in the network, and try to find a QoS route based on B-nodes. If even the above scheme fails, S will flood the RR packet to all nodes in the network, which is similar to the QoS routing protocol based on AODV in [1].

7. If $B_d$ does not receive Ack from node D for a certain time, it means D is no longer in the neighbor cells of $C_d$. Node $B_d$ will obtain the current location information of node D (described in step 10), then it will forward the RR packet to a B-node close to node D, and that B-node will process as above. Because we mainly consider MANETs without high mobility, most of the time, node D will not be far away from $B_d$. In the case where node D moves to a new location far away from $B_d$, after having D's new location, $B_d$ will send a route-failure packet (with D's new location) to source node S, and S will start a new QoS route discovery process.

8. If there is no B-node in the destination cell $C_d$ that can be detected by a B-node (say $B_{d-1}$) in a neighbor cell of the destination cell (i.e., $B_{d-1}$ does not overhear the transmission of a probe packet from $B_d$ after it sends the RR packet to the destination cell for a certain time), then $B_{d-1}$ will flood the RR packet in its cell and the destination cell, and try to find a QoS route via general nodes to destination D. The general nodes (possibly including D) will compute the available bandwidth and slot allocation according to the FA algorithm. If a QoS route is found, D will send the Ack via the reverse route to $B_{d-1}$, and then node $B_{d-1}$ will send the Ack to source S.

9. QoS route discovery from general nodes. If the source node S is not a B-node, S will first find a route to a nearby B-node with enough bandwidth. Node S floods a Route Discovery (RD) packet to all the nodes in its cell $C_s$. The RD packet includes the following fields: source, source-cell, sequence-$n$, path, slot-set-list, destination, and RB, where source-cell is the cell in which source stays, and RB is the required bandwidth. Only nodes in the same cell as S will process and forward the RD packet. This reduces the routing overhead from route discovery. When other general nodes receive the RD packet, they will calculate the available bandwidth from the source to itself and compare it with the required bandwidth. If QoS is satisfied, the node stores the slot allocation, updates the slot-set-list, and forwards the RD packet to its neighbors. Otherwise, the RD packet is dropped. When the B-node in cell $C_s$ receives the first RD packet and if it has enough bandwidth, it will flood the route request packet to all B-nodes in the routing cells, and proceed as in step 4. Because B-nodes have much larger bandwidth than general nodes, most of the time the B-node will have enough bandwidth. In the case where the B-node in cell $C_s$ does not have enough bandwidth, this B-node will send a route-failure packet to the source node directly. (Recall a B-node can directly reach all nodes in its cell.) It is also possible that there is no B-node in cell $C_s$ when S wants to discover a QoS route. The source node S can detect no B-node in $C_s$ if S does not overhead the transmission of RR from a B-node in $C_s$ after S sending out RD for a certain time. If the B-node in cell $C_s$ does not have enough bandwidth, or if there is no B-node in cell $C_s$, source node S will flood the RD packet to all neighbor cells and find a nearby B-node with enough bandwidth. In the worst case, if all the nearby B-nodes do not have enough bandwidth to continue the QoS route discovery, source node S will flood RD packets to all nodes in the network, and find a QoS route if possible. In the worst case, B-QoS is similar to the QoS routing protocol in [1], which combines AODV with the FA algorithm.

10. Dissemination of node location information. As mentioned in step 4, in B-QoS routing, a starting B-node needs to know the current location of the destination node D. Because nodes move around, an algorithm is needed to disseminate updated node location information. We propose an efficient dissemination scheme, and it is described in the following. If a node moves within the same cell, there is no need to update its

location information. When a node moves out of its previous cell, it sends a location update packet (with its new location) to the B-node in the new cell (or the nearest B-node). The location update packet can be sent out via broadcast within a small hop count. And all B-nodes periodically send aggregated node location information to a special B-node $B_0$, for example, $B_0$ could be the command headquarters in a battlefield. The period of updating location information should not be too long, because this will cause the location information to not be accurate. Also the period should not be too short, because updating the location information too often will cause large overhead. The special B-node $B_0$ is preferred to be a fixed B-node, or a B-node only moving within one cell. If $B_0$ is fixed or within one cell, the dissemination algorithm is very simple. When a starting B-node S needs to know the location of a node D, S sends a location request packet to $B_0$; then $B_0$ sends the location of D to S. Because both S and $B_0$ are B-nodes, they know how to communicate with each other (step 3). If $B_0$ also moves around, then if needs to multicast its current location to all B-nodes when it moves from one cell to another. Then all B-nodes know the current location of $B_0$, and they are able to request location information from $B_0$. In many MANETs, it is possible to choose a static or slowly moving B-node as $B_0$. And in many (as with military) MANETs, it worthwhile to deploy a static B-node as $B_0$.

## 4.2 More Protocol Details

Route maintenance and election of B-nodes is discussed here.

### 4.2.1 QoS Route Maintenance

Node mobility can cause an established QoS route to be broken. Route maintenance is very important for QoS routing in MANETs. Assume a QoS route $R: S \rightarrow g_1 \rightarrow B_1 \rightarrow B_2, \ldots, \rightarrow B_k \rightarrow g_2 \rightarrow D, \ldots, n_1 \rightarrow n_0)$ is set up between source node S and destination node D, where $g_1$ and $g_2$ are general nodes, and $B_j (j = 1, \ldots, k)$ are B-nodes. Usually B-node $B_k$ can send a packet directly to destination D. We add node $g_2$ to cover a more general case when there is no B-node in the cell of destination node D, and general nodes are used to form the QoS route.

**Detection of Broken Route**

Moving away (or failure) of any node in the route can cause the route to be broken, and the broken route is detected by the upstream node (closer to the source). That is, after node i sends a packet to its downstream node j, if node i does not overhead a transmission of the packet from node j for a certain time, node i assumes node j moves away or fails. And node i will start the route repairing process; if route repairing does not work, node i will notify source node S to discover a new route. If source S is the node moving away, S itself will detect the broken link. The route repairing and rerouting processes are discussed in the sequel. We refer to the node that moves away and causes a broken link as the leaving node L. The upstream node and downstream node of L are denoted as up-L and down-L respectively.

**Route Repairing and Rerouting**

There are two different cases of broken routes, depending on whether the leaving node is source node S.

1. The leaving node is source S. If S is still in the same cell, or in a nearby cell, S will flood RE (Route rEpair) packets to nodes in the cell (or plus the nearby cell) and try to find a new QoS path to the starting B-node; the RE packet includes the bandwidth requirement and the slot assignment at the starting B-node $B_1$. The slot assignment at the upstream node of $B_1$ must not conflict with the slot assignment at $B_1$. So if a new QoS path is found between S and $B_1$, the slot assignments from $B_1$ to D do not need to change. If S moves far away from its previous cell, or if the route repairing fails, rerouting will be used: S will use the B-QoS routing protocol to discover a new QoS route to destination D.

2. The leaving node is a node other than the source. The upstream node up-L broadcasts a RE packet, which includes its slot assignment and address, with a TTL (Time-To-Live) set as two hops. The address of up-L is used to solve the unidirectional problem. It is possible to have the unidirectional link problem during route repairing. We solve the problem using the similar approach as found in step 6 of Section 4.1. A general node will use limited-hop, small-area directional flooding to send a packet back to a B-node. When down-L receives a RE packet, there are two cases depending on whether the RE packet comes directly from up-L or via an intermediate node.

   1. If the RE packet is directly from up-L, because the original slot assignments at up-L and down-L are conflict-free with each other, a repaired route is found. Node down-L will send a route-repaired packet

to up-L, and the repaired route is: up-L →down-L.

2. If the RE packet is from an intermediate node K, K will add its free slots and address to the RE packet, and down-L will try to find a slot assignment at the intermediate nodes that satisfies the QoS requirement and does not conflict with the slot assignments at nodes up-L and down-L. If found, down-L will send a route-repaired packet to up-L via node K. Node K will reserve the slots for the QoS flow, and route repairing is done. Otherwise, the RE packet is discarded. For example, assume the bandwidth requirement is two slots, and assume the slot assignments at nodes up-L, L and down-L are 1,2; 3,4; 7,8 respectively. When node L moves away, there is another path that connects up-L to down-L, for example, up-L down-L. And node K has free slots 5, 6, which satisfy the QoS requirement and do not conflict with the slot assignments at nodes up-L and down-L. Then a repaired route is found.

If up-L does not receive any route-repaired packet for a certain time, it assumes the route repairing failed. And up-L will send a route-failure packet to source S. Then S will start a new QoS route discovery process. In the above route repairing process, the TTL can be set to a value larger than 2, that is, allow more than one intermediate nodes to relay the QoS flow between up-L and down-L. This will increase the chance of successful route repairing, but it will increase the routing overhead.

### 4.2.2 Election of B-node

Initially, one B-node is elected in each cell if there are BC-nodes available in the cell. Because B-nodes also move around, an algorithm is needed to elect a new B-node. When a B-node moves out of its current cell, it initiates the B-node election process. When a general node discovers there is no B-node in the cell (as stated in step 8 of the B-QoS routing protocol), for example, because the B-node failed, it initiates the B-node election process. Initially nodes know which node is the B-node in the cell. The election process works as follows. The leaving B-node or the general node floods an election message to all the nodes in the cell. When a BC-node receives the election message, it broadcasts a claim message that claims it will become the B-nodes to all nodes in the cell. Because there is a delay in propagating the claim message to neighbor nodes, several BC-nodes may broadcast during this period. To reduce such concurrent broadcasts, a random timer is used.

Each BC-node defers a random time before its B-node claim. If it hears a claim message during this random time, it then gives up its broadcast. And then one of the BC-nodes T becomes the new B-node in the cell, and T will start using the second address, which is the same as the cell ID. Because all nodes in the cell can hear the claim message, they know that T is the new B-node. This idea is similar to the cluster-head election scheme proposed in [7].

#### 4.2.3 A Routing Example

We present a routing example by using the B-QoS routing protocol in Figure 4, where the routing area is divided into nine cells, and the black boldface number is the cell ID. In Figure 4, the larger grey nodes are B-nodes, and the smaller green nodes are general nodes. In the example, node 2 (in cell 7) wants to set up a QoS session and sends packets to node 41 (in cell 3).

1. Node 2 floods RD packets (black arrows) to all the nodes in its cell. When a general node receives the RD packet, it forwards the RD packet to its neighbors. Only nodes in the same cell as node 2 will process the RD packet as in step 9 of B-QoS routing.

2. When B-node B7 receives the RD packet, first it checks if it has enough bandwidth. In the example assume B7 has enough bandwidth. Then B7 requests the location of destination node 41 from (B4 in the example), and then B7 knows that node 41 is in cell 3 with B-node B3. B7 determines the width of the routing cells to be $a\sqrt{2}/2$ , where a is the side length of a cell. The routing cells are the cells between the two blue border lines, that is, cells 4, 2; 7, 5, 3; and 8, 6. Then B7 floods RR packets (red arrows) to all B-nodes in the routing cells, and tries to find a QoS route to destination.

3. When B3 receives a RR packet, it sends a probe packet (blue arrow) to find destination node 41. In this example, node 41 is close to B3, and node 41 sends the Ack back to node B3. Then B3 sends RP back to B7, and to source node 2. A QoS route is set up, and it is $2 \rightarrow 1 \rightarrow B7 \rightarrow B5 \rightarrow B6 \rightarrow B3 \rightarrow 41$.

**Routing Latency and Overhead**

An important feature in B-QoS routing is to permit most of the transmissions based on B-nodes. Because B-nodes have large bandwidth, it increases

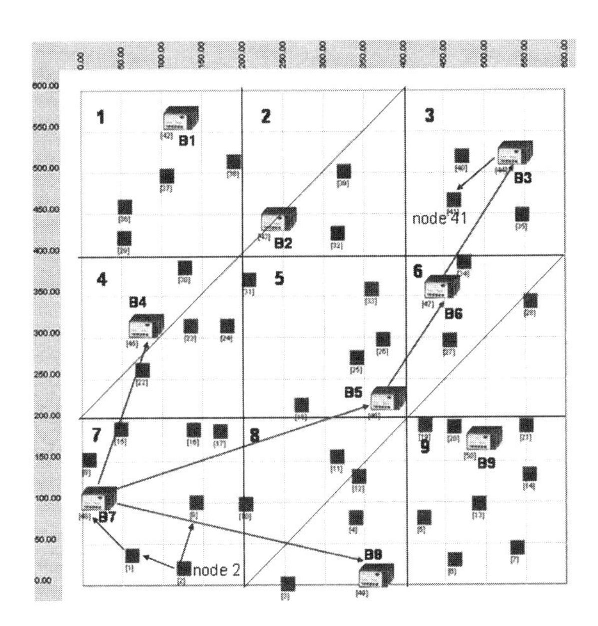


Figure 4: A routing example.

the chance of finding the route that satisfies the QoS requirement. Also B-nodes have a long transmission range, which greatly reduces the hop number in the route.

Based on node location information and cell structure, routing among B-nodes is very efficient (step 3 in Section 4.1). Small hop number and efficient B-node routing ensure B-QoS has low routing latency. Low latency is very important for routing in MANETs, because nodes in MANETs are constantly moving. Low routing latency means the intermediate nodes will not move far away from previous locations when the data packet comes, and this reduces the chance of a broken link. It also means the destination node will not be far away from its previous location when the data packet arrives, which also reduces routing overhead.

The routing overhead in B-QoS routing includes a small-area flooding from the source (or destination) to a nearby B-node, plus B-node route discovery among routing cells. And usually the nearby B-node is close to the source (or destination), for example, in the same or neighbor cell. So the overhead from small-area flooding is not large. We want to point out that usually the number of B-nodes is small (although the number of BC-nodes may be large). Because the transmission range of a B-node, R, is large, the side length of a cell $a = R/2\sqrt{2}$ is also large. Then the number of cells in a fixed routing area is small. Recall that only one B-node is maintained in each cell, thus the number of B-nodes is small. So the overhead from B-node route discovery is limited. The routing overhead from disseminating node location information is also not large. Providing the location of all nodes to a B-node $B_0$ does not incur much overhead. And other B-nodes request node location information from $B_0$ only when a QoS route needs to be discovered.

## 5 Performance Evaluation

The B-QoS routing protocol is implemented in QualNet, a scalable packet-level simulator with an accurate radio model. TDMA is used as the MAC protocol. The transmission rate of the general node and the B-node are 1 Mbps and 4 Mbps, respectively. There are 50 slots in a TDMA frame. For B-nodes, each TDMA slot is further divided into 4 subslots. So there are 200 subslots in the TDMA frame of B-nodes. For the transmission between two B-nodes, a subslot can handle the data transmitted by one slot of a general node. Note: for the transmission from a B-node to a general node, 1 Mbps data rate and slot (not subslot) should be used to avoid overflow at the general node. A subslot is only used between two B-nodes. The simulation testbed that we used consists of 35 general nodes and 15 BC nodes uniformly distributed at random in an area of 600 m $\times$ 600 m, which is divided into 9 cells. The radio transmission ranges of the general node and the B-node are 80 m and 320 m respectively. The side length of a cell is set as $a = R/1.6 = 200$ m. The detailed discussion of cell size is given in Section 5.6. Each simulation was run for 600 simulated seconds. The mobility in the environment was simulated using a random-waypoint mobility model. In our simulations, the pause time was set to 1 millisecond (close to 0 second), which correspondsto constant motion. We control the node mobility by varying the maximum node velocities. The maximum velocities range from 0 m/s to 20 m/s. In all simulations, B-nodes have the same mobility as general nodes, also the special B-node $B_0$ has the same mobility as general nodes.

User traffic is generated with CBR sources, where the source and the destination of a session are chosen randomly among the nodes. The default parameter settings are given below. A particular parameter is varied when we test the QoS routing performance according to the parameter. During its lifetime of 100 seconds, a CBR source generates 20 packets per second. A CBR source does not adjust its transmission according to the network congestion, and all 2000 packets are always transmitted irrespective of how many of them get through. The size of a CBR packet is 256 bytes. The starting time of a session is randomly chosen between 0 and 500 seconds, so a session always ends naturally by the end of the simulation. The offered traffic load is varied by increasing the number of CBR sessions generated during the simulation from 20 to 300. For each simulation configuration, we generate 20 different traffic patterns and get the average results.

We compare our B-QoS routing protocol with the QoS routing protocol proposed in [1], which is referred to as A-QoS. A-QoS has also been implemented in QualNet. We chose A-QoS as the routing protocol for comparison because it has a similar route discovery mechanism as B-QoS. One of the performance metrics is the "serviced session," which is used in [1]. A session is called "serviced" if at least 90% of the packets are received by the destination. This is an approximate measurement of the quality-of-service provided to the end-user. AODV [17] is also used in the performance comparison because it is a widely used benchmark for MANET routing protocols. AODV is an on-demand best-effort routing protocol that uses flooding to discover the route. The following metrics are used to compare routing performances.

1. *Routing overhead.* Routing overhead is the number of routing-related packets (RR, RE, RP, RF packets, etc.) for each QoS session request. This metric is used to measure the efficiency of the routing protocols. In all the tests, the routing overhead of B-QoS includes the overhead of disseminating node location information. Section 5.1 presents the result of routing overhead comparison.

2. *Success ratio.* The success ratio is the ratio between the number of accepted sessions and the number of session requests. This metric measures the effectiveness of finding QoS routes, and the result is discussed in Section 5.2.

3. *Session good-put.* Session good-put is the number of sessions that are serviced. The session good-put is not the same as the success ratio. Even after a QoS route has been set up, it may become broken during the session because some intermediate nodes move away. Such a session can be counted as an accepted session for the success ratio, but it cannot be counted as a serviced session for good-put if less than 90% of the packets are delivered. Whether the session can be serviced depends on the route repairing or rerouting. Also, a session may be serviced by a best-effort route even if a QoS route is not found. Session good-put is discussed in Section 5.3.

For the simulations presented in Section 5.3, we modified the B-QoS and A-QoS routing protocols so that if B-QoS (or A-QoS) cannot find a QoS route on the first try, a best-effort route will be discovered and used to deliver packets.

4. *Throughput and delay.* These two metrics are used to measure the effectiveness of the routing protocols. Sections 5.4 and 5.5 present the results of throughput and delay comparison.

## 5.1 Routing Overhead

Routing overhead is the number of routing-related control messages (RR, RE, RP, RF packets, etc.) per QoS connection request. Sending a control packet over one link is counted as one message. If a control packet traverses a route of $k$ hops, $k$ messages are counted. The control packet from route repairing is also included in the overhead. We compare the routing overhead of B-QoS, A-QoS, and AODV for different node mobility. Figure 5 is the average routing overhead per QoS connection request when node maximum speed varies from 0 m/s (the actual value is a small number close to 0) to 20 m/s. Although AODV does not process QoS request, the routing overhead from best-effort AODV routing is still presented for comparison purposes. Figure 5 shows that all the routing overheads increase as node speed increases. Higher mobility causes more broken links, and thus increases

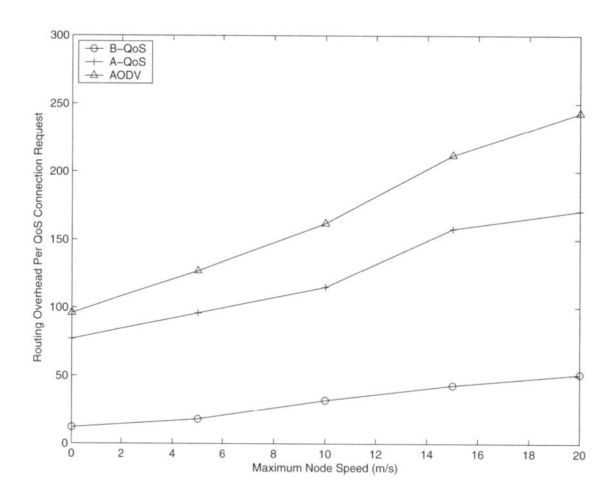


Figure 5: Routing overhead versus mobility.

routing overhead. B-QoS has much smaller routing overhead than both A-QoS and AODV. Recall that B-QoS tries to find a route based on B-nodes first. Large bandwidth of B-nodes increases the chance of satisfying the QoS requirement. In addition, a route based on B-nodes has a small hop number, which also increases the chance of satisfying the QoS requirement, because a small hop number reduces the chance of conflicting slot assignments. So in B-QoS routing, most QoS requests can find a route mainly based on B-nodes. Thus for most route discoveries in B-QoS, the route request packets are only propagated among B-nodes in the routing cells. Usually the number of B-nodes is small, and the routing cells further reduce the number of involved B-nodes. Thus the average routing overhead of B-QoS is small. On the other hand, both A-QoS and AODV discover a route by flooding route request packets to all nodes in the network, which causes large routing overhead. We also observe that A-QoS has less routing overhead than AODV. The reason is stated below. During the route discovery of A-QoS, if the required bandwidth cannot be met, the route request packet will be dropped. So the number of control messages used in A-QoS is smaller than in AODV.

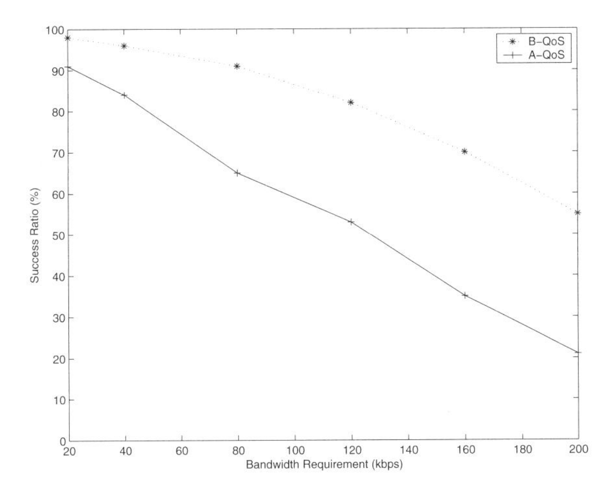


Figure 6: Success ratio.

## 5.2 Success Ratio

The success ratio represents the chance of finding a route for a given QoS requirement. We only compare B-QoS with A-QoS for the success ratio, because AODV does not provide QoS routing. Figure 6 plots the success ratios of B-QoS and A-QoS for different bandwidth requirements. In the test, the number of CBR sessions is fixed to be 50. We change the bandwidth requirement by varying the data rate (packet per second) of each CBR. And the node maximum speed is set as 10 m/s. Figure 6 shows that B-QoS has a much higher success ratio than A-QoS, especially when the bandwidth requirement is large. B-QoS builds QoS routes mainly by B-nodes, which have large bandwidth and long transmission range. The large bandwidth of B-nodes increases the chance of finding a QoS route. The long transmission range of B-nodes reduces the hop number in a QoS route. As stated in Section 5.1, a small hop number increases the chance of satisfying QoS requirements.

## 5.3 Session Good-Put

Then we compare session good-puts, the number of sessions that are serviced in B-QoS, A-QoS, and AODV. A session may be serviced by a best-effort route even if a QoS route is not found. The best-effort AODV is also used in the comparison. In this test, the data rate of each CBR session is fixed to be 50 kbps. The offered traffic load is varied by increasing the number of CBR sessions generated during the simulation from 20 to 300. The node maximum speed is set as 10 m/s. The session good-puts for B-QoS, A-QoS, and AODV are reported in Figure 7. As we can see, B-QoS has much larger session good-put than both A-QoS and AODV. This is because the QoS route in B-QoS is mainly based on B-nodes, which have larger bandwidth and a longer transmission range than general nodes. Also routing in B-QoS has small latency and overhead. So B-QoS has a higher delivery ratio than A-QoS and AODV.

We also compare the session good-puts of the three routing protocols for different node mobility, and the results are plotted in Figure 8, where the number of transmitted sessions is 150. The node maximum speed varies from 0 m/s to 20 m/s. Figure 8 shows the session good-puts of the three routing protocols are close when node speed is zero (static). When node mobility increases, all the session good-puts decrease. High mobility causes more broken links, and thus causes good-put drop. However, the session good-put of

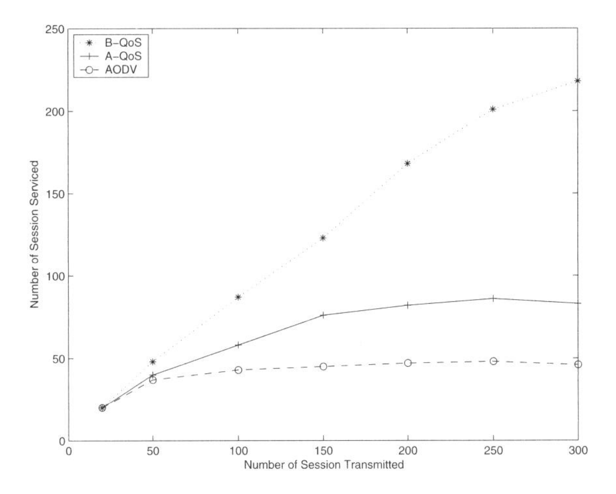


Figure 7: Session good-put.

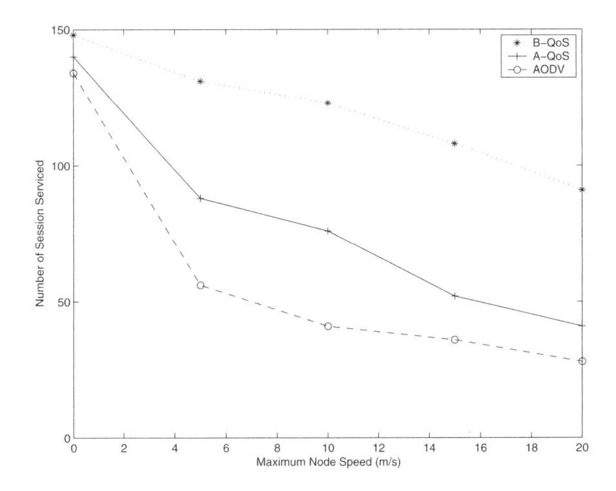


Figure 8: Session good-put versus mobility.

B-QoS is always higher than both A-QoS and AODV. The reason is similar to that above. In addition, the same mobility causes fewer broken links for B-nodes than for general nodes, because the transmission range of B-nodes is large. Even for relatively high node mobility (e.g., 20 m/s), the number of sessions serviced in B-QoS is still large, about 95 of 150 sessions, whereas the number is lower than 50 in both A-QoS and AODV.

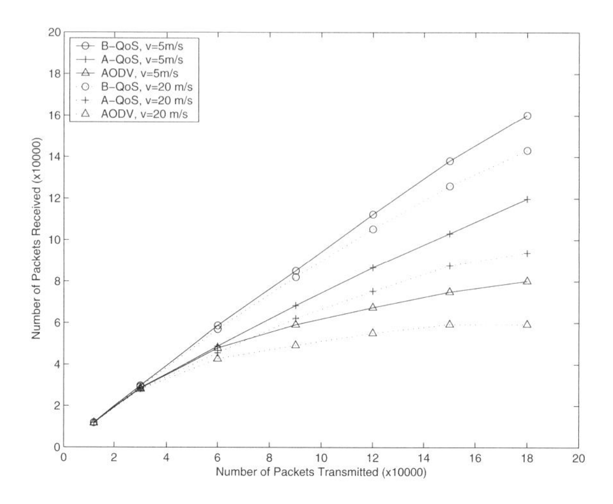


Figure 9: Comparison of throughput.

## 5.4 Throughput Comparison

We compare the throughput of B-QoS with A-QOS and AODV under different network traffic loads. The network traffic load is varied by changing the number of CBR sessions generated during the simulation. The number of packets transmitted varies from $1.2 \times 10^4$ to $1.8 \times 10^5$. Node mobility also affects the throughput. We measure the throughput under both low mobility (maximum node speed = 5 m/s) and relatively high mobility (maximum node speed = 20 m/s). Figure 9 reports the throughputs of the three routing protocols. We observe that B-QoS always has higher throughput than A-QoS and AODV. Under a light traffic load, the throughputs are close for B-QoS, A-QoS, and AODV. The difference becomes larger when traffic becomes heavy. High throughput in B-QoS comes from the efficient QoS routing based on B-nodes. Because B-nodes have a large bandwidth and transmission range, most packets can be successfully delivered even under heavy traffic. Figure 9 also shows that both the two QoS routing protocols (B-QoS and A-QoS) have higher throughput than AODV. The QoS routing protocols try to find and use routes satisfying bandwidth constraints for different flows, and distribute traffic load over different routes. The throughputs of the QoS routing protocols do not decrease much as the network load becomes heavy. On the contrary, the best-effort AODV only uses one route for multiple flows, therefore as the network load gets heavy, the throughput drops significantly. Also, we observe that for each routing protocol, the

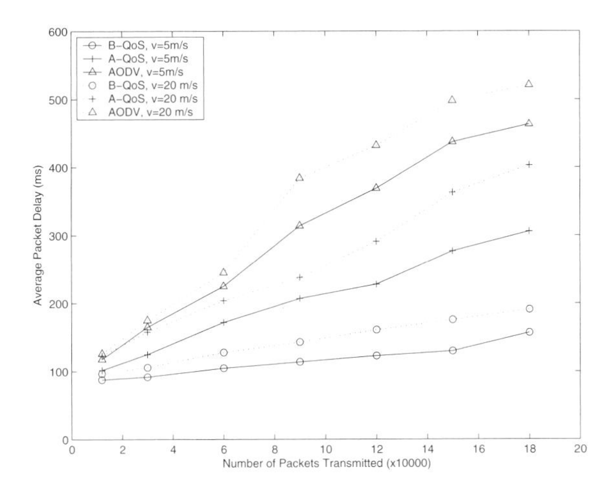


Figure 10: Average packet delay.

throughput under low mobility is larger than when the throughput is under high mobility. High mobility causes more broken links and thus reduces throughput. From Figure 9, we also find out that the throughput performance for the B-QoS routing protocol at high mobility (20 m/s) is better than A-QoS (and AODV) at low mobility (5 m/s). This demonstrates that B-QoS performs much better than A-QoS (and AODV).

## 5.5 Delay Comparison

Figure 10 presents the average (end-to-end) packet delay for the three routing protocols under different network loads. The packet delay also depends on the node mobility. We measure the delay under both low mobility (5 m/s) and relatively high mobility (20 m/s). When the traffic is light, the average packet delays are close for B-QoS, A-QoS, and AODV. When network traffic increases, the packet delay of B-QoS increases slowly, and the packet delays of A-QoS and AODV increase much faster. The packet delay of B-QoS remains low for all tested mobility. The low packet delay of B-QoS mainly comes from the small hop number in B-QoS routing. Also B-QoS has less routing latency and overhead than A-QoS and AODV (as shown in Section 5.1). Less routing overhead reduces the chance of congestion and thus reduces packet delay. Figure 10 also shows that both B-QoS and A-QoS have less delay than AODV. For the best-effort AODV routing protocol, one source-destination pair usually uses only one active route for all the packet flows. And when the network traffic is heavy, this single route becomes heavily loaded, causing packets to be delayed or dropped, and hence increases the packet delay a lot. On the other hand, the QoS routing protocols try to find and use routes satisfying bandwidth constraints for different flows, even between the same pair of source and destination. So when a particular route does not have enough bandwidth, a new route is discovered. Thus, the network load is balanced among different routes by the QoS routing protocols.

The above simulation tests demonstrate that the B-QoS routing protocol has much better performance than both A-QoS and AODV. B-QoS provides a high success ratio, high delivery rate, high throughput, and low delay for QoS routing, and it incurs small routing overhead. B-QoS achieves such good performance by taking advantage of the powerful B-nodes, which have large bandwidth and transmission range.

## 5.6 The Size of the Cell

An important parameter in B-QoS routing is the size of the cell. For a fixed B-node transmission range, we want the size of a cell to be as large as possible. The larger the cell, the smaller the number of cells, and hence the smaller the number of B-nodes needed. Recall in Section 3, for a given transmission range of a B-node, R, the side length of a cell is set as $a = R/2\sqrt{2}$. This is to ensure a B-node can directly communicate with B-nodes in all neighbor cells, even under the worst conditions. However, more often, two nearby B-nodes are not as far as $2a\sqrt{2}$ (i.e., at the two opposite corners). This suggests that we can set the side length of a cell to be larger than $R/2\sqrt{2}$, and the connections between two nearby B-nodes are still valid for most of the time. Increasing the size of the cell is especially useful when the number of backbone-capable nodes is small. We run simulations to measure the impact of different cell sizes on routing overhead. For a fixed R = 320 m, we changed the value of a from $R/2\sqrt{2} = 113$ m to 250 m. The results are plotted in Figure 11. In the simulation, the maximum node speed is 10 m/s.

Figure 11 shows the routing overhead increases when the size of the cell increases. When the cell becomes larger, sometimes a B-node may not be able to connect to a nearby B-node directly, and this increases the routing overhead. We observe from Figure 11 that the routing overhead does not increase much when the side length is less than 200 m. If the size of the cell is only a little bit larger than $R/2\sqrt{2}$, B-nodes can still directly connect with nearby B-nodes most of the time. The simulation results suggest that we can

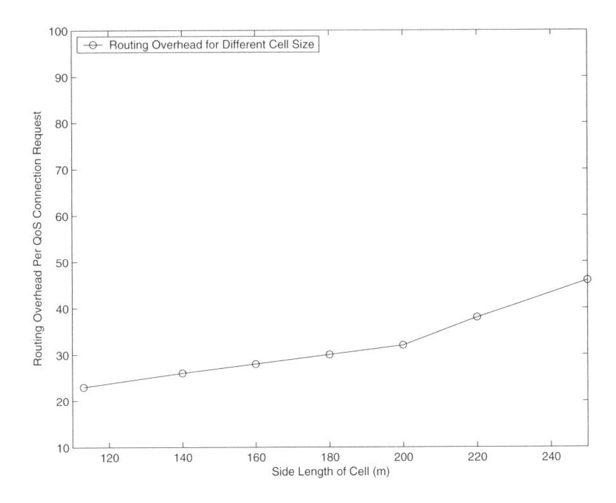


Figure 11: Routing overhead for different cell sizes.

choose the value of a to be a little larger than $R/2\sqrt{2}$, and still achieve good routing performance. In the next section, we study the probability of having B-nodes in one cell.

# 6 The Probability of Having B-Nodes

To ensure good performance of the B-QoS routing protocol, it is important to have B-nodes in most cells. If there is no B-node in a cell, then small-area flooding or flooding among B-nodes (or even among all nodes) in the network may be used, which increases routing overhead and degrades routing performance. We design simulations to measure the probability of having a B-node in a cell. Later, we use a computation approach to estimate the probability of having a B-node.

The simulation is run in the same testbed used in Section 5, with an area of 600 m $\times$ 600 m divided into 9 cells. We measure the probability in the following way. We record the time $t_j$ of each cell j having a B-node during the whole simulation process. Then we add all the $t_j$ together, and divide by the number of cells and the simulation time, that is, P (having B-nodes) $= \sum t_j/(MT)$, where $M$ is the number of cells, and $T$ is the simulation time. We measure the average probability of having a B-node in a cell when the number of backbone-capable nodes varies from $M$ (the number of cells) to $3M$.

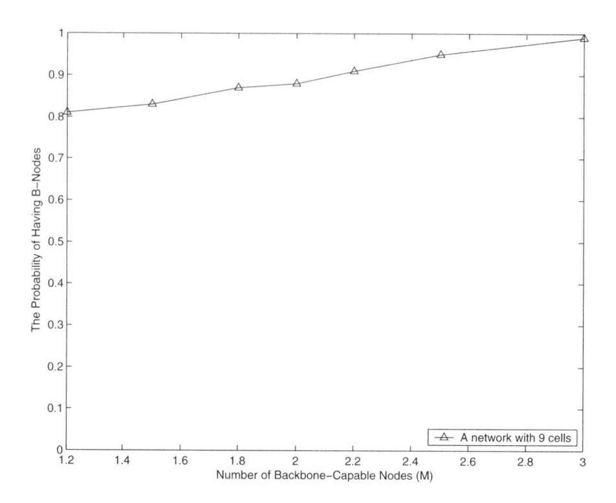


Figure 12: The probability of having B-nodes.

The test results are given in Figure 12. The maximum node speed is set as 10 m/s. The pause time is set to zero, which means the nodes continuously move during the simulation. In Figure 12, the $x$-axis is the number of backbone-capable nodes in terms of cell number $M$. Figure 12 shows that the probability of having a B-node in each cell is very high when there are a reasonable number of B-nodes. The probability is always larger than 0.80, and it is larger than 0.90 when the number of backbone-capable nodes is larger than twice the cell number ($2M$).

We also estimate the probability of having a B-node by the computational method. To simplify the computation, we assume nodes move in all directions with equal probability. Assume there are $M$ cells and a total $N$ BC-nodes in the network. For each cell, the probability of having a particular BC-node in the cell is $1/M$, and the probability of this BC-node not in the cell is $1 - 1/M$. The probability of having no BC-nodes in the cell is $(1 - 1/M)^N$. So the probability of having at least one B-node in the cell is

$$P = 1 - (1 - \tfrac{1}{M})^N. \qquad (1)$$

| M | 9 |
|---|---|
| N | 15 |
| P | 0.829 |

Based on Equation (1), we compute the probability P for the testbed used in our simulations. The result is listed in the table above. As we can see, the probability is very high (0.829). The high probability of having B-nodes in each cell guarantees the good performance of B-QoS routing protocols. Also, the simulation result in Figure 11 is close to the computational result in the table, where the number of BC nodes is $15/9 = 1.67$ in terms of M. Equation (1) can also be used to determine the value of W, the width of routing cells. If P is large, then W can be small; and if P is small, then W should be large.

## 7 Conclusions

In this chapter, we proposed a new QoS routing protocol, B-QoS for heterogeneous mobile ad hoc networks. B-QoS takes advantage of the different transmission capability of multiple types of nodes. Most of the links of the QoS route are based on B-nodes, which have large bandwidth and transmission range. Large bandwidth increases the chance to meet QoS requirements. And a large transmission range reduces the number of hops in the route. A small hop number not only increases the chance of satisfying the QoS requirement, but also reduces routing latency. We implemented B-QoS in QualNet, and compared routing performance of B-QoS with another QoS routing protocol — A-QoS— and with a popular routing protocol — AODV. Extensive simulations show that the B-QoS routing protocol performs much better than A-QoS and AODV. B-QoS provides a high success ratio, high delivery rate, high throughput, and low delay for QoS routing, and it incurs small routing overhead. B-QoS obtains such good performance by utilizing powerful B-nodes, which have large bandwidth and transmission range. In addition, we show by simulation and computation that the probability of having B-nodes in each cell is very high, given a reasonable number of backbone-capable nodes.

## References


[1] C. Zhu and M. Corson, *QoS routing for mobile ad hoc networks.* In IEEE INFOCOM 2002.

[2] C. R. Lin, *On-demand QoS routing in multihop mobile netowrks.* In IEEE INFOCOM 2001.

[3] C. R. Lin and J.-S. Liu, *QoS routing in ad hoc wireless networks*, IEEE JSAC, 17(8), 1999.

[4] S. Chen and K. Nahrstedt, *Distributed quality-of-service routing in ad hoc networks*, IEEE JSAC, 17(8), 1488–1505, August, 1999.

[5] G Xue, *Minimum cost QoS multicast and unicast routing in communication networks*, IEEE Transactions on Communications, 51, 817–824, 2003.

[6] J. Li et al., *A scalable location service for geographic ad hoc routing.* In ACM Mobicom 2000, Boston.

[7] K. Xu, X. Hong, M. Gerla, *An ad hoc network with mobile backbones.* In IEEE ICC 2002.

[8] S. Butenko, X. Cheng, D.-Z. Du, and P. Pardalos, *On the construction of virtual backbone for ad hoc wireless networks.* In Proceedings of the second Conference on Cooperative Control and Optimization.

[9] Z. Ye, S. Krishnamurthy, S. Tripathi, *A framework for reliable routing in mobile ad hoc networks.* In IEEE INFOCOM 2003, San Francisco, April, 2003.

[10] Y.-B. Ko and N. H. Vaidya, *Location-aided routing (LAR) in mobile ad hoc networks.* In ACM/IEEE International Conference. Mobile Computer Networks,1998.

[11] S. Chakarabarti, A. Mishra, *QoS issues in ad hoc wireless networks*, IEEE Comm., February, 2001.

[12] E. Elmallah, et al., *Supporting QoS routing in mobile ad hoc networks using probabilistic locality and load balancing.* In IEEE GLOBECOM 2001.

[13] R. Sivakumar, et al., *CEDAR: A core-extraction distributed ad hoc routing algorithm.* In INFOCOM 1999.

[14] J. Tsai, T. Chen, and M. Gerla, *QoS routing performance in multihop, multimedia, wireless networks.* In Proceedings of IEEE ICUPC, 1997.

[15] Y.-C. Hsu and T.-C. Tsai, *Bandwidth routing in multihop packet radio environment.* In Proceedings of the third International Mobile Computing Workship, 1997.

[16] F. Cao, H. Fang, and M. Conlon, *Performance analysis of measurement-based call admission control on voice gateways*. In Internet Telephony Workshop 2001, April 2001, New York.

[17] C. Perkins and E. Royer, *Ad hoc on-demand distance vector routing*, Proceedings of the secnd IEEE Workshop on Mobile Computing Systems and Applications, New Orleans, February, 1999.

[18] S. Yi, et al., *Security aware ad hoc routing for wireless networks*. In Proceedings of the 2001 ACM International Symposium on Mobile Ad Hoc Networking Computing, Long Beach, Ca, 2001.

[19] M. Gerla, X. Hong, and G. Pei, *Landmark routing for large ad hoc wireless networks*, In Proceedings of IEEE Globecom 2000, San Francisco, CA, November, 2000.

[20] W.H. Liao, Y.C. Tseng, and J.P. Sheu, *GRID: A fully location-aware routing protocol for mobile ad hoc networks*, Telecommunication Systems, 18, 2001, 37–60.

[21] Z. Li, and P. Mohapatra, *QRON: QoS-Aware Routing in Overlay Networks*, IEEE Journal on Selected Areas in Communications, 22(1), 29 – 40, January, 2004.

[22] C.R. Lin and M. Gerla, *Adaptive Clustering for Mobile Wireless Networks*, IEEE Journal on Selected Areas in Communications, 15(7), 1265–1275, September, 1997.

[23] Y. Xu, J. Heidemann, and D. Estrin, *Geography-informed energy conservation for ad hoc routing*. In Proceedings of the ACM/IEEE International Conference on Mobile Computing and Networking, pp. 70–84, Rome, Italy, ACM. July, 2001.

[24] S. Lee and A. T. Campbell, *INSIGNIA: In-band signalling support for QoS in mobile ad hoc networks*. In Proceedings of the Fifth International Workshop on Mobile Multimedia Communication, 1998.

[25] A. Orda and A. Sprintson, *Precomputation schemes for QoS routing*. IEEE/ACM Transactions on Networking. 11(4) 578–591, August, 2003.

[26] B. Karp and H. T. Kung, *GPSR: Greedy perimeter stateless routing for wireless networks*. In Proceedings of the Sixth Annual International

Conference. Mobile Computing and Networking (MobiCom 2000), Boston, 2000, pp. 243–54.

[27] S. Basagni et al., *A distance routing effect algorithm for mobility (DREAM).* In ACM/IEEE International Conference. Mobile Computing and Networks, 1998, pp. 76–84.

[28] V. Ramasubramanian, Z. J. Haas, and E. Sirer, *SHARP: A hybrid adaptive routing protocol for mobile ad hoc networks.* In MobiHoc 2003, Annapolis, MD, June, 2003.

[29] P. Sinha and S. Krishnamurthy, *Scalable unidirectional routing with zone routing protocol (ZRP) extenstions for mobile ad hoc networks.* In WCNC, Chicago, September, 2000.

# Chapter 2

## Optimal Path Selection in Ad Hoc Networks Based on Multiple Metrics: Bandwidth and Delay

Hakim Badis
*Computer Science Research Laboratory (LRI)*
*University of Paris-Sud, 91405 Orsay, France*
E-mail: badis@lri.fr

Khaldoun Al Agha
*Computer Science Research Laboratory (LRI)*
*University of Paris-Sud, 91405 Orsay, France*
E-mail: alagha@lri.fr

## 1 Introduction

A link state routing approach makes available detailed information about the connectivity and the topology found in the network. Moreover, it increases the chances that a node will be able to generate a route that meets a specified set of requirement constraints. The OLSR protocol [1] is an optimization over the classical link state protocol, proposed for the Mobile Ad hoc NETworks (MANET) [2]. It performs hop-by-hop routing; that is, each node uses its most recent information to route a packet. Therefore, each node selects a set of its neighbor nodes as MultiPoint Relays (MPRs) [3]. In the OLSR protocol, only nodes, selected as such MPRs, are responsible for forwarding control traffic, intended for diffusion into the entire network. MPRs provide an efficient mechanism for flooding control traffic by reducing

M.X. Cheng, D. Li (eds.) *Advances in Wireless Ad Hoc and Sensor Networks.*
Signals and Communication Technology, doi: 10.1007/978-0-387-68567-0_2.

the number of transmissions required. Nodes, selected as MPRs, also have a special responsibility when declaring link state information in the network. Nevertheless, no QoS information is taken into account.

We have presented in [4, 5] the QOLSR protocol, which is an enhancement of the OLSR routing protocol to support multiple-metric routing criteria. This work details how the metric measurements and the routing table calculation are achieved by the QOLSR protocol. We include quality information in *TC* messages about each link, between a node and its MPR selectors. The heuristic for the selection of MPRs limits its number in the network and ensures that the overhead is as low as possible. However, there is no guarantee that OLSR finds the optimal path in terms of QoS requirements. With this heuristic, the good quality links may be hidden from other nodes in the network. In this chapter, we present two heuristics for the selection of MPRs based on QoS measurements. We show that the proposed heuristics find optimal widest paths (paths with maximum bandwidth) based on the QoS over the links. The idea is to introduce more appropriate metrics, such as bandwidth and delay. The QOLSR protocol calculates these metrics between each node and its neighbors having a direct and symmetric link. This metric information is stored in the neighbor table and used to calculate the MPRs. Afterwards, each MPR node generates a topology control message including its MPR selector's bandwidth and delay. Based on this information, the routing table is deduced to find the optimal routes in terms of bandwidth and delay. The QOLSR protocol is presented, including the routing table calculation and MPR computation.

The remainder of this chapter is organized as follows. Section 2 presents the QoS routing in ad hoc networks, describing some related work and our QOLSR protocol. The routing table calculation based on multiple metrics is presented in Section 4. Section 5 presents the disadvantages of the standard MPR selection heuristic and introduces the new heuristics for the selection of MPR algorithms including bandwidth and delay information. Finally, we conclude in Section 7.

## 2 QoS Routing in Ad Hoc Networks

### 2.1 Related Work

The existing research on QoS routing for ad hoc networks can be divided into two categories: QoS route information and QoS route computation. QoS route information provides the QoS information over the paths using traditional best-effort routing algorithms. Such information helps the source node to fulfill the call admission task. QoS route computation calculates feasible routes based on various QoS requirements.

For the QoS route information, Chen et al. [6] propose a bandwidth-constrained routing algorithm. Each node calculates the available bandwidth over the wireless links to the destination. Such bandwidth information is piggybacked in the Destination Sequence Distance Vector (DSDV) routing algorithm [7]. Thus, each node knows the bottleneck bandwidth over the paths calculated by DSDV to all known destinations.

In [8] a similar approach using DSDV is presented. Focusing on bandwidth control, bandwidth information is embedded in the nodes' routing tables and sent to the neighbors. Upon receiving a routing table from a neighbor, a node updates its own routing table and the path bandwidth information. With the bandwidth information, a node can decide whether it should accept a new connection request based on the bandwidth requirement of that connection.

In [9], the authors have developed an architecture that supports accurate permissible throughput explicit feedback to multimedia transports and call admission applications. The motivation is for the architecture to base a cost/benefit analysis of end-to-end measurements versus lower-level explicit feedback in 802.11 access networks. They use the fragment acknowledgment and link-fail message used for unicast in DCF to produce a link-by-link (source, destination) pair throughput measurement. Link layer queue utilization measurements are then combined to produce a permissible throughput measurement, in a suitable way for a wireless multihop network.

For the QoS route computation, [10] considers a number of issues in QoS routing. The authors first examine the basic problem of QoS routing, namely, finding a path that satisfies multiple constraints, and its implications on routing metric selection. They present a centralized algorithm that is suitable for source routing and two distributed algorithms that are suitable for hop-by-hop routing based on bandwidth and delay constraints.

Chen [11] proposes a heuristic algorithm for the delay-cost constrained routing problem which is NP-complete. The main idea of this algorithm is to first reduce the NP-complete problem to a simpler one that can be solved in polynomial time, and then solve the new problem by either using an extended Dijkstra's algorithm or extended Bellman–Ford algorithm.

In [12], the authors propose a routing algorithm to find the shortest path between one source and one destination node while considering the criteria of multiple metric constraints. They have several goals to achieve: (1) sorting the QoS metrics according to the requested service; (2) finding the possible routes between the source and the destination; and (3) speeding up the path determination by using a sliding window and hierarchical clustering technique.

In [13] Costa et al. propose and analyze the performance of a distance-vector QoS routing algorithm that takes into account three metrics: propagation delay, available bandwidth, and loss probability. Classes of service and metric combination are used to make the algorithm scalable, as on a two-metric scheme.

Cheng and Nahrstedt [14] give an algorithm to find a path that meets two requirements in polynomial time. The algorithm reduces the original problem to a simpler one by modifying the cost function, based on which the problem can be solved using an extended shortest-path algorithm. The shortcoming of this method is that the algorithm has to use high granularity in approximating the metrics, so it can be very costly in both time and space, and it cannot guarantee that the simpler problem has a solution if the original one has a solution.

The algorithms in [15] and [16] are based on calculating a simple metric from the multiple requirements. By doing so, we can use a simple shortest-path algorithm based on a single cost aggregated as a combination of weighted QoS parameters. The main drawback of this solution is that the result is quite sensitive to the selected aggregating weights.

Efficient heuristics for K-constrained optimal path selection are proposed in [18, 17]. They show that when the link weights are positively correlated, the path weights are also positively correlated and thus the linear approximation algorithm often succeeds in finding feasible paths. However, when link weights are negatively correlated, the path weights also tend to be negatively correlated, degrading the performance of linear approximation algorithms.

## 2.2 QOLSR: QoS in OLSR Protocol

OLSR [1] is a proactive routing protocol, which inherits the stability of a link state algorithm and has the advantage of having the routes immediately available when needed due to its proactive nature. In a pure link state protocol, all the links with neighbor nodes are declared and are flooded in the whole network. The OLSR protocol is an optimization of the pure link state protocol for mobile ad hoc networks. First, it reduces the size of the control packets: instead of all links, it declares only a subset of links with its neighbors that are its multipoint relay selectors. Second, it minimizes the flooding of its control traffic by using only the selected nodes, called multipoint relays, to broadcast its messages. Therefore, only the multipoint relays of a node retransmit the packets. This technique significantly reduces the number of retransmissions in a flooding or broadcast procedure. OLSR protocol performs hop-by-hop routing; that is, each node uses its most recent information to route a packet.

### 2.2.1 Multipoint Relay

The idea of multipoint relays is to minimize the flooding of broadcast packets in the network by reducing duplicate retransmissions in the same region. Each node $m$ of the network independently selects a set of nodes in its one-hop neighbors, which retransmits its packets. This set of selected neighbor nodes, called the MultiPoint Relay (MPR) of $m$ and denoted MPR(m), is computed in the following manner: it is the smaller subset of one-hop neighbors with a symmetric link, such that all two-hop neighbors of $m$ have symmetric links with MPR(m). This means that the multipoint relays cover (in terms of radio range) all the two-hop neighbors. Figure 1 shows the multipoint relay selection by node $m$.

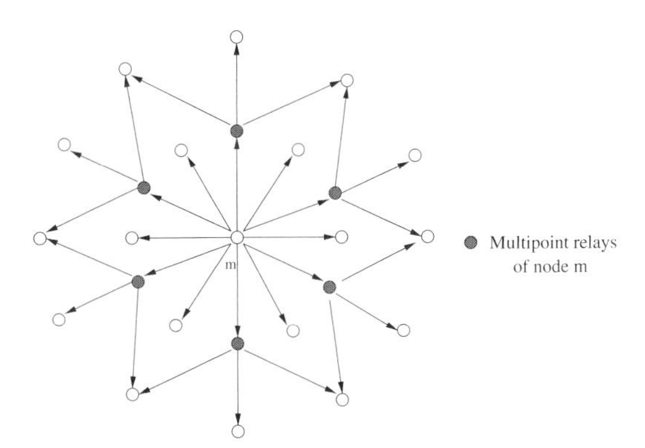


Figure 1: Multipoint relays of node $m$.

Only MPR nodes forward broadcast messages received from one of their MPR selectors.

### 2.2.2 Neighbor Sensing

Each node must detect the neighbor nodes with which it has a direct and bidirectional link. The uncertainties over radio propagation may make some links unidirectional. Consequently, all links must be checked in both directions in order to be considered valid. For this, each node periodically broadcasts its *hello* messages, containing the list of neighbors known to the node and their link status. The *hello* messages are received by all one-hop neighbors, but are not forwarded. They are broadcast at a low frequency determined by the refreshing period *Hello Interval* (the default value is two seconds).

These *hello* messages permit each node to absorb the knowledge of its neighbors up to two hops. On the basis of this information, each node performs the selection of its multipoint relays. These selected multipoint relays are indicated in the *hello* messages with link status MPR. On the reception of *hello* messages, each node can construct its MPR selectors table.

### 2.2.3 Topology Information

Each node with a nonempty MPR selector set periodically generates a Topology Control message ($TC$ message). This $TC$ message is diffused to all nodes in the network at least every $TC$ Interval. A $TC$ message contains the list of neighbors that have selected the sender node as a multipoint relay. The information diffused in the network by these $TC$ messages will help each node to build its topology table. Based on this information, the routing table is calculated. The route entries in the routing table are computed with Dijkstra's shortest-path algorithm [19]. Hence, they are optimal as concerns the number of hops.

The routing table is based on the information contained in the neighbor and the topology tables. Therefore, if any of these tables is changed, the routing table is recalculated to update the route information about each known destination in the network.

### 2.2.4 QoS-Enhanced OLSR

The QOLSR protocol includes quality parameters in a link state routing protocol. In such a case, a novel approach is applied because the QoS constraints are available before the routing table calculation. The QOLSR is an extension introduced to the OLSR protocol without breaking backwards compatibility. No additional control traffic is generated (only *hello* and $TC$

messages). As in the standard OLSR, link state information is generated only by nodes selected as MPRs. This information is then used by the route calculation. QOLSR requires only a partial link state to be flooded in order to provide optimal paths in terms of bandwidth and delay. The QOLSR does not require any changes to the format of IP packets. Thus any existing IP stack can be used and the protocol only interacts with routing table management.

## 3 Metrics Measurements

The delay and bandwidth metrics are taken into account as QoS constraints for the proposed QOLSR protocol. Such metrics are included on each routing table entry corresponding to each destination.

### 3.1 Delay Metric

With the QOLSR protocol, a route is immediately available when needed, satisfying the QoS requirements. So, before sending data traffic we must inform each node about the delay information between any node and its MPRs. The only possibility to calculate the delay is to use the control traffic messages. Each node in the ad hoc network periodically broadcasts its *hello* messages locally. These control messages are transmitted in broadcast mode without acknowledgments in response. Consequently, to measure the round-trip time *rtt* (time passed from sending the message in the sender to arrival of the acknowledgment in the sender), each node calculates the one-way time (from generation of the *hello* message in the source node to arrival in the neighbor node) from its neighbors. Now, there are two cases to consider: synchronized and asynchronized ad hoc networks.

In synchronized networks, the measured delay computing is very simple. Each node includes in the *hello* message, during the neighbor discovery performed by the QOLSR, the creation time of this message. When a neighbor node receives this message, it calculates the difference between such time and the current time, which represents the measured delay. For instance, if we use the IEEE 802.11b as the medium access control, the measured delay is represented by the following formula:

$$\mathrm{Measured\ delay} = \mathrm{t_q} + (\mathrm{t_S} + \mathrm{t_{CA}} + \mathrm{t_{Overhead}}) \times \mathrm{R} + \textstyle\sum_{\mathrm{r}=1}^{\mathrm{R}} \mathrm{B_T},$$

where: $t_q$ = Mac queueing time; $t_S$ = transmission on time of S bits; $t_{CA}$ = collision avoidance phase time; $t_{Overhead}$ = control overhead time (e.g.,

RTS, CTS, etc.); $R$ = number of necessary transmissions; and $B_T$ = backoff time for $r$.

In asynchronized networks, due to the characteristics of sparse ad hoc networks, classical clock synchronization algorithms are not applicable. Time synchronization in ad hoc networks is a wide subject of research. Here, we propose a simple algorithm to estimate an average of measured delays without synchronizing the local computer clocks of the devices. Figure 2 shows the *hello* messages exchanged between a node $a$ and its neighbor $b$ and how to measure the measured delay.

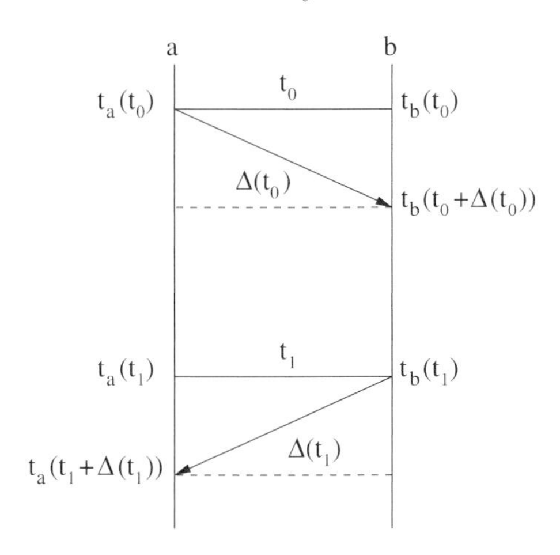


Figure 2: Hello message delay estimation.

We suppose that the absolute time when the node $a$ is generating its *hello* message is $t_0$. So, the local time in $a$ is $t_a(t_0) = t_0 + \delta_a(t_0)$ and $t_b(t_0) = t_0 + \delta_b(t_0)$ in $b$, where $\delta_a(t_0)$ and $\delta_b(t_0)$ are the differences between the local times in $a$ and $b$, respectively, and the absolute time $t_0$. The node $a$ includes $t_a(t_0)$ in its *hello* message. When the neighbor node $b$ receives this message, it has a local time $t_b(t_0 + \Delta(t_0)) = t_0 + \delta_b(t_0 + \Delta(t_0)) + \Delta(t_0)$; it calculates the difference between its local time and the attached time $t_a(t_0)$, which represents the $\Delta_{\mathrm{pars(b,a)}} = \delta_b(t_0 + \Delta(t_0)) + \Delta(t_0) - \delta_a(t_0)$. We suppose that the absolute time when node $b$ is generating its *hello* message is $t_1$. At this time, the local times in $a$ and $b$ are, respectively, $t_a(t_1) = t_1 + \delta_a(t_1)$ and $t_b(t_1) = t_1 + \delta_b(t_1)$. The node $b$ includes $t_b(t_1)$ and its $\Delta_{\mathrm{pars(b,a)}}$ in the *hello* message. When the neighbor node $a$ receives this message, it has a local time $t_a(t_1+\Delta(t_1)) = t_1+\delta_a(t_1+\Delta(t_1))+\Delta(t_1)$; it calculates the difference between its local time and the attached time $t_b(t_1)$, which represents the $\Delta_{\mathrm{pars(a,b)}} =$

$\delta_a(t_1 + \Delta(t_1)) + \Delta(t_1) - \delta_b(t_1)$. Then, node $a$ gets the average between its $\Delta_{\text{pars(a,b)}}$ and the attached $\Delta_{\text{pars(b,a)}}$ which represents the measured delay. By the assumption that $\delta_b(t_0 + \Delta(t_0)) = \delta_b(t_1)$ and $\delta_a(t_0) = \delta_a(t_1 + \Delta(t_1))$ (we suppose that the hardware clocks in $a$ and $b$ are perfect; i.e., there is no clock drift in $a$ and $b$) $(\Delta(t_0) + \Delta(t_1))/2$ represents the measured delay. It is clear that this measured delay has no information about $\Delta_a$ and $\Delta_b$ individually.

After calculating the measured delay, each node must smooth the measured delay due to variations of delay in the ad hoc networks. The average delay is calculated as follows.

$$\text{Average delay} = \alpha \times \text{average delay} + (1 - \alpha) \times \text{measured delay},$$

where $\alpha$ is a constant filter gain. In [4], we have shown that the best performance is achieved using $\alpha = 0.4$. The average delay is updated every time a new measurement is made. Forty percent of each new estimate is from the previous estimate and sixty percent is from the new measurement to put more weight on the actual topology.

## 3.2 Bandwidth Metric

The bandwidth measurement in the on-demand protocols is made when a new route is established between a source and a destination host. A remarkable work is presented in Kazantzidas and Gerla [9], considering the acknowledgment time from the data packets. We have started the bandwidth measurements based on their statements.

We consider in our proposal a linkstate paradigm. For our approach, the bandwidth is calculated between a node and its neighbors having a direct and symmetric link. For our analysis we consider data packets and signaling traffic that also use the available bandwidth and then must be taken into account (e.g., *hello* messages and traffic control messages in the OLSR protocol).

Measuring the throughput over the link in multihop networks, as presented in [9], demands the analysis of three aspects: the unique contention for each source–destination link pair, the distributed knowledge of the contention, and the rapid changes due to mobility.

For our purpose, as we have taken into account a linkstate paradigm, afterward the considered parameters are in a source node function and its neighbor, such as a hop-by-hop model, differing as to the final destination considered in the on-demand approach.

Available bandwidth on link $(i,j)$ is given by the following formula.

$$Bw_{(i,j)} = (1-u) \times Throughput_{(i,j)},$$

where $u$ is the link utilization.

The source–neighbor pair throughput is measured for a window of packets using existing traffic. Each successful packet transmission contributes its bits to the numerator of the throughput measurement and the time from when it was ready for transmission at the head of the link queues, to the acknowledgment receipt. Using IEEE 802.11b as the medium access control, this interval is packet size dependent as shown in the next equation:

$$Throughput_{packet} = \frac{S}{t_q + (t_S + t_{CA} + t_{Overhead}) \times R + \sum_{r=1}^{R} B_T},$$

where $S$ = packet size. In order to filter another measurement issue, we use a packet window. As performed in [9], we also use a packet window to increase statistical robustness of the measurements. We have seen in such work, that a packet window of 16 or 32 samples (packets) is adequate to produce fast enough, noise-immune measurements. The idle time and window duration are calculated to produce the link utilization factor and the permissible throughput measurement as:

$$\frac{\text{idle\_time\_in\_window}}{\text{window\_duration}} \times \text{Throughput\_measured}.$$

# 4 Routing Table Calculation

Let $G = (V,E)$ be the network with $|V| = n$ nodes and $|E| = m$ arcs and $\text{met}_{\text{ij}}$ a metric for link $(i,j)$. The value of a metric over any directed path $p = (i,j,k,\ldots,q,r)$ can be one of the following compositions.

- *Additive metrics:* We say metric met is additive if

  $$\text{met(p)} = \text{met}_{\text{ij}} + \text{met}_{\text{jk}} + \cdots + \text{met}_{\text{qr}}.$$

  It is obvious that delay (del), delay jitter (dej), hop-count (hop), and cost (co) follow the additive composition rule.

- *Multiplicative metrics:* We say metric met is multiplicative if

  $$\text{met(p)} = \text{met}_{\text{ij}} \times \text{met}_{\text{jk}} \times \cdots \times \text{met}_{\text{qr}}.$$

The probability of successful transmission (pst) follows the multiplicative composition rule. The composition rule for loss probability (Lp) is more complicated. It can be transformed to an equivalent metric pst,

$$\mathrm{Lp(p)} = 1 - ((1 - \mathrm{Lp_{ij}}) \times (1 - \mathrm{LP_{jk}}) \times \cdots \times (1 - \mathrm{Lp_{qr}})).$$

- *Concave metrics:* We say metric met is concave if $\mathrm{met(p)} = \min\{\mathrm{met_{ij}}, \mathrm{met_{jk}}, \ldots, \mathrm{met_{qr}}\}$. It is obvious that Bandwidth (Bw) follows the concave composition rule.

## 4.1 Single Metric Approach

In traditional data networks, routing protocols usually characterize the network with a single metric such as hop-count or delay, and use the shortest-path algorithms for path computation. For a delay metric, each arc $(i, j)$ in the path $p$ is assigned a real number $\mathrm{del_{ij}}$. When the arc $(i, j)$ is nonexistent or $j$ is not a MPR of $i$ (referring to the OLSR routing mechanism), then $\mathrm{del_{ij}} = \infty$. Let $\mathrm{del(p)} = \mathrm{del_{ij}} + \mathrm{del_{jk}} + \cdots + \mathrm{del_{qr}}$. The routing problem is to find a path $p*$ between $i$ and $r$ so that del(p∗) is the minimum. In such a case, we use the well-known Dijkstra routing algorithm. For the bandwidth metric, each arc $(i, j)$ in the path is assigned a real number $\mathrm{Bw_{ij}}$. When the arc $(i, j)$ is nonexistent or $j$ is not a MPR of $i$, $\mathrm{Bw_{ij}} = 0$. Let $\mathrm{Bw(p)} = \min\{\mathrm{Bw_{ij}}, \mathrm{Bw_{jk}}, \ldots, \mathrm{Bw_{qr}}\}$. The routing problem is to find a path $p*$ between $i$ and $r$ that maximizes Bw(p∗). In order to implement such a metric, we can use a variant Dijkstra algorithm.

The worst-case complexity of Dijkstra's algorithm on networks with nonnegative arc length depends on the way of finding the labeled node with the smallest distance label. A naive implementation that examines all labeled nodes to find the minimum runs in $O(n^2)$ time [19]. The implementation using $k$-ary heaps runs on $O(m \log n)$ time (for a constant $k$). The implementation using Fibonacci heaps runs in $O(m+n \log n)$ time. The implementation using one-level R-hrapsb runs in $O(m + (n \log C))$ time and the one using two-level R-heaps together with Fibonacci heaps, in $O(m + n\sqrt{\log C})$ time.

## 4.2 Multiple Metrics Approach

The second approach treats each metric individually. Such an approach is not feasible due to algorithm complexity. The problem of finding a path with

$n$ additive and $m$ multiplicative metrics is NP-complete if $n + m \geq 2$ [20]. Including a single metric, the best path can be easily defined. Otherwise including multiple metrics, the best path with all parameters at their optimal values may not exist.

We consider the bandwidth and delay routing problem. A path with both maximum bandwidth and minimum delay may not necessarily exist. Thus, we must decide the precedence among the metrics in order to define the best path. The delay has two basic components: queueing delay and propagation delay. The queueing delay is more dynamic and traffic-sensitive, thus bandwidth is often more critical for most multimedia applications. If there is no sufficient bandwidth, queueing delay and probably the loss rate will be very high. So, we define the precedence as bandwidth and then the propagation delay. Our strategy is to find a path with maximum bandwidth (a widest path), and when there is more than one widest path, we choose the one with the shortest delay. We refer to such a path as the shortest-widest path. The widest path problem is to find a path $p*$ between $i$ and $j$ that maximizes $\mathrm{Bw(p*)}$. For a given topology, there are usually many widest paths with equal width, and loops can be formed as a result. However, a shortest-widest path is always free of loops. Intuitively, the delay metric eliminates the loops.

**Theorem 4.1** *Shortest-widest paths are loop-free in a distributed computation.*

*Proof.* See [4]. □

Each arc $(i, j)$ in the path is assigned the following values: $\mathrm{del_{ij}}$, which is the propagation delay and $\mathrm{Bw_{ij}}$, which is the available bandwidth. When the arc $(i, j)$ is nonexistent or $j$ is not a MPR of $i$ (due to the OLSR routing mechanism), $\mathrm{del_{ij}} = \infty$ and $\mathrm{Bw_{ij}} = 0$. Let

$$\mathrm{del(p)} = \mathrm{del_{ij}} + \mathrm{del_{jk}} + \ldots + \mathrm{del_{qr}} \text{ and } \mathrm{Bw(p)} = \min\{\mathrm{Bw_{ij}}, \mathrm{Bw_{jk}}, \ldots, \mathrm{Bw_{qr}}\}.$$

The shortest-widest path algorithm [4, 5] based on the Dijkstra algorithm finds at each iteration a node with maximum width from a source $s$; if there is more than one widest path found, it chooses the one with minimum length. The time complexity of the shortest-widest path algorithm is equal to that of Dijkstra's shortest-path algorithm.

# 5 Heuristics for the Selection of Multipoint Relays

Finding a MPR set with minimal size falls in the category of the dominating set problem, which is known to be NP-complete [3]. The information needed to calculate the MPRs is the set of one-hop neighbors and two-hop neighbors. To select the MPRs for the node $x$, the following terminology is used in describing the heuristics.

- MPR(x): the multipoint relay set of node $x$ that is running this algorithm.
- N(x): the one-hop neighbor set of node $x$ containing only symmetric neighbors.
- N2(x): the two-hop neighbor set of node $x$ containing only symmetric neighbors in N(x). The two-hop neighbor set N2(x) of node $x$ does not contain any one-hop neighbor of node $x$.
- D(x, y): degree of one-hop neighbor node $y$ (where $y$ is a member of N(x)), is defined as the number of symmetric one-hop neighbors of node $y$ excluding the node $x$ and all the symmetric one-hop neighbors of node $x$; that is, $\mathrm{D}(x, y)$ = number of elements of $\mathrm{N}(y) - x - \mathrm{N}(x)$.
- Widest path: is a path with maximum bandwidth, calculated by the source node with its known partial network topology. In the widest path, any intermediate node is MPR of its previous node.
- Shortest-widest path: is the widest path, and with shortest delay when there is more than one widest path.
- Optimal widest path: is the widest path between two nodes in the whole network topology. Any node in the network can be selected as an intermediate node in the optimal widest path.
- Optimal shortest-widest path: is the shortest-widest path between two nodes in the whole network topology. Any optimal shortest-widest path is an optimal widest path.

The heuristic used in the standard OLSR protocol computes a MPR set of cardinality at most log$n$ times the optimal multipoint relay number, where $n$ is the number of nodes in the network. The approximation factor of the upper bound can be given by $\log\Delta$ where $\Delta$ is the maximum number of two-hop nodes a one-hop node may cover.

The standard OLSR heuristic limits the number of MPRs in the network and ensures that the overhead is as low as possible. However, in QoS routing by such a MPR selection mechanism, the good quality links may be hidden from other nodes in the network.

**Theorem 5.1** *There is no guarantee that OLSR finds the optimal shortest-widest or optimal widest path.*

*Proof.* (1) By construction. The heuristic for the selection of multipoint relays in the standard OLSR does not take into account the bandwidth and delay information. It computes a multipoint relay set of minimal cardinality. So, the links with high bandwidth and low delay can be omitted. After, the path calculated between two nodes using the shortest-widest path algorithm has no guarantee that it is the optimal widest path or shortest-widest path in the whole network. (2) By example. From Figure 3 and Table 1 we have: when $g$ is building its routing table, for destination $a$, it will select the route $(g, f, b, a)$ whose bandwidth is 5. The optimal widest path between $g$ and $a$ is $(g, f, d, a)$. It has 100 as bandwidth. □

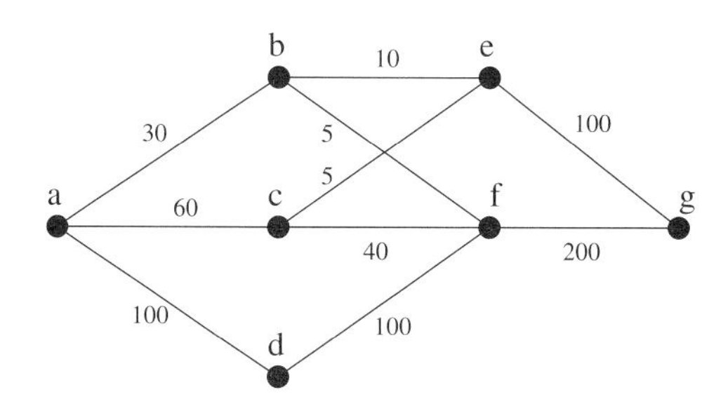


Figure 3: Network example for MPR selection.

The decision of how each node selects its MPRs is essential for determining the optimal bandwidth and delay route in the network. In the MPR selection, the links with high bandwidth and low delay should not be omitted.

Table 1: MPR selected in the standard OLSR.

| Node | 1-Hop Neighbors | 2-Hop Neighbors | MPRs |
|---|---|---|---|
| a | b, c, d | e, f | b |
| b | a, e, f | c, d, g | f |
| c | a, e, f | b, d, g | f |
| d | a, f | b, c, g | f |
| e | b, c, g | a, f | b |
| f | b, c, d, g | a, e | c |
| g | e, f | b, c, d | f |

## 5.1 QOLSR_MPR1

In this protocol, MPR selection is almost the same as that of the standard OLSR. However, when there is more than one one-hop neighbor covering the same number of uncovered two-hop neighbors, the one with maximum bandwidth link (a widest link) to the current node is selected as the MPR. If there is more than one widest link, we choose the one with the shortest delay.

This heuristic has the same time complexity of the standard OLSR heuristic. It computes a MPR set of cardinality at most log $n$ times the optimal multipoint relay number where $n$ is the number of nodes in the network.

**Theorem 5.2** *There is no guarantee that QOLSR_MPR1 finds the optimal shortest-widest or optimal widest path.*

*Proof.* (1) By construction. The heuristic for the selection of multipoint relays in the QOLSR_MPR1 is almost the same as that of the standard OLSR. We use the bandwidth and delay information when there is more than one one-hop neighbor covering the same number of uncovered two-hop neighbors. So, the links with high bandwidth and low delay can be omitted. (2) By example. From Figure 3 and Table 2, we have: between $b$ and $c$, $c$ is selected as $a$'s MPR because it has the larger bandwidth. When $g$ is building its routing table, for destination $a$, it will select the route $(g, f, c, a)$ whose bandwidth is 40. The optimal widest path between $g$ and $a$ is $(g, f, d, a)$. It has 100 as bandwidth.

Table 2: MPR selected in the QOLSR_MPR1.

| Node | 1-Hop Neighbors M | 2-Hop Neighbors | MPRs |
|---|---|---|---|
| a | b, c, d | e, f | c |

□

## 5.2 QOLSR_MPR2

In this protocol, neighbors that guarantee maximum bandwidth and minimum delay among two-hop neighbors are selected as MPRs. The heuristic used in QOLSR_MPR2 protocol is as follows.

**Step 1:** Start with an empty multipoint relay set MPR(x).

**Step 2:** Calculate D(x, y), $\forall$ nodes $y \in$ N(x).

**Step 3:** First, select those one-hop neighbor nodes in N(x) as the multipoint relays that provide the only path to reach some nodes in N2(x), and add these one-hop neighbor nodes to the multipoint relay set MPR(x).

**Step 4:** While there still exist some nodes in N2(x) that are not covered by the multipoint relay set MPR(x):

- **Step 4.a:** For each node in N(x) that is not in MPR(x), calculate the number of nodes that are reachable through it among the nodes in N2(x) and which are not yet covered by MPR(x).

- **Step 4.b:** Select that node of N(x) with the maximum bandwidth and minimum delay as a MPR.

- **Step 4.c:** In case of a tie in the above step, select that node which reaches the maximum number of uncovered nodes in N2(x).

**Claim 5.3** *Let $p = (a_1, \ldots, a_{i-1}, a_i, a_{i+1}, \ldots, a_k)$ an optimal widest path, $k \geq 3$. For any intermediate node $a_i$ $(i \neq 1)$ in $p$ that is not selected as MPR by its previous node $a_{i-1}$, we can find a node $b_i$ selected as MPR by $a_{i-1}$ such that the path $(a_1, \ldots, a_{i-1}, b_i, a_{i+1}, \ldots, a_k)$ has the same bandwidth performance.*

*Proof.* Let $p = (a_1, \ldots, a_{i-1}, a_i, a_{i+1}, \ldots, a_k)$, $k \geq 3$ an optimal widest path from $a_i$ to $a_k$ (Figure 4).

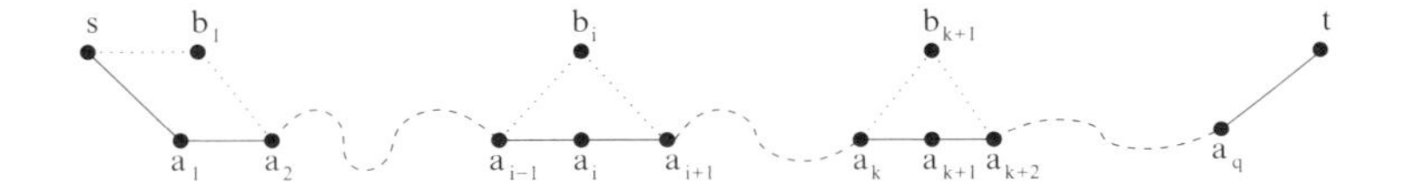


Figure 4: Optimal widest path from $s$ to $t$.

Suppose that on the optimal widest path, the node $a_i$ is not selected as MPR by its previous node $a_{i-1}$. We can assume that for each node on the path, its next node in the path is its one-hop neighbor, and the node two hops away from it is its two-hop neighbor. For example, $a_i$ is $a_{i-1}$'s one-hop neighbor, and $a_{i+1}$ is $a_{i-1}$'s two-hop neighbor. Based on the basic idea of the MPR selection that all the two-hop neighbors of a node should be covered by this node's MPR set, $a_{i-1}$ must have another neighbor $b_i$, which is selected as its MPR, and is connected to $a_{i+1}$. Let $p' = (a_1, \ldots, a_{i-1}, b_i, a_{i+1}, \ldots, a_k)$, $k \geq 3$. According to the criteria of MPR selection specified in QOLSR_MPR2, $a_{i-1}$ selects $b_i$ instead of $a_i$ as its MPR because:

$$\mathrm{Bw}_{\mathrm{a_{i-1}b_ia_{i+1}}} > \mathrm{Bw}_{\mathrm{a_{i-1}a_ia_{i+1}}}. \tag{1}$$

Or

$$\mathrm{Bw}_{\mathrm{a_{i-1}b_ia_{i+1}}} = \mathrm{Bw}_{\mathrm{a_{i-1}a_ia_{i+1}}} \mathrm{del}_{\mathrm{a_{i-1}b_ia_{i+1}}} < \mathrm{del}_{\mathrm{a_{i-1}a_ia_{i+1}}}. \tag{2}$$

From (1) we have $\mathrm{Bw(p')} \geq \mathrm{Bw(p)}$ and there is no guarantee about $\mathrm{del(p')} \geq \mathrm{del(p)}$.

From (2) we have

$$\mathrm{Bw(p')} = \mathrm{Bw(p)} \mathrm{del(p')} < \mathrm{del(p)}. \tag{3}$$

In both cases, $\mathrm{Bw(p')} \geq \mathrm{Bw(p)}$. Based on our assumption, path $p$ is an optimal widest path. So, path $p'$ is also optimal widest. This completes the proof. □

**Claim 5.4** *There is an optimal widest path in the whole network such that all the intermediate nodes are selected as MPR by their previous nodes.*

*Proof.* By a recurrence. Let $p = (s, a_1, \ldots, a_{i-1}, a_i, a_{i+1}, \ldots, a_k, \ldots, a_q, t)$, $k < q$ an optimal widest path (Figure 4).

(a) We demonstrate that the first intermediate node $a_1$ is selected as MPR by source $s$. By using claim 1, we can find a node $b_1$ selected as MPR by $s$ such as the path $p' = (s, b_1, \ldots, a_{i-1}, a_i, a_{i+1}, \ldots, a_k, \ldots, a_q, t)$ has the same bandwidth performance of the optimal path ($p'$ is also an optimal widest path). So, the source's MPRs are on the optimal widest path.

(b) We assume that all the nodes $\{a_1, \ldots, a_{i-1}, a_i, a_{i+1}, \ldots, a_k\}$ are selected as MPR by their previous node in the path $p$. We prove that the next hop node of $a_k$ on $p$ is $a_k$'s MPR. Suppose that $a_{k+1}$ is not an MPR of $a_k$. The same as above, by using claim 1, we can find a node $b_{k+1}$ selected as MPR by $a_k$ such as the path $p' = (s, a_1, \ldots, a_{i-1}, a_i, a_{i+1}, \ldots, a_k, b_{k+1}, \ldots, a_q, t)$ has the same bandwidth performance of the optimal widest path ($p'$ is also an optimal widest path). So, in an optimal widest route, the $(k+1)$th intermediate node is the MPR of the $(k)$th intermediate node.

Based on (a) and (b), all the intermediate nodes of an optimal widest path are the MPRs of the previous nodes. □

By claim 5.4, there is an optimal widest path such that all the intermediate nodes are the MPR of the previous nodes on the same path. So the optimal widest path for the whole network topology is included in the partial topology the node knows. And by using the shortest-widest path algorithm, we can compute the optimal widest path in the partial network topology. We can conclude that the QOLSR_MPR2 finds the optimal widest path.

**Theorem 5.5** *QOLSR_MPR2 finds optimal widest paths using only the known partial network topology.*

The heuristic used in the QOLSR_MPR2 finds exactly the optimal MPRs that guarantee maximum bandwidth and minimum delay. So, this heuristic is an algorithm. The upper bound of the time complexity of this algorithm is $O(\alpha)$ where $\alpha$ is the maximum number of two-hop nodes.

# 6 Performance Evaluation

In this section, we evaluate the performance of the QOLSR protocol and its extensions applying multiple metrics (bandwidth and delay) and the different MPR selection algorithms in different configurations and scenarios. We use the OPNET simulator for our evaluation.

## 6.1 Heuristic Evaluation in Static Networks

### 6.1.1 MPR Selection Simulation Model and Results

We assume that the ad hoc network topology is stable (a wireless network consisting of desktops, laptops, and printers for home business may keep its original topology for a long time until someone moves one of the laptops to another room). We generate 100 random networks of 100 nodes. Each node is placed in an area of 1000 $m \times 1000\ m$ randomly selecting its $x$ and $y$ coordinates. Each node is randomly assigned an idle_time ranging from 0 to 1. Each link has the same bandwidth, 2 Mb/s. The available link bandwidth between two nodes is equal to the minimum of their idle_time × max_bandwidth.

In our simulated scenarios, we collect results over three values of range transmission (100 m, 200 m, 300 m). Table 3 shows the average number of one-hop neighbors and two-hop neighbors. We can see that when the range transmission decreases, the number of one- or two-hop neighbors decreases. These values affect the MPR number in the network. By assuming high connectivity of the network: (1) the more one-hop neighbors a node has, the fewer MPRs it may select, because with a high probability a small subset of its one-hop neighbor can reach a high number of the two-hop neighbors; (2) the more two-hop neighbors a node has, the more MPRs may be needed to cover them all.

Table 3: Average number of (1, 2)-hop neighbors.

| Transmission Range | 300 m | 200 m | 100 m |
|---|---|---|---|
| 1-hop neighbors | 21 | 10 | 2 |
| 2-hop neighbors | 33 | 15 | 4 |

The next results show the performance of the routes found by the implemented algorithms (standard OLSR, QOLSR_MPR1, QOLSR_MPR2, pure link state algorithm: each node floods its link state information into the entire network). The results are given in two categories: performance and cost. Performance is characterized by: (a) error rate: the percentage of the bad routes (bandwidth not optimal), and (b) average difference: the average of the difference between the optimal bandwidth and current bandwidth found in routing algorithms in percentage. The larger the value is, the worse is the result. Cost is measured by: (a) overhead: average number of the TC messages transmitted in the network, and (b) MPR number: average number of the MPRs in the network.

Table 4: Performance and cost.

| Algorithm | Transmission | Performance | | Cost | |
|---|---|---|---|---|---|
| | Range (m) | Error rate (%) | Average Difference (%) | Overhead | Nb MPR |
| Standard OLSR | 300 | 28 | 46 | 12 | 65 |
| | 200 | 41 | 51 | 24 | 68 |
| | 100 | 12 | 45 | 5 | 42 |
| QOLSR_MPR1 | 300 | 14 | 22 | 12 | 65 |
| | 200 | 21 | 26 | 24 | 68 |
| | 100 | 8 | 44 | 5 | 42 |
| QOLSR_MPR2 | 300 | 0 | 0 | 26 | 71 |
| | 200 | 0 | 0 | 38 | 73 |
| | 100 | 0 | 0 | 5.7 | 44 |
| Pure link state | 300 | 0 | 0 | 1245 | 100 |
| | 200 | 0 | 0 | 979 | 100 |
| | 100 | 0 | 0 | 28 | 100 |

Table 4 shows that for each transmission range the standard OLSR has the worst performance (it has the highest error rate and average difference). The bandwidth difference between the paths found by the standard OLSR and the optimal paths is large. QOLSR_MPR1 achieves a large improvement in performance compared to the standard OLSR. The explanation is that the shortest-widest path algorithm enhances the bandwidth of the found paths. However, QOLSR_MPR1 does not always find an optimal path, in as much as its MPR selection heuristic may omit the optimal bandwidth link from the partial network topology the node learned. QOLSR_MPR2 achieves the best performance each time (it finds the optimal bandwidth route).

The cost is directly related to the number of the retransmitting nodes. If the number of the retransmitting nodes increases, the cost increases. The pure link state algorithm has the highest overhead, because each node retransmits the messages it receives. As the MPR selection heuristic in the standard OLSR and QOLSR_MPR1 emphasizes reducing the number of MPRs in the network, the standard OLSR and QOLSR_MPR1 have the same and the lowest MPR number, and therefore the lowest overhead compared with QOLSR_MPR2 and the pure link state algorithm. QOLSR_MPR2 selects more MPRs, so there is more overhead than the standard OLSR and QOLSR_MPR1.

We can see that the standard OLSR and QOLSR_MPR1 and QOLSR_MPR2 have more MPRs and thus more overhead with a transmission range of 200 $m$. In a higher density network (such as for a node transmission range of 300 $m$), node connectivity is also high (see Table 3), so a node may need fewer MPRs to cover its 2-hop neighbors. In a lower density network (such as for a node transmission range of 100 $m$), because of the lower connectivity, a node may have fewer two-hop neighbors; therefore, it also needs fewer MPRs. However, the transmission range of 200 $m$ falls within these two extremes, so it may well result in the largest number of MPRs to produce the highest overhead. This situation is not found in the pure link state algorithm, where a node's entire neighbor set is its MPR set.

### 6.1.2 Performances in Varying Load Conditions

In this simulation, we study the behavior of the QOLSR, QOLSR_MPR1, and QOLSR_MPR2 protocols and the maximum utilization of the bandwidth by varying the load in the network. We have taken a static network composed of 50 nodes without unidirectional links. This network operates in a stable state when each node has complete and correct information about the network. All nodes are packet-generating sources. We have taken the mean packet size as 1 K bytes, 200 packets for each transmit buffer, and the same size as the receive buffer. We increase the data packet arrival rate from 100 packets per second (which represents approximately 100 k bytes per second) up to 1400 per second. With the arrival rate equal to 100 packets per second, each node generates 2 packets per anasecond in the average.

Figure 5 plots the data load delivered to the destination. As the average route length is three hops, the maximum throughput of data that we can obtain is therefore

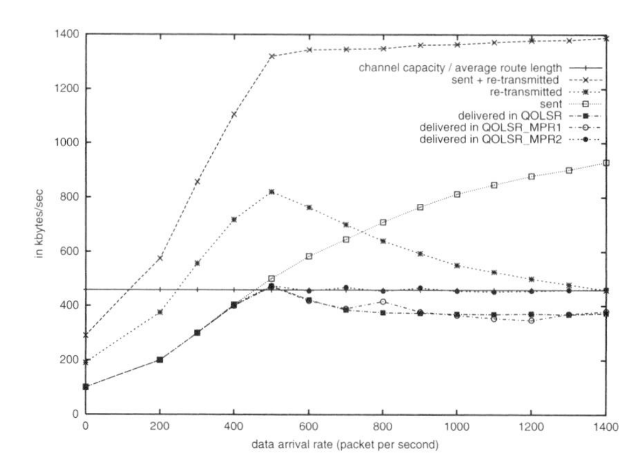


Figure 5: Data load transmitted in varying load conditions.

$$\begin{aligned} \text{max throughput} &= \frac{\text{channel capacity}}{\text{average route length}} \\ &= \frac{(11000 \text{ kbps}/8)}{3} = 458.3 \text{ K bytes/s.} \end{aligned}$$

We can see that the network attains almost the maximum throughput before saturation for both protocols. The drop in throughput after the saturation point using QOLSR and QOLSR_MPR1 is due to the absence of any congestion control mechanism as each node continues to generate data packets at a high rate. However, the throughput after saturation using the QOLSR_MPR2 remains stable in the maximum utilization bandwidth because all the paths founded are optimal widest paths and are between nodes that have enough space in their buffers.

## 6.2 Heuristic Evaluation in Mobile Networks

The simulation model introduced in [21] is very close to real ad-hoc network operations. At each time, we can detect the position of mobiles by our mobility model. Each node is represented by a subqueue and placed in the region by randomly selecting its $x$ and $y$ coordinates. The number of nodes can reach 100,000. With our method, the simulation model is very optimized, enabling us to reduce the CPU time and consequently to increase the time of simulation.

The random mobility model proposed is a continuous-time stochastic process. Each node's movement consists of a sequence of random length intervals, during which a node moves in a constant direction at a constant speed. A detailed description can be found in [21].

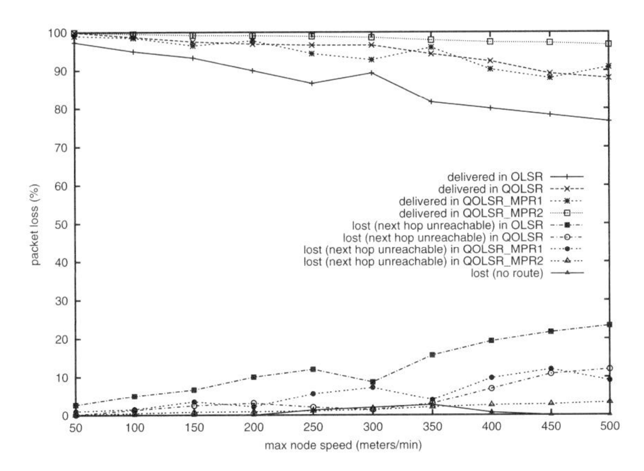


Figure 6: Data load transmitted at varying node speeds.

Figure 6 shows the results of our simulation in which the data packets sent and successfully delivered are plotted against the increasing speed. The speed is increased from 50 m/min (3 Km/hr) up to 500 m/min (30 Km/hr). In this simulation, 50 nodes constitute the network in a region of $1000^2$ $m^2$, and all 50 nodes are packet-generating sources. We also keep the movement probability as 0.3 (i.e., only 20% of nodes are mobile and the rest are stationary). Each mobile node selects its speed and direction which remains valid for the next 60 seconds. We can see that when the mobility (or speed) increases, the number of packets delivered to the destinations decreases. This can be explained by the fact that when a node moves, it goes out of the neighborhood of a node which may be sending it the data packets. There are about 99.92% of packets delivered for QOLSR at a mobility of 2 meters/minute (99.01% for QOLSR_MPR1 and 99.99% for QOLSR_MPR2). At a mobility of 500 meters/minute, 88% of packets are delivered for QOLSR (90.9% for QOLSR_MPR1 and about 97% for OLSR_MPR2). QOLSR_MPR2 has the highest number of packets delivered because the routes are optimal and chosen with minimal interference. The data packets are lost because the next-hop node is unreachable. QOLSR with the classic MPR selection algorithm and QOLSR_MPR1 have the same performance in terms of lost packets. A node keeps an entry about its neighbor in its neighbor table for about 6 seconds. If a neighbor moves, which is the next-hop node in a route, the node continues to forward it the data packets considering it as a neighbor. Also, the next-hop is unreachable if there is interference. A few of the packets are also lost because of unavailability of routes and it is the same for OLSR with or without QoS. This happens when a node movement causes the node to be disconnected from the network temporarily, until it rejoins the network again.

## 7 Conclusions

This chapter presented a link state QoS routing protocol for ad hoc networks. In order to improve quality requirements in routing information, delay and bandwidth measurements are applied. The implications of routing metrics on path computation are examined and the rationales behind the selection of bandwidth and delay metrics are discussed. The heuristic used in the standard OLSR finds a MPR set with minimal size. There is no guarantee that OLSR finds the optimal widest path. We have proposed two heuristics that allow OLSR to find the maximum bandwidth path. In order to improve quality requirements in MPR selection and also in routing information, delay and bandwidth measurements are applied. Delay and bandwidth are calculated between each node and its neighbors having a direct and symmetric link. We have demonstrated and also shown by simulations that QOLSR_MPR2 finds optimal widest paths using only the known partial network topology. From the analysis of the static and mobile network simulation, QOLSR and QOLSR_MPR1 present the same performance.

## References

[1] T. Clausen and P. Jacquet, *Optimized Link State Routing Protocol*, RFC 3626, October 2003.

[2] *http://www.ietf.org/html.charters/manet-charter.html.*

[3] A. Qayyum, L. Viennot, and A. Laouiti, *Multipoint relaying technique for flooding broadcast message in mobile wireless networks.* In Hawaii International Conference on System Sciences, Hawaii, January 2002.

[4] A. Munaretto, H. Badis, K. Al Agha, and G. Pujolle, *QoS for ad hoc networking based on multiple metrics: bandwidth and delay.* In IEEE MWCN2003, Singapore, October 2003.

[5] H. Badis, A. Munaretto, K. Al Agha and G. Pujolle, *Optimal path selection in a link state QoS routing protocol.* In IEEE VTC2004 Spring, May 2004.

[6] W. Chen, J. T. Tsai and M. Gerla, *QoS routing performance in multi-hop, multimedia, wireless networks.* In IEEE International Conference on Universal Personal Communications, October 1997, pp. 451–557.

[7] C. Perkins and P. Bhagwat, *Highly dynamic destination-sequenced distance-vector routing (DSDV) for mobile computers.* In Association for Computing Machinery's Special Interest Group on Data Communication, 1994, pp 234–244.

[8] C. R. Lin and J. S. Liu, *QoS routing in ad-hoc wireless networks*, IEEE Journal on Selected Areas in Communications, 17(8) 1999 pp. 1426–1438.

[9] M. Kazantzidis and M. Gerla, *End-to-end versus explicit feedback measurement in 802.11 networks.* In IEEE Symposium on Computers and Communications, 2002.

[10] Z. Wang and J. Crowcroft, *Quality of service routing for supporting multimedia applications*, IEEE Journal on Selected Areas in Communications 14(7) 1996 pp. 1228–1234.

[11] S. Chen, *Routing support for providing guaranteed end-to-end quality of service.* PhD, Engineering College, Urbana, IL, 1999.

[12] M. Al-Fawaz and M. E. Woodward, *Fast quality of service routing algorithms with multiple constraints.* In IFIP Workshop on ATM&IP, Ilkely, 2000.

[13] L. H. Costa, S. Fdida, and M. B. Duarte, *Distance-vector QoS-based routing with three metrics*, NETWORKING 2000, pp. 847–858.

[14] S. Cheng and K. Nahrstedt, *On finding multi-constrained paths.* In ICC'98, Atlanta, 1998.

[15] H. Neve and P. Mieghem, *A multiple quality of service routing algorithm for PNNI.* In IEEE ATM'98 Workshop, Fairfax, VA, 1998, pp. 306–314.

[16] L. Guo and I. Matta, *Search space reduction in QoS routing*, Technical Report NU-CCS-98-09, 1998.

[17] X. Yuan and X. Liu, *Heuristic algorithms for multi-constrained quality of service routing.* In Infocom 2001, pp. 844–853.

[18] T. Korkmaz and M. Krunz, *Multi-constrained optimal path selection.* In Infocom 2001, pp. 834–843.

[19] A. S. Tanenbaum, *Computer Networks*, Prentice Hall, 1996.

[20] F. Kuipers, P. Van Mieghem, T. Korkmaz and M. Krunz, *An overview of constraint-based path selection algorithms for QoS routing*, IEEE Communications 40(12), 2002.

[21] H. Badis and K. Al Agha, *Scalable model for the simulation of OLSR and Fast-OLSR protocol*, Med-hoc-Net, June 2003.

Chapter 3

# Analyzing Voice Transmission Between Ad Hoc Networks and Fixed IP Networks Providing End-to-End Quality of Service

Mari Carmen Domingo
*Department of Telematics Engineering, Technical University of Catalonia (UPC)*
*Av. del Canal Olímpic s/n. 08860 Castelldefels (Barcelona), SPAIN*
E-mail: cdomingo@mat.upc.es

David Remondo
*Department of Telematics Engineering, Technical University of Catalonia (UPC)*
*Av. del Canal Olímpic s/n. 08860 Castelldefels (Barcelona), SPAIN*
E-mail: remondo@mat.upc.es

## 1 Introduction

In the last several years there has been growing interest in the integration of wireless networks and the Internet. Making VoIP available over Wireless Local Area Networks (WLANs) has made it possible to develop some applications such as voice-enabled Personal Digital Assistants (PDAs) or wireless phones in the enterprise. These wireless VoIP systems are ideal for hospitals, retail stores, and other corporations that want to keep their employees connected. However, as VoIP moves into the wireless world, performance issues arise because some Quality of Service (QoS) parameters such as delay

M.X. Cheng, D. Li (eds.) *Advances in Wireless Ad Hoc and Sensor Networks.*
Signals and Communication Technology, doi: 10.1007/978-0-387-68567-0_3.

are magnified in an IEEE 802.11 WLAN environment due to access point congestion in infrastructure-based WLANs and link quality in ad hoc as well as infrastructure-based wireless networks. The three most important factors that affect speech quality are packet loss, delay, and jitter. Packet loss issues encountered in IP applications can reduce the quality of voice calls below acceptable levels. Dropped or excessively delayed packets in real-time voice communication result in unsatisfactory voice quality. All these requirements make the Distributed Coordination Function (DCF) scheme an infeasible option to support quality of service for real-time traffic. Therefore the goal would be to improve the current standards with additions that take into account the different requirements of regular data traffic and time-sensitive voice traffic.

We focus our research on providing service differentiation in wireless ad hoc networks. Ad hoc wireless networks are generally mobile networks, with a dynamic topology, where the nodes are connected through radio links and configure themselves on the fly without the intervention of a system administrator.

In an ad hoc network it is possible that two nodes communicate even when they are outside their radio ranges because the intermediate nodes can function as routers, forwarding data packets from the source to the destination node. This is also known as a multihop wireless network.

Some authors have presented several proposals to support QoS in wireless ad hoc networks including QoS MAC protocols [30–36], QoS routing protocols [1–4], and resource reservation protocols [5]. Moreover, in [6] a Flexible QoS Model for Mobile ad hoc networks (FQMM) is proposed.

Our objective is to study the interworking between a mobile ad hoc network and the Internet, extending the Differentiated Services (DiffServ) model to a wireless environment. For this reason a scenario where an ad hoc network is connected via a single gateway to a fixed IP network has been chosen. Moreover, to provide QoS and differentiate the service level between applications the fixed IP network supports the DiffServ architecture. We consider a single DS (DiffServ) domain covering the whole network between the wired corresponding hosts and the gateway. The ad hoc network incorporates the SWAN model to provide QoS. In [26] and [27] the behavior of Const Bit Rate (CBR) voice traffic is studied, but voice transmission of Variable Bit Rate (VBR) VoIP has not been yet analyzed. There is some work related to voice transmission in IEEE 802.11 but only very few in the ad hoc mode [7]. To our knowledge, there has been little or no prior work on analyzing the voice transmission capacity between an ad hoc network and a fixed IP network providing end-to-end QoS.

The chapter is structured as follows. Section 2 explains how to provide Internet access to mobile ad hoc networks modifying the ad hoc on-demand distance vector routing protocol. Section 3 introduces the DiffServ architecture. Section 4 describes related work about how to apply QoS in mobile ad hoc networks modifying the DiffServ model, introducing the SWAN model, or analyzing some service differentiation mechanisms that modify the IEEE 802.11 DCF protocol. Section 5 analyzes voice transmission in ad hoc networks. Section 6 presents the modified version from the SWAN model named DS-SWAN (Differentiated-Services SWAN). Section 7 shows our simulation results. Finally, Section 8 concludes this chapter.

## 2 Internet Connectivity with Mobile Ad Hoc Networks

The aim is for a fixed IP and an ad hoc network to communicate. The Internet draft Global Connectivity for IPv6 Mobile Ad Hoc Networks [8] describes how to provide Internet access to mobile ad hoc networks modifying the Ad Hoc On-demand Distance Vector (AODV) routing protocol [9]. IPv6 enables a wider range of Internet-connected devices because it provides several advantages over IPv4 such as much larger address space and support for stateless address autoconfiguration.

In order to communicate, the ad hoc and the Internet network packets must be transmitted to a gateway as illustrated in Figure 1. This device must be able to implement both the ad hoc network protocol stack and the protocols of the fixed Internet, routing the packets from one network to the other.

The protocol stacks used by mobile nodes, gateways, and Internet nodes are shown in Figure 2.

Consider that a mobile node needs to send data to the Internet. It needs a global address to send these data. Furthermore, it needs to discover the location and address of the gateway and use the route to this device as the default route for sending packets to the Internet. We have chosen a hybrid gateway discovery method [10] to find a gateway. This method combines proactive and reactive gateway discovery minimizing the disadvantages of using these two methods separately. The hybrid gateway discovery defines a transmission range where the gateway periodically sends a RREP_I message (an extended Route Reply where the I-flag (Internet-Global Information

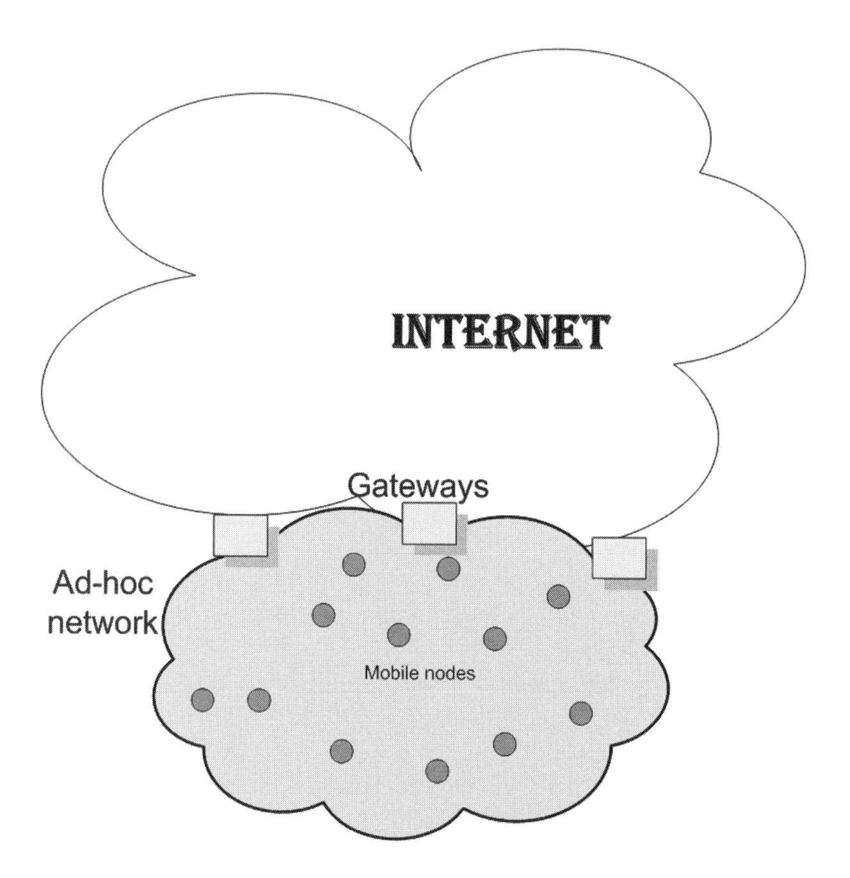


Figure 1: Interworking scenario.

Flag) is used for global address resolution). The RREP_I contains gateway information and it is propagated around a limited zone. A mobile node receiving a RREP_I must use the information about the global prefix length and the IPv6 address from the gateway carried in this message to discover the global prefix. Afterwards this mobile node autoconfigures a new routable IPv6 address and selects the gateway address as the default route.

If a mobile node outside the gateway transmission range wants Internet connectivity, it broadcasts a RREQ_I message (an extended Route Request where the I-flag is used for global address resolution) to the Internet gateway multicast group, that is, the IP address for the group of gateways in the ad hoc network. The mobile node can use any available global address as source address (e.g., its Mobile IPv6 home address) or it can create a new

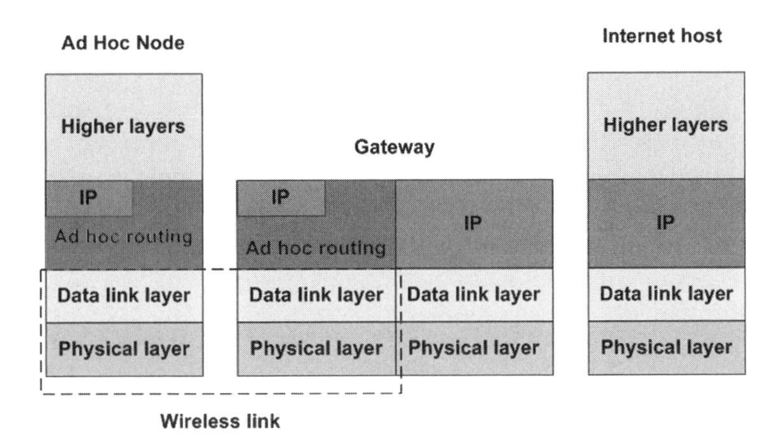


Figure 2: Protocol architecture.

temporary one using the MANET_INITIAL_PREFIX as described in the IP Address Autoconfiguration for Ad Hoc Networks [11]. If another mobile node receives this RREQ_I, it rebroadcasts it until the RREQ_I arrives at a gateway that responds, sending back a RREP_I. Then the source node deletes the temporary address and obtains the globally routable IPv6 address from the gateway.

Now we assume that a source node S in the ad hoc network wants to send packets to a destination node D and does not know whether this node is located in the ad hoc network or on the Internet [12]. Thus, the node S first consults its routing table and uses a route to this destination if it exists. Otherwise the node sends a RREQ waiting for a RREP. However, node S will not receive any RREP if node D is a fixed node. In this case the source node must send the packet using the default route (if it exists; otherwise the node must obtain it using the already explained method) and rely on the gateway to deliver the packet.

# 3 The Differentiated Services Architecture

The objective is to provide Internet access to an ad hoc network. The fixed IP network supports DiffServ and we want to extend the DiffServ model to a wireless environment. The DiffServ architecture has been defined by the IETF (Internet Engineering Task Force) [13,14] to provide scalable services differentiation in the Internet and to offer different services to customers and applications according to their requirements.

This architecture defines the DS-region, an area that is composed of one or more DS-domains (perhaps under different administrative authorities). A DS-domain (see Figure 3) can be defined as a contiguous portion of the Internet over which a consistent set of DS policies is administered. It is made up of two parts: the core network formed by core routers and the access network where an edge router (a DS ingress or DS egress node) connects one DS-domain to a node either in another DS-domain or in a domain that is not DS-capable.

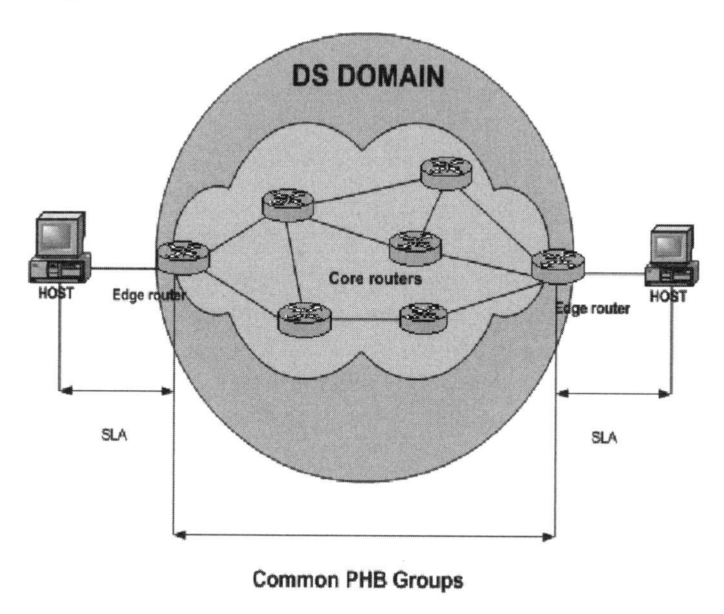


Figure 3: Basic elements of the Differentiated Services network.

The main elements of traffic handling in the edge routers at the boundary of the network are classifier, meter, marker, shaper, and dropper (see Figure 4). The edge routers classify the incoming packets according to a policy specified by the Service Level Agreement (SLA). This document is a kind of contract between the customer and its Internet Service Provider (ISP) where the customers communicate their requirements and it addresses not only technical issues such as traffic conditioning or service availability but also business aspects such as pricing. A Traffic Conditioning Agreement (TCA) is a subset of a SLA and it addresses traffic classification and conditioning rules to be applied to a traffic stream entering the network. A flow [15] can be designated by source, destination, maximum rate, and any combination of header fields (port number, etc.). A meter checks that the incoming packets from a flow accomplish certain traffic parameters. Then the flows are aggregated with a small number of different traffic classes by marking the Differentiated Services Code Point (DSCP) field in the IP header so that this state reflects the desired level of service. The policing and shaping mechanisms ensure that the traffic fulfills the specifications from the SLAs.

A particular marking on a packet indicates a PHB (Per-Hop Behavior) that is applied when the packet is forwarded. For this reason the core routers examine the incoming packet's codepoint and decide to forward the packet according to the Per-Hop Behavior associated with it because the PHB field defines the priority class of the packet. These routers do not keep per-flow state information and process the incoming packets very quickly because there are a limited number of service classes defined by the DSCP field.

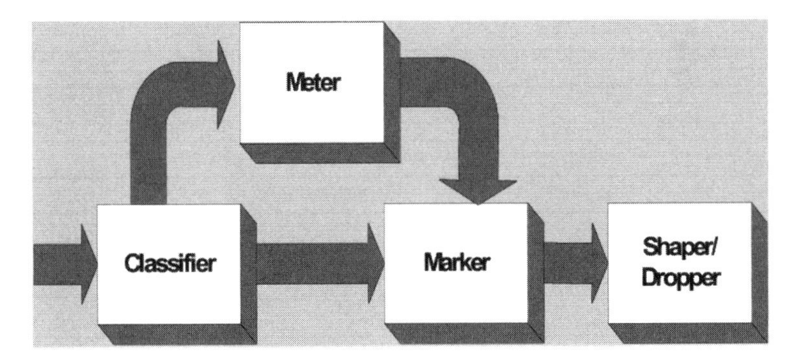


Figure 4: The main blocks of traffic handling.

One service class is the EF (Expedited Forwarding) PHB [16,17], that provides low loss, low latency, low jitter, and end-to-end assured bandwidth service. It provides a premium service. On the other hand, the AF (Assured Forwarding) PHB [18] defines a group of codepoints that can be used to define four classes of traffic, each of which has three drop precedences (see Table 1). An IP packet that belongs to an AF class i and has drop precedence j is marked with the AF codepoint AFij, where $1 <= i <= N$ and $1 <= j <= M$. Currently, four classes ($N = 4$) with three levels of drop precedence in each class ($M = 3$) are defined for general use but it is possible to define more AF classes and levels of drop precedence for local use. Within each AF class an IP packet can be assigned one of the three levels of drop precedence. The drop precedence value is used in the case of congestion to determine

Table 1: DiffServ AF codepoint table.

| DROP PRECEDENCE | Class #1 | Class #2 | Class#3 | Class #4 |
|---|---|---|---|---|
| Low Drop Prec | (AF11) 001010 | (AF21) 010010 | (AF31) 011010 | (AF41) 100010 |
| Medium Drop Prec | (AF12) 001100 | (AF22) 010100 | (AF32) 011100 | (AF42) 100100 |
| High Drop Prec | (AF13) 001110 | (AF23) 010110 | (AF33) 011110 | (AF43) 100110 |

which packets must be first dropped (those with a higher value), protecting packets with a lower drop precedence from being lost. Each class is assigned a certain amount of resources such as buffer space and bandwidth. These classes provide Assured Service (AS) [19,20], which means that the user of such a service receives the assurance that such traffic is unlikely to be dropped as long as it stays within the expected capacity profile. Behavior aggregates can be given different forwarding assurances. Three of the four currently defined classes of the AF PHB group are used for Olympic Service [21], so that traffic can be divided into gold (which has a large share of the link), silver (intermediate share), and bronze (lowest share) classes. Finally, BE (Best-Effort) defines a class where no QoS is guaranteed.

# 4 QoS in Mobile Ad Hoc Networks

The support of quality of service in mobile ad hoc networks is an important challenge. We focus on three active research lines: the DiffServ model applied to ad hoc networks, the SWAN model, and the service differentiation mechanisms in wireless networks modifying the IEEE 802.11 DCF protocol.

## 4.1 The DiffServ Model Applied to Ad Hoc Networks

There has been some research trying to adapt DiffServ (originally designed for wired and relatively high-speed networks) to mobile wireless ad hoc networks. Some problems have been found:

- Every node [22] should be able to act as an ingress node (when it sends data as a source node) and to act as core router (when it forwards packets from others as an intermediate node). These functioning modes have a heavy storage cost.

- The service level agreement where a service profile for aggregated flows is specified does not exist in wireless ad hoc networks and it is complicated to establish traffic rules between mobile nodes in such a kind of network.

- In a mobile environment [23] it is difficult to provide a certain QoS because the network topology changes dynamically and in wireless networks the packet loss rates are much higher and more variable than in wired networks.

Some authors [24] have adapted the DiffServ model for mobile ad hoc networks modifying the interface queues between the LL and the Mac Layer, setting two different scheduling disciplines (priority and round-robin scheduling) and distinguishing between two different classes of traffic. The model achieved is not exactly the DiffServ model because this architecture has been explicitly designed for wired networks and it cannot be strictly applicable to wireless networks; however, the results shown are quite satisfactory and a certain degree of service differentiation is maintained. Nevertheless, when the DiffServ model applied to ad hoc networks is compared with the SWAN Model [25], then SWAN clearly outperforms DiffServ in terms of throughput and delay requirements. For this reason, some attention must be given to the SWAN Model.

## 4.2 SWAN

SWAN (Stateless Wireless Ad Hoc Networks) [26] is a stateless network model that has been specifically designed to provide service differentiation in wireless ad hoc networks employing a best-effort distributed wireless MAC. It distinguishes between two traffic classes: real-time UDP traffic and best-effort UDP and TCP traffic. A classifier (see Figure 5) [27] differentiates between real-time and best-effort traffic; then a leaky bucket traffic shaper delays best-effort packets at a rate previously calculated applying an AIMD (Additive Increase Multiplicative Decrease) rate control algorithm. Every node measures the per-hop MAC delays from packet transmission locally and this information is used as feedback to the rate controller. Rate control restricts the bandwidth for best-effort traffic so that real-time applications can use the required bandwidth. On the other hand the bandwidth not used

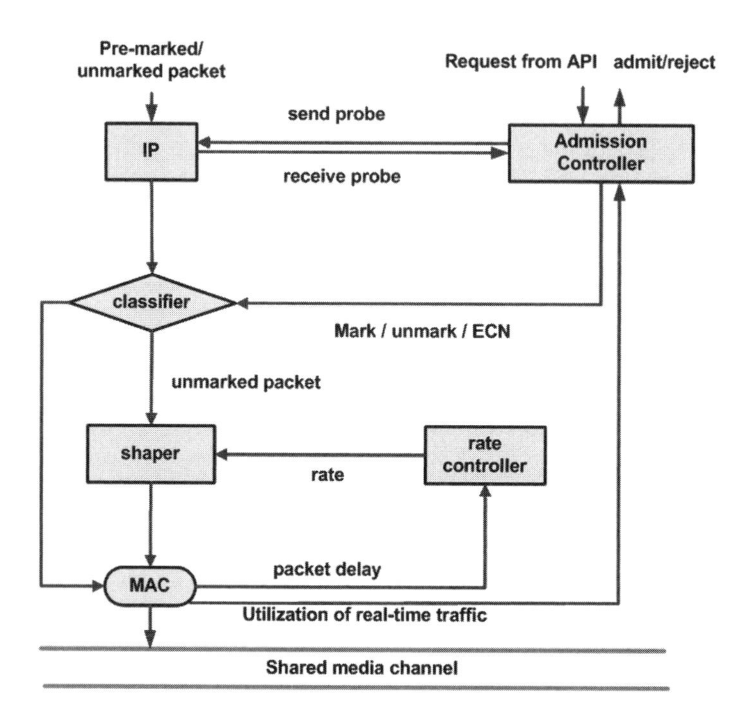


Figure 5: SWAN model.

by real-time applications can be efficiently used by best-effort traffic. The total best-effort and real-time traffic transported over a local shared channel is limited below a certain threshold rate to avoid excessive delays.

SWAN also uses sender-based admission control for real-time UDP traffic [28]. The rate measurements from aggregated real-time traffic at each node are employed as feedback. This mechanism sends an end-to-end request/response probe to estimate the local bandwidth availability and then determine if a new real-time session should be admitted. The source node is responsible for sending a probing request packet toward the destination node. This request is a UDP packet containing a bottleneck bandwidth field. All intermediate nodes between the source and destination must process this packet, check their bandwidth availability, and update the bottleneck bandwidth field in the case where their own bandwidth is less than the current value in the field. The available bandwidth can be calculated as the difference between an admission threshold and the current rate of real-time traffic. The admission threshold is set below the maximum available resources to enable real-time and best-effort traffic to share the channel efficiently. Finally the destination node receives the packet and returns a probing response packet with a copy of the bottleneck bandwidth found along the path back to the source. When the source receives the probing response it compares the end-to-end bandwidth availability and the bandwidth requirement and decides whether to admit a real-time flow accordingly. In the first case the Real-Time packets are marked as RT and they bypass the shaper mechanism and are thus not regulated.

The traffic load conditions and network topology change dynamically so that real-time sessions might not be able to maintain the bandwidth and delay bound requirements and they must be rejected or readmitted. For this reason it is said that SWAN offers soft QoS. The Explicit Congestion Notification (ECN) regulates real-time sessions as follows. When a mobile node detects congestion or overload conditions, it starts marking the ECN bits in the IP header of the real-time packets. The destination monitors the packets with the marked ECN bits and informs the source, sending a regulate message. Then the source node tries to reestablish the real-time session with its bandwidth needs accordingly.

Intermediate nodes do not keep any information per flow thus avoiding complex signaling and state control mechanisms and making the system simpler and more scalable.

## 4.3 Service Differentiation Mechanisms in Wireless Networks Modifying the IEEE 802.11 DCF Protocol

IEEE 802.11 is a standard for wireless communication that has received very good acceptance. It defines two possible operation modes: the Distributed Coordination Function (DCF) and the Point Coordination Function (PCF). The latter aims at providing some service differentiation for supporting real-time services, but it is rather inefficient. Besides, PCF requires a central point of control (the access point), which prevents the user from working in an ad hoc fashion. Given this context, researchers have introduced many modifications to the standard IEEE 802.11 DCF to provide service differentiation for QoS support.

The basic scheme for DCF is the Carrier Sense Multiple Access with Collision Avoidance (CSMA/CA) [29]. When a station wants to send a packet, it must first sense the medium. If the channel is sensed idle for a time interval equal to or greater than a Distributed InterFrame Space (DIFS), then the station can send the packet; otherwise the station waits a random time after the channel being idle for DIFS, named the backoff time, and then it sends the packet. The backoff time interval can be computed as follows: Tbackoff = Rand(0,CW) * Tslot, where Tslot represents the time slot selected by the physical layer. This formula means that the node chooses a backoff time equal to a random number of time slots between 0 and a Contention Window, CW. The backoff interval is used to initialize a backoff timer, which will be decreased for each station while the medium is idle and will be frozen when the transmission from another station is detected. If the medium remains idle for a period greater than DIFS, the backoff timer is periodically decremented by one for every time slot. When the backoff timer expires the station immediately accesses the medium. It can happen that two or more stations start transmission at the same time slot and so a collision occurs.

To reduce the probability of consecutive collisions, the CW is doubled after each unsuccessful transmission attempt until a maximum value (CW-max) is reached. After a successful transmission, the CW is reset to CWmin.

The hidden terminal effect can be reduced through the use of RTS-CTS handshake packets [29].

Several authors have proposed different modifications of the IEEE 802.11 DCF protocol that introduce service differentiation.

The DENG scheme [30] is an access method where the CSMA/CA protocol has been modified to support four priority classes. The priority access to the wireless medium can be provided due to the InterFrame Space (IFS) and giving different backoff windows.

The IEEE 802.11e or enhanced DCF [31] is an access scheme that provides differentiated DCF access to the wireless medium for prioritized traffic categories. There are eight traffic categories and each one has a different minimum contention window value to assure service differentiation. Besides, the different traffic classes use different interframe spaces named Arbitration InterFrame Space (AIFS), where the DIFS plus some (possibly zero) time slots are added. The Adaptative Enhanced Distributed Coordination Function (AEDCF) [32] is a modification and improvement of the EDCF protocol that aims at establishing different priority classes access the wireless medium. The EDCF mechanism resets the contention window of the corresponding class i to CWmin[i] after a successful transmission. It would be better to update the contention window more slowly. AEDCF ensures different priorities by adjusting the size of the contention window of each traffic class according to applications requirements and network conditions.

In the Distributed Fair Scheduling (DFS) [33] access scheme fair queueing is applied to the wireless domain.

Black burst [34] is a distributed MAC scheme whose main goal is to minimize the delay for real-time traffic. Stations with low priority can access the medium using the standard IEEE 802.11 MAC DCF. Nodes that transmit real-time packets (stations with higher priority) schedule their next transmission attempt for a time tsch. Real-time nodes have the ability of jamming the channel with pulses of energy named black burst. If a node with high-priority traffic schedules an access attempt for the present time and notices that the channel has been idle during a medium interframe interval, the node starts transmitting a black burst. Otherwise, it waits until the channel becomes idle and then enters a black burst contention period by jamming the channel for a period of time. The length of the black burst depends on how much time the station has been waiting to access the medium and thus incorporating a first-come first-served mechanism with the high-priority stations. After transmitting the black burst, the node waits for an observation interval to see if another node transmitted a longer black burst, implying that it would have been waiting longer for access to the channel. Afterwards, if the medium is perceived idle, the node transmits its packet; otherwise, it will wait until the channel becomes idle again and will enter another black burst contention period. After the successful transmission of a frame, the station schedules the next transmission attempt for a time tsch. By doing this, real-time flows will synchronize.

ARME (Assured Rate MAC Extension) is an extension of the IEEE 802.11b MAC protocol that aims at supporting differentiated services. Two types of service (assured rate service and best-effort) access the channel with the DCF mode but they use different contention windows [35]. DIME (DiffServ MAC Extension) is an extension of ARME where expedited forwarding traffic is also considered and reuses the IFS of the point coordination function in a distributed manner [36].

## 5 Voice Transmission in Ad Hoc Networks

There is some work related to voice transmission in IEEE 802.11 networks but most of them are in infrastructure-based mode [37,38].

In [39] the authors study the capability of the enhanced point coordination function and enhanced distributed coordination function to support voice over IP applications in infrastructure-based IEEE 802.11 WLANs. The results show that the end-to-end delays are maintained low and best-effort background traffic does not alter the performance of VoIP connections.

The delay, jitter, and packet loss in VoIP applications are higher in an infrastructure-based wireless network than in a wired one because when the number of users connected to the same access point increases then congestion problems in the AP occur and the system deteriorates. In ad hoc networks the problem is reduced because the VoIP packets should not be sent through an access point but if we try to develop voice over IP applications in wired-cum-wireless scenarios through a gateway the problem persists and in this case the gateway will be the bottleneck.

On the other hand, in IEEE 802.11 WLANs (working in infrastructure-based as well as in ad hoc mode) the limited radio coverage, interference from other devices, and mobility favor that the link quality is degraded and consequently the available bandwidth is reduced. In this case the number of retransmissions increases and therefore the latency and jitter of such sensitive applications as VoIP are affected.

Voice over IP is susceptible to delay, jitter, and traffic loss. For this reason the evaluation has been done according to these metrics:

- Delay can be measured as one-way delay and it represents the time taken from point to point in the network.
- VoIP jitter is the variation in delay over time from point to point. It can be defined as a smoothed function of the delay differences between consecutive packets over time. The voice call quality can be seriously degraded if this parameter is too wide. The amount of jitter tolerable is affected by the depth of the jitter buffer on the network equipment in the voice path. The more jitter buffer available, the more the network can reduce the effects of jitter.
- Packet loss along the data path can severely degrade the voice application too.

The ITU-T (International Telecommunication Union) recommends in standard G.114 that the one-way delay should be kept below 150 ms to maintain an acceptable conversation quality [40].

# 6 DS-SWAN (Differentiated Services SWAN)

Real-time VoIP traffic has some special QoS requirements such as bounded end-to-end delay, low delay jitter, and limited loss rate. Therefore the classical IEEE 802.11 MAC DCF is not able to offer these special requirements and other solutions must be found and analyzed. Thus we have run simulations where a service differentiation SWAN model is introduced in the ad hoc network The authors in [26] and [27] studied the behavior of CBR voice traffic but VBR VoIP transmission had not been analyzed yet and we have done it. We have found that the SWAN model is not totally appropriate for such kind of applications. The problem is that a rate control is used for UDP (real-time) and TCP (best-effort) traffic [26]. Voice flows will not be rate controlled only once admitted through the source-based admission control. Therefore, some VoIP packets will not be accepted as real-time traffic when the source starts sending traffic because the admission control mechanism has not finished and they are shaped as best-effort traffic at a rate determined by the rate controller. Then these packets will be delayed at a rate according to the shaper and it will take longer for them to access the MAC layer. This means that the voice application will be severely degraded because the one-way delay will not be kept under 150 ms during some period of time [41].

We have modified this algorithm and created a new version named DS-SWAN (Differentiated Services-SWAN). In our algorithm we have assumed that all real-time flows should bypass the best-effort traffic shaper during the admission control process. Only if the admission control is rejected will the VoIP packets be considered as best-effort traffic and they will have to be rate controlled just as the others. In this way we prevent some VoIP packets from suffering unacceptable one-way and jitter delays, and intolerable packet losses, and the VoIP applications can function properly.

# 7 Simulations

## 7.1 Scenario

The simulator used in this work for providing end-to-end QoS in wired-cum-wireless environments is ns-2 [41].

The objective is to study the interworking between a mobile ad hoc network and the Internet, extending the DiffServ model to a wireless environment. For this reason a scenario where an ad hoc network is connected via a single gateway to a fixed IP network has been chosen. We have used an ns-2 package based on [8] that enables us to use AODV as an ad hoc routing protocol for simulations of wired-cum-wireless scenarios. The system framework is shown in Figure 6.

The addressing architecture helps us to connect the ad hoc network with the Internet. The ad hoc network can be considered as a subnet within the hierarchy of the wired IP network [42]. Wired and wireless nodes are segregated by placing them in different hierarchical domains. Hierarchical routing is used to route packets between these hierarchical domains. A gateway connects nodes in the wireless ad hoc network with nodes outside the domain.

Moreover, to provide QoS and differentiate the service level between applications the fixed IP network supports the DiffServ architecture. We consider a single DS-domain covering the whole network between the wired corresponding hosts and the gateway. The ad hoc network also supports QoS. Therefore we have run simulations where SWAN has been selected to provide QoS. The chosen scenario consists of 20 mobile nodes, one gateway, three routers and three fixed hosts. The nodes are distributed in a square region of 500 m by 500 m. The gateway is placed close to the center of the area with $x$, $y$ coordinates (200, 200). Simulation runs last for 200 seconds.

In our scenario we assume that two traffic classes are transmitted: best-effort CBR traffic and real-time VBR VoIP traffic. The mobile nodes communicate with one of the three fixed hosts located in the Internet through the gateway. Therefore the destination of all the CBR and VBR VoIP traffic is one of the three hosts in the wired network, although some nodes in the ad hoc network will act as intermediate nodes or routers forwarding the packets from other nodes. In order to represent best-effort background traffic, 13 of the 20 mobile nodes were selected to act as CBR sources (each node establishes two CBR connections) and one node is selected to send VBR VoIP traffic. All these mobile hosts use IEEE 802.11b. Initially, they are randomly distributed according to a uniform distribution within the ad hoc network. The simulations use the random waypoint model. Each node selects a random destination within the area and moves toward it at a velocity uniformly distributed between 0 and 3 m/s. The size of the area and the number of nodes have been chosen so as to have several hops from the source to the gateway in the wireless part on average.

Upon reaching the destination the node pauses a fixed time period of 20 seconds, selects another destination, and repeats the process. To avoid synchronization problems due to deterministic start time, background traffic is generated with CBR traffic sources whose starting time is chosen from a uniform random distribution in the range [15 s, 20 s] for the first source, [20 s, 25 s] for the second one, and so on up to [140 s, 145 s] for the last one. They have a rate of 32 Kbit/s with a packet size of 80 bytes. The sources continue sending data until the end of the simulation.

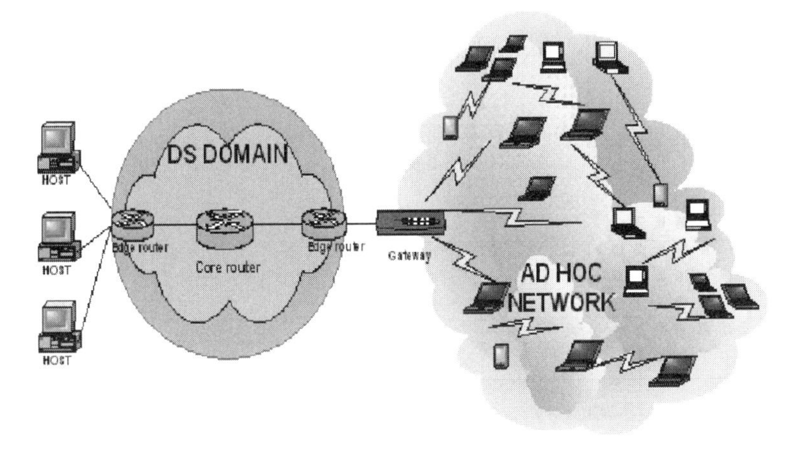


Figure 6: Differentiated services for wireless ad hoc networks.

The VoIP traffic is modeled as an on/off source with exponentially distributed on and off periods of 312.5 ms and 325 ms average each. Packets are generated during on periods at a constant bit rate of 50.536 Kbit/s and no packets are sent during off periods. One VoIP connection is activated at a starting time chosen from a uniform distribution in the range [10 s, 15 s]. The source sends data until the end of the simulation. Packets have a constant size of 128 bytes, thus the interpacket time is 20.3 ms.

In our first setting the fixed Internet network uses DiffServ as the QoS mechanism. The edge router functions are presented in Figure 7. Incoming packets are classified and marked with a DSCP. The recommended DSCP values for EF are 46 [17] and for BE, 0 [14] is used. Shaping of EF (VoIP) and BE (CBR) traffic is done in two different drop tail queues of size 30 and 100 packets, respectively. The EF and BE aggregates are policed with the token bucket meters and droppers described in Table 2. Some bursts are tolerated but the traffic that exceeds the profile is marked with a different codepoint and then it is dropped. Token bucket policer codepoint 46 is policed to codepoint 51 and token policer codepoint 0 is policed to codepoint 50. Accepted packets are served using a round-robin scheduler. The Traffic Conditioning Specification (TCS) describes some values of certain parameters that specify a set of classifier rules and a traffic profile for DiffServ. The TCS for the edge router/core router interface is explained in Table 2. CBS (Committed Burst Size) refers to the maximum size of the token bucket and it is measured in bytes. CIR (Commited Information Rate) refers to the rate at which tokens are generated and it is specified in bits per second.

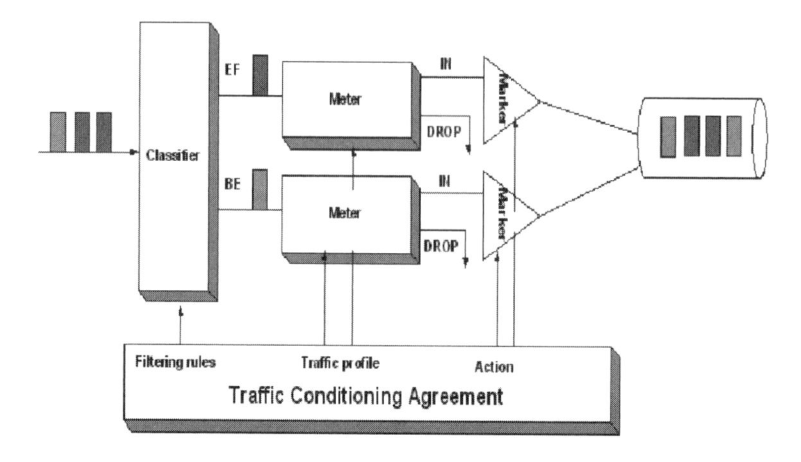


Figure 7: Edge router functions.

Table 2: TCS for the edge router/core router interface.

| Traffic class | Traffic type | PHB | DSCP value | Assigned bandwidth | Meter | Dropper |
|---|---|---|---|---|---|---|
| Premium | VoIP | EF | 46 | 100 Kbps | Token Bucket CIR 100 Kbps CBS 1000 bytes | Drop out of profile |
| Best Effort | other | BE | 0 | 200 Kbps | Token Bucket CIR 200 Kbps CBS 1000 bytes | Drop out of profile |

The architecture of the core router is composed of one queue for each class of traffic. Packets are scheduled using a round-robin discipline.

## 7.2 Simulation Results

We have run simulations to assess the one-way delay, jitter, and packet loss of VoIP traffic in the network. We evaluate and compare the performance of the already explained scenario in three cases: when there is not any service differentiation mechanism in the ad hoc network and the IEEE 802.11

DCF protocol is used (original system), when the SWAN model has been applied in the ad hoc network (SWAN), and when the DS-SWAN has been introduced in the ad hoc network (DS-SWAN).

At the beginning of the simulation, data packets waiting for the AODV routing protocol to find a route for them should be buffered at the source; but sensitive traffic such as VBR VoIP is severely affected. For this reason we have decided to exchange just a single CBR packet before running the VoIP application so that the routing tables are set up in advance.

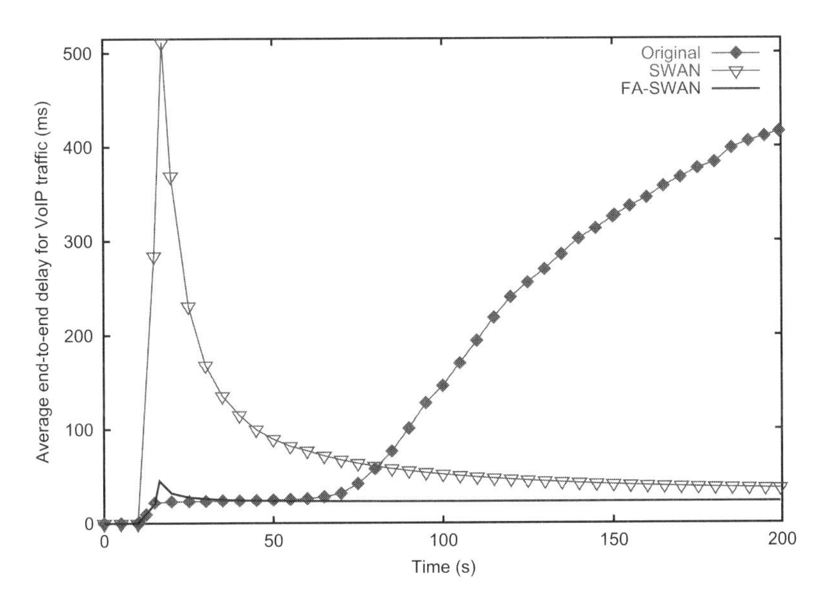


Figure 8: Average one-way delay for EF traffic.

Figure 8 shows the average one-way delay for EF (VoIP) traffic in all systems. It has been measured at the arrival times of the packets to the receiver. We have run 80 simulations. The average one-way delay is unacceptable for the original system (IEEE 802.11 DCF) because there is no service differentiation when the background traffic increases and the impact of this kind of traffic on VoIP performance is enormous; users lose interactivity and the voice is distorted, becoming unintelligible. With the SWAN model, VoIP undergoes very high one-way delays at the beginning of the simulation. The reason is that this service differentiation model supposes that all traffic (VoIP and CBR) is considered as best-effort traffic until a new session of real-time traffic is admitted. The problem therefore is that

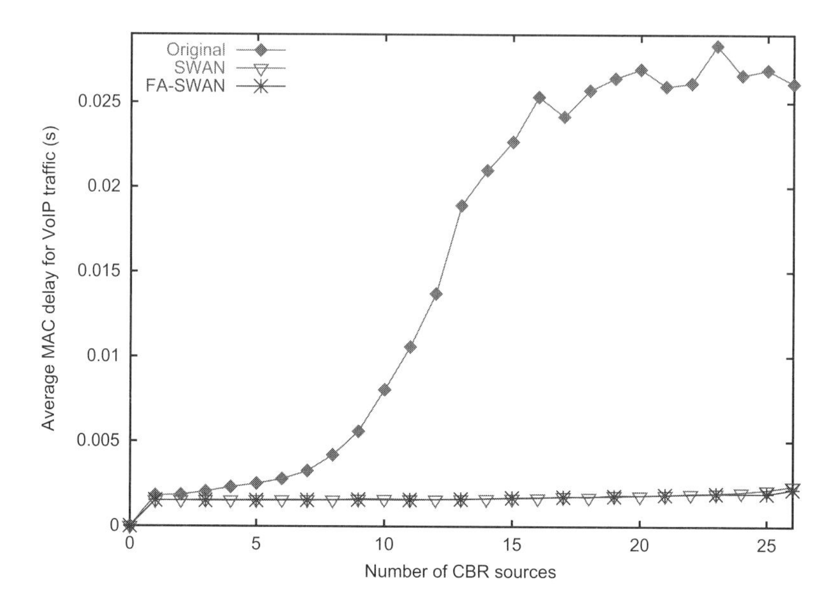


Figure 9: Average MAC delay for EF traffic.

the first VoIP packets are shaped at a rate determined by the rate controller and this means that the voice application will be severely degraded because the one-way delay goes beyond the 500 ms and the ITU recommendation [40] has not been fulfilled. After its peak the curve decreases slowly in comparison with the DS-SWAN model because it takes a certain period of time until the real-time packets delayed by the shaper arrive at the destination.

On the other hand, using the DS-SWAN model the one-way delays remain around 23 ms so that they are always kept under 150 ms [40] and the VoIP quality will be satisfactory. This way we can check that the traffic prioritization has been correctly done using the DS-SWAN model. VoIP traffic may be effectively controlled and is not sensitive to the best-effort traffic offered. Figure 9 shows the average delay of VoIP traffic with a growing number of CBR flows. We can see that the average MAC delay increases in the original system (IEEE 802.11 DCF) from 4 to 28 ms when the number of CBR sources increases from 8 to 23. In contrast, the average delay of the real-time traffic remains around 15–20 ms applying the SWAN or the DS-SWAN model. Therefore we can observe that with the SWAN and DS-SWAN models the real-time flow experiences low and stable MAC delays for an increasing amount of CBR traffic by controlling the rate of

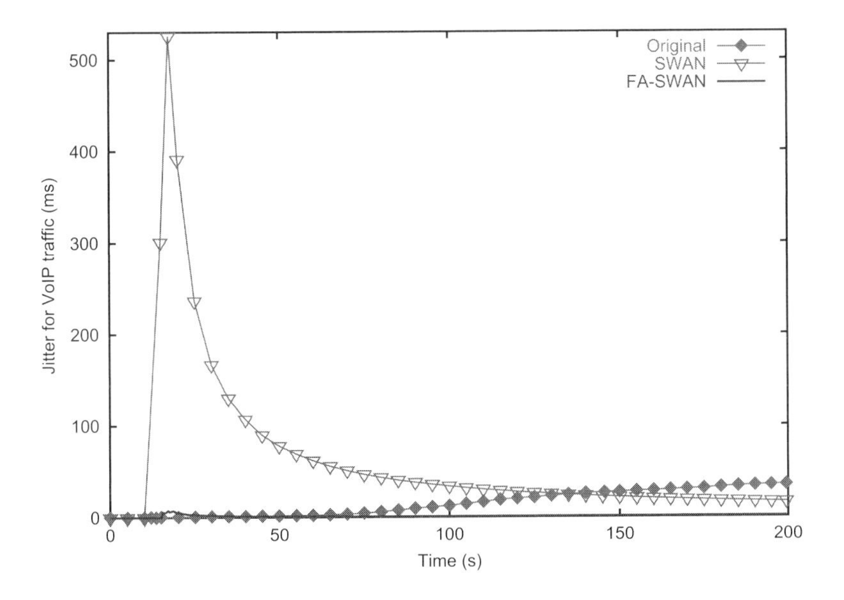


Figure 10: Jitter for EF traffic.

best-effort CBR.

The jitter for VoIP traffic is illustrated in Figure 10. We can appreciate that the traffic shaper conditioning best-effort and real-time traffic when the real-time session has not been yet admitted has a great impact on jitter in the SWAN model with delays higher than 500 ms. In the original system (IEEE 802.11 DCF) jitter increases slowly from 0 to around 35 ms. The DS-SWAN model shows the best results and the jitter is practically negligible with delays around 2–3 ms.

Figure 11 shows the packet loss in the ad hoc network for VoIP traffic. Packet loss may occur due to mobility and congestion. When a source or an intermediate node must send a packet the routing protocol at the network layer consults if there is a valid route to the destination and uses it if it exists. Otherwise, if the nodes have moved and there is no route available to the destination at the moment, the packet is buffered until a route is discovered. Packets are dropped due to mobility at the network layer because either the buffer is full and does not accept any more packets or the time that the packet has been buffered exceeds the limit. On the other hand, if a packet at the MAC layer manages to access the channel and mobility is high the next hop may be out of range at the moment and there

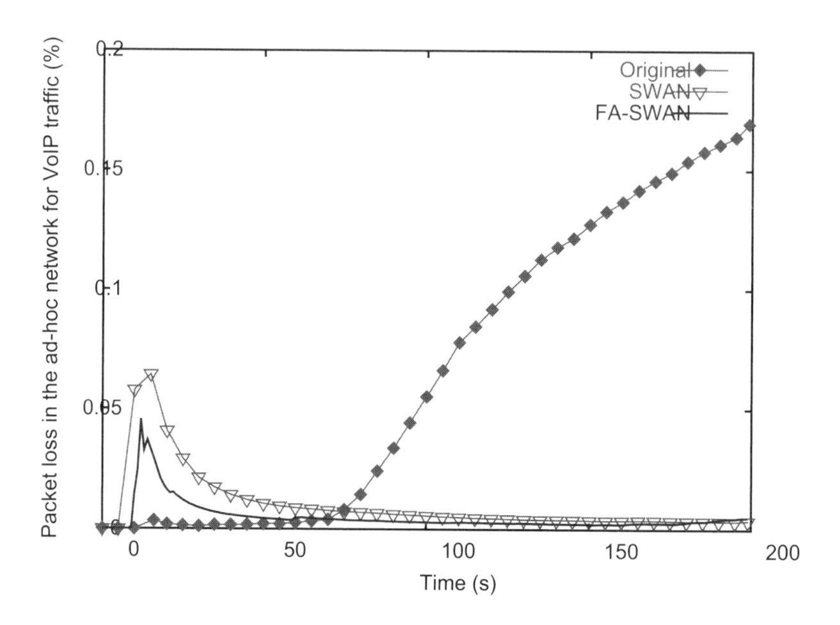


Figure 11: Packet loss for VoIP traffic in the ad hoc network.

would be packet loss at the MAC layer due to mobility, too. Packet loss due to congestion happens for several reasons. When the maximum amount of time allowed for a backoff interval is exceeded because the wireless channel is busy the packet is dropped at the MAC layer. Furthermore, if the queue that buffers packets waiting to access the medium is full due to congestion, packets are also dropped at the MAC layer. We can appreciate in Figure 11 that the number of lost packets is negligible in all systems. Two different causes for packet loss in the ad hoc network are ARP drops and MAC callback. If a node does not have the hardware address for the destination, it broadcasts an ARP query and caches the packet temporarily. For each unknown destination hardware address, there is a buffer for a single packet. When an additional packet for the same destination is sent to the node and the ARP reply has not arrived at the node due to congestion, the earlier buffered packet is dropped (ARP drop).

On the other hand, MAC callback means that the MAC layer is not able to transmit the packet and hence it informs the upper layer about the transmission failure. The cause for transmission failure is link failure due to mobility. In SWAN the number of lost packets is increased in relation to FA-SWAN to 0.4% in the peak because some VoIP packets in SWAN are

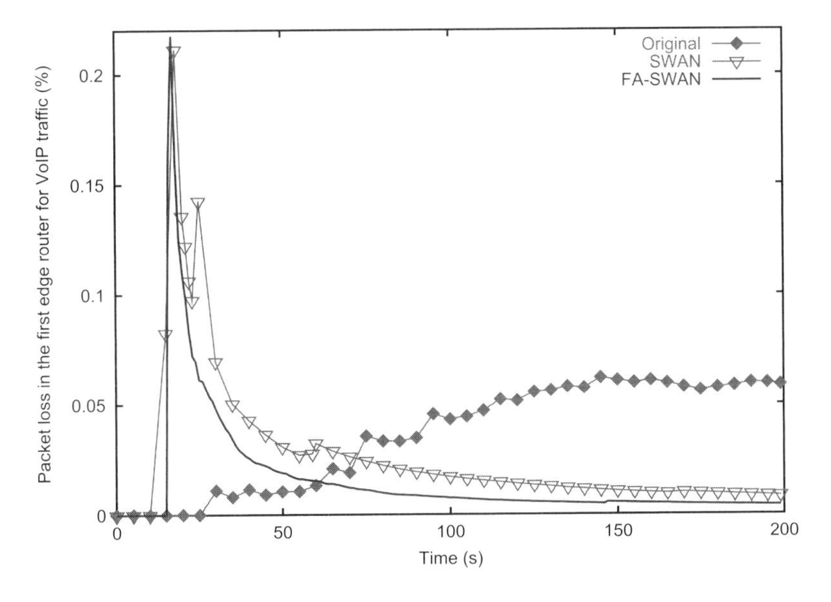


Figure 12: Packet loss in the first edge router for EF traffic.

delayed by the best-effort traffic shaper and when they finally access the medium after a certain period of time the nodes have moved and there are link failures due to node mobility and consequently the packets are dropped. Other VoIP packets are dropped in SWAN because they are delayed by the best-effort traffic shaper and its buffer is full or due to ARP drops. Packet loss in FA-SWAN is motivated basically due to mobility and ARP drops.

In the original system the number of lost VoIP packets at wireless nodes is augmented with a growing number of CBR flows because there is no service differentiation and in case of congestion VoIP packets are not prioritized over the best-effort ones. Hence, when VoIP packets finally access the congested medium the next hop will probably be out of range due to mobility and packets will be dropped. Some packets are lost due to ARP drops, too. However, most of them are dropped at the MAC layer due to congestion either when their backoff intervals are exceeded or at the buffers waiting for medium access. Figure 12 shows the number of lost packets in the first edge router for EF traffic. In the original system the number of lost packets increases slowly and afterwards it is maintained around 6%. In SWAN there is a peak at second 18 of 0.2% and another at second 25 of 0.14%; afterwards this number is strongly reduced. In DS-SWAN there is a peak at second 17 of 0.21% and afterwards the number of lost packets is also strongly reduced. We can observe that prioritization of voice packets can be an effective way of getting around congestion problems, by penalizing data traffic. However, when voice traffic is high it will still experience congestion and there are losses because the traffic exceeds the profile and it is marked with a different codepoint and then dropped.

## 8 Conclusions

We have evaluated the performance of the DCF IEEE 802.11 protocol, the SWAN model, and the DS-SWAN model in carrying VoIP applications with QoS between a mobile ad hoc network and a wired network. Our results show that in the scenario where VoIP calls are made between wired and wireless networks the DS-SWAN model provides low end-to-end delays for voice calls, and its performance is not sensitive to background best-effort traffic. The system is stable even under heavy-load background traffic and still meets the QoS requirements posed by VoIP connections. Moreover, DS-SWAN significantly reduces the effects of packet loss, delay, and jitter. Through our simulation study we conclude that DS-SWAN can provide differentiated channel access for different traffic types.

## 9 Acknowledgments

This work was partially supported by the Ministerio de Ciencia y Tecnología of Spain under the project TIC2003-08129-C02 and under the Spanish Ramón y Cajal program.

## References

[1] C. R. Lin and J. S. Liu, QoS routing in ad hoc wireless networks, *IEEE Journal on Selected Areas in Communications*, 17(8), pp. 1426–1438, 1999.

[2] P. Sinha, R. Sivakumar, and V. Bharghavan, CEDAR: A core-extraction distributed ad hoc routing algorithm. *In Proceedings of IEEE INFOCOM'99* (New York, 1999) pp. 202–209.

[3] S. Chen and K. Nahrstedt, Distributed quality-of-service routing in ad hoc networks, *IEEE Journal on Selected Areas in Communications*, 17(8), pp. 1488-1505, 1999.

[4] A. Iwata, C-C. Chiang, G. Yu, M. Gerla, and T-W. Chen, Scalable routing strategies for ad hoc wireless networks, *IEEE Journal on Selected Areas in Communications*, 17(8), pp. 1369–1379, 1999.

[5] S. B. Lee and A. T. Campbell, INSIGNIA: In-band signaling support for QoS in mobile ad hoc networks. *In Fifth International Workshop on Mobile Multimedia Communications. (MoMuc'98)* (Berlin, 1998).

[6] H. Xiao, K.G. Seah, A. Lo, and K.C. Chua, A flexible quality of service model for mobile ad hoc networks. *In IEEE Vehicular Technology Conference (VTC Spring 2000)* (Tokyo, 2000), pp. 445-449.

[7] P. B., Velloso, M. G., Rubinstein, and O. C. M. B., Duarte, Analyzing Voice Transmission Capacity on Ad Hoc Networks. *In International Conference on Communication Technology — ICCT 2003* (Beijing, 2003).

[8] R. Wakikawa, J. T. Malinen, C. E. Perkins, A. Nilsson, and A. J. Tuominen, Global connectivity for IPv6 mobile ad hoc networks, *Internet Engineering Task Force* Internet Draft, July 2002 (Work in Progress).

[9] C. E. Perkins, E. M. Belding-Royer, and I. Chakeres, Ad Hoc On Demand Distance Vector (AODV) Routing, *IETF Internet draft, draft-perkins-manet-aodvbis-00.txt,* Oct 2003 (Work in Progress).

[10] J. Xi and C. Bettstetter, Wireless multi-hop internet access: Gateway discovery, routing, and addressing. *In Proceedings of the International Conference on Third Generation Wireless and Beyond (3Gwireless'02)* (San Francisco, 2002).

[11] J. P. Jeong, J.-S. Park, K. Mase, Y.-H. Han, B. Hakim, and J.-M. Orset, Requirements for ad hoc IP address autoconfiguration, *IETF Internet draft, draft-jeong-manet-addr-autoconf-reqts-00.txt,* August 2003 (Work in Progress).

[12] A. Nilsson, C. E. Perkins, A. Tuominen, R. Wakikawa, and J. T. Malinen, AODV and IPv6 Internet Access for Ad Hoc networks, *ACM Mobile Computing and Communications Review* 6 (2002).

[13] S. Blake, D. Black, M. Carlson, E. Davies, Z. Wang, and W. Weiss, An architecture for differentiated service, *Request for Comments (Informational) 2475* Internet Engineering Task Force, December 1998.

[14] K. Nichols, S. Blake, F. Baker, and D. Black, Definition of the differentiated services field (DS field) in the IPv4 and IPv6 headers, *Request for Comments (Proposed Standard) 2474* Internet Engineering Task Force, December 1998.

[15] K. Nichols, V. Jacobson, and L. Zhang, A two-bit differentiated services architecture for the Internet, *RFC-2638* July 1999.

[16] V. Jacobson, K. Nichols, and K. Poduri, An expedited forwarding PHB, *Request for Comments (Proposed Standard) 2598* Internet Engineering Task Force, June 1999.

[17] B. Davie, et al., An expedited forwarding PHB, *RFC 3246* (2002).

[18] J. Heinanen, F. Baker, W. Weiss, and J. Wroclawski, Assured Forwarding PHB Group, *RFC 2597* June 1999.

[19] D. D. Clark and W. Fang, Explicit allocation of best-effort packet delivery service, *IEEE/ACM Transactions on Networking* 6(4), pp. 362–373, 1998.

[20] J. Ibanez and K. Nichols, Preliminary Simulation Evaluation of an Assured Service, *Internet Draft, draft-ibanez-diffserv-assured-eval-00.txt* August 1998.

[21] Cisco Systems, DiffServ-The Scalable End-to-End QoS Model (White Paper, March 2001).

[22] K. Wu and J. Harms, QoS support in mobile ad hoc networks, *Crossing Boundaries — the GSA Journal of University of Alberta* 1(1), pp. 92–106, 2001.

[23] T. Braun, C. Castelluccia, and G. Stattenberger, An analysis of the DiffServ approach in mobile environments. *In IQWiM-Workshop'99*, 1999.

[24] H. Arora and H. Sethu, A simulation study of the impact of mobility on performance in mobile ad hoc networks. *In Applied Telecommunications Symposium* (San Diego, 2002).

[25] H. Arora, L. Greenwald, U. Rao, and J. Novatnack, Performance comparison and analysis of two QoS schemes: SWAN and Diffserv, *Drexel Research Day Honorable Mention* (April, 2003).

[26] G.-S. Ahn, A. T. Campbell, A. Veres, and L.-H. Sun, SWAN *draft-ahn-swan-manet-00.txt,* February 2003.

[27] G.-S. Ahn, A. T. Campbell, A. Veres, and L.-H. Sun, Supporting service differentiation for real-time and best effort traffic in stateless wireless ad hoc networks (SWAN), *IEEE Transactions on Mobile Computing (TMC)* July-September 2002.

[28] G.-S. Ahn, A. T. Campbell, A. Veres, and L.-H. Sun, SWAN: Service differentiation in stateless wireless ad hoc networks, *In Proceedings of IEEE INFOCOM'2002* New York, June 2002.

[29] J.H. Schiller, *Mobile Communications.* Reading, MA Addison Wesley Professional, 2000.

[30] D-J. Deng and R-S. Chang, A priority scheme for IEEE 802.11 DCF access method, *IEICE Transactions on Communications* E82-B(1), January 1999.

[31] M. Benveniste, G. Chesson, M. Hoeben, A. Singla, H. Teunissen, and M. Wentink, EDCF proposed draft text, *IEEE working document 802.11-01/131r1* March 2001.

[32] L. Romdhani, Q. Ni, and T. Turletti, AEDCF: Enhanced service differentiation for IEEE 802.11 wireless ad hoc networks, *INRIA Research Report* No. 4544 (2002).

[33] N. H. Vaidya, P. Bahl, and S. Gupta, Distributed fair scheduling in a wireless LAN. *In Proceedings of the Sixth Annual International Conference on Mobile Computing and Networking* Boston (2000).

[34] J. L. Sobrinho and A. S. Krishnakumar, Quality-of-service in ad hoc carrier sense multiple access networks, *IEEE Journal on Selected Areas in Communications*, 17(8), pp. 1353–1368, 1999.

[35] A. Banchs and X. Pérez, Providing throughout guarantees in IEEE 802.11 Wireless LAN. *In Proceedings of IEEE Wireless Communications and Networking Conference (WCNC 2002)* Orlando, FL, USA (2002).

[36] A. Banchs, M. Radimirsch, and X. Pérez, Assured and expedited forwarding extensions for IEEE 802.11 Wireless LAN. *In Quality of Service, Tenth IEEE International Workshop on* (2002).

[37] A. Köpsel, J.-P Ebert, and A. Wolisz, A performance comparison of point and distributed coordination function of an IEEE 802.11 WLAN in the presence of real-time requirements. *In Workshop on Mobile Multimedia Communications (MoMuC2000)* Tokyo, Japan (2000).

[38] A. Köpsel and A. Wolisz, Voice transmission in an IEEE 802.11 WLAN based access network. *In Workshop on Wireless Mobile Multimedia (WoWMoM2001)* Rome, Italy (2001).

[39] D. Chen, S. Garg, M. Kappes, and K. S. Trivedi, Supporting VoIP traffic in IEEE 802.11 WLAN with enhanced medium access control (MAC) for quality of service, *www.research.avayalabs.com/techreport/ALR-2002-025-paper.pdf.*

[40] ITU-T Recommendation G. 114, One way transmission time, (2000).

[41] Ns-2: Network Simulator, *http://www.isi.edu/nsnam/ns*

[42] J. Broch, D. A. Maltz, and D. B. Johnson, Supporting hierarchy and heterogeneous interfaces in multi-hop wireless ad hoc networks. *In Proceedings of the Workshop on Mobile Computing held in conjunction with the International Symposium on Parallel Architectures, Algorithms, and Networks, IEEE* Perth, Australia (1999).

# Chapter 4

## Separable Threshold Decryption for Ad Hoc Wireless Networks

Willy Susilo
*School of IT and Computer Science*
*University of Wollongong, Wollongong, NSW 2522, Australia*
Email: wsusilo@uow.edu.au

Fangguo Zhang
*Department of Electronics and Communication Engineering*
*Sun Yat-Sen University, Guangzhou 510275, P.R. China*
Email: isdzhfg@zsu.edu.cn

Yi Mu
*School of IT and Computer Science*
*University of Wollongong, Wollongong, NSW 2522, Australia*
Email: ymu@uow.edu.au

## 1 Introduction

Let us consider the following scenario. Suppose a sender $\mathcal{S}$ would like to send a message to a group that consists of $n$ mobile nodes. $\mathcal{S}$ would like to ask $t$ out of $n$ mobile nodes to work together to reveal and read the message, but less than $t$ nodes cannot perform this task. This problem is essential especially in mobile networks, because it is easy to compromise a single node. We note that this problem can be solved by employing threshold encryption

M.X. Cheng, D. Li (eds.) *Advances in Wireless Ad Hoc and Sensor Networks.* Signals and Communication Technology, doi: 10.1007/978-0-387-68567-0_4.

schemes (e.g., [11, 15]), if the mobile nodes use a uniform cryptosystem, for example, RSA, and publish his or her public key beforehand. However, it is unknown whether it is possible to allow each mobile node to have his or her own selected cryptosystem and publish his or her public key accordingly. We also note that in an ad hoc system, we do not set the group of receivers beforehand, and therefore, having some restrictions to "force" each mobile node to use a uniform cryptosystem is somewhat unrealistic.

Threshold decryption is particularly useful where there is a concern about centralizing the power to decrypt. On the other hand, the idea of separable threshold decryption is to allow the sender to select the group of receivers in an ad hoc manner without the need to preselect the members of the group. A major advantage of such a system is that it allows each group member (or each mobile node, vice versa) to use his own public key cryptosystem, and publish his public key accordingly.

The important concept of separable threshold decryption is to allow "separability" and "threshold decryption" among the receivers. A successful combination of these two concepts will allow one to build a scheme that is very useful in an ad hoc wireless network. As another example, consider a situation where Alice, who is a news reporter, would like to send an encrypted message to the parliament members, such that $t$ of them can decrypt the message. In this scenario, Alice can collect the public keys of the parliament members, regardless of the cryptosystems that they use, and form her encrypted message. By sending her encrypted message, any $t$ members in the parliament can obtain and read the message. The notion of separability has recently attracted many researchers, because it can provide more attractive applications of the scheme (e.g., a separable group signature scheme [7], a separable ring signature [1], a separable threshold ring signature [14], etc.).

In this chapter, we take a fresh approach to separable threshold decryption. We focus our work by defining the requirements for such a scheme and follow up our idea by constructing a secure scheme. Our scheme can be used in many different public-key cryptosystems, including the one based on RSA, the discrete logarithm assumption, the Elliptic Curve Discrete Logarithm Problem (ECDLP), bilinear pairings, and so on. Our scheme can also be easily "upgraded" to cater to new schemes that might appear in the near future.

#### Related Work

Ghodosi et al. proposed a threshold cryptosystem in [11, 15] that allows a threshold of receivers to reveal the encrypted message. They also extended their work to allow decryption without any trusted authority involved [15]. A similar scheme has also been proposed in [18] that is secure against chosen ciphertext attack. However, these schemes do not allow separability among the receivers.

The term separability was originated in [13] and was diversified in [7]. An application is said to be *separable* if all participants can freely select their keys independently with different parameter domains and for different types of cryptosystems. There exist weaker forms of separability, for example, partial separability [13] and weak separability [7]. Partial separability allows only a subset of participants to select their keys independently and weak separability requires all participants to use some common system parameters. Another type of separability is a type-restricted separability [14]. In type-restricted separability, all participants are allowed to select their keys independently but requires that all the keys be from one single type of cryptosystem. The size of the keys can vary.

The secret sharing scheme allows a subset of the group to reveal the secret on behalf the group. The secret sharing scheme was independently proposed in [16, 3]. The secret sharing scheme has been widely integrated with several different primitives to create a threshold system that has many applications in threshold cryptography.

Threshold cryptography, and in particular threshold signature, was independently invented by Desmedt [9], Boyd [6], and Croft and Harris [8]. The main goal of threshold cryptography is to replace a system entity — such as a transmitter — in a classical cryptosystem with a group of entities sharing the same power. A threshold cryptosystem must remain secure not only under the attacks on the original cryptosystem, but also new types of attacks that are introduced because of the distributed structure of the system.

In a $(t, n)$ threshold signature scheme [10], signature generation requires collaboration of at least $t$ members of a set of $n$ signers. Although construction of threshold signature schemes generally uses a combination of secret sharing schemes and signature schemes, as noted in [10], a simplistic combination of the two primitives could result in a completely insecure system that allows the members of an authorized group to recover the secret key of the signature scheme. In a secure threshold signature scheme the power of signature generation must be shared among $n$ signers in such a way that $t$ signers can collaborate to produce a valid signature for any given message and no subset of fewer than $t$ participants can forge a signature.

**Our Contributions**

We present a model of a separable threshold decryption scheme. This scheme can be used in an ad hoc wireless network. Each mobile node can freely select the public key from different cryptosystems. The sender can send an encrypted message and a subset of these mobile nodes ($t$ out of $n$) can always decrypt the message. We also provide a secure scheme that satisfies this model. We provide a complete security analysis for our scheme.

The rest of this chapter is organized as follows. In Section 2, we review some building blocks that are used in our construction. In Section 3, we describe our model of a separable threshold decryption scheme. In Section 4, we present a secure scheme that satisfies our model. In Section 5, we provide some security analysis for our scheme. Section 6 concludes the chapter.

# 2 Building Blocks

In this section, we briefly review some building blocks that are used to construct our scheme.

## 2.1 Shamir's Secret Sharing Scheme

In [16] Shamir proposed a way to split a secret among participants. A $(t, n)$ threshold scheme is a method of sharing a key $\mathcal{K}$ among a set of $n$ participants, in such a way that any $t$ participants can compute the value of $\mathcal{K}$, but no group of $t-1$ participants can do so. The value of $\mathcal{K}$ is determined by a special entity called a dealer, which is denoted by $\mathcal{D}$. The set of $n$ participants is denoted by $\mathcal{P}$. We assume $\mathcal{D} \notin \mathcal{P}$. When $\mathcal{D}$ wants to share $\mathcal{K}$ among the participants in $\mathcal{P}$, he generates some partial information for each participant called a *share*. The shares must be distributed securely via an authenticated channel, so no participant knows the share given to another participant. At the end of this process, each participant holds his or her own share as his or her secret key. At a later time, a subset of participants $\hat{\mathcal{P}} \subseteq \mathcal{P}$ will pool their shares to compute the key $\mathcal{K}$. If $|\hat{\mathcal{P}}| \geq t$, then the value $\mathcal{K}$ can be recomputed as a function of the shares, but if $|\hat{\mathcal{P}}| < t$, then they should not be able to compute $\mathcal{K}$.

In Shamir's threshold scheme, $\mathcal{K} \in Z_p$, where $p \geq n+1$ is a prime number. The scheme consists of two stages as follows.

*Initialization Phase.* $\mathcal{D}$ selects $n$ distinct, nonzero elements of $Z_p$, denoted by $x_i$, $1 \leq i \leq n$. For each $i$, $\mathcal{D}$ sends $x_i$ to $\mathcal{P}_i$ via an authenticated channel. The values $x_i$ are publicly available.
*Share Distribution Phase.* When $\mathcal{D}$ would like to share a key $\mathcal{K} \in Z_p$, $\mathcal{D}$ will perform the following steps.

1. Select $t-1$ random elements of $Z_p$, denoted by $a_1, \ldots, a_{t-1}$.
2. For $1 \leq i \leq n$, $\mathcal{D}$ computes $y_i = f(x_i)$, where
$$f(x) = \mathcal{K} + \sum_{j=1}^{t-1} a_j x^j \bmod p.$$
3. For $1 \leq i \leq n$, $\mathcal{D}$ sends the share $y_i$ to $\mathcal{P}_i$.

When $t$ out of $n$ participants would like to reveal $\mathcal{K}$, they can jointly compute
$$\mathcal{K} = \sum_{j=1}^{t} y_{i_j} \prod_{1 \leq k \leq t, k \neq j} \frac{x_{i_k}}{x_{i_k} - x_{i_j}}.$$

## 2.2 Bilinear Pairings

In recent years, bilinear pairings have been used to construct numerous new cryptographic primitives. We recall the basic concept and properties of bilinear pairings.

Let $\mathbb{G}_1$ be a cyclic additive group generated by $P$, whose order is a prime $q$, and $\mathbb{G}_2$ be a cyclic multiplicative group with the same order $q$. Let $e : \mathbb{G}_1 \times \mathbb{G}_1 \to \mathbb{G}_2$ be a bilinear pairing with the following properties.

1. *Bilinearity:* $e(aP, bQ) = e(P,Q)^{ab}$ for all $P, Q \in \mathbb{G}_1, a, b \in Z_q$.
2. *Nondegeneracy:* There exist $P, Q \in \mathbb{G}_1$ such that $e(P,Q) \neq 1$; in other words, the map does not send all pairs in $\mathbb{G}_1 \times \mathbb{G}_1$ to the identity in $\mathbb{G}_2$.
3. *Computability:* There is an efficient algorithm to compute $e(P,Q)$ for all $P, Q \in \mathbb{G}_1$.

Throughout this chapter, we define the system parameters in all schemes as follows. Let $P$, $Q$ be two generators of $\mathbb{G}_1$ with order $q$ (it is assumed that the discrete logarithm of $Q$ to the base $P$ is unknown); the bilinear pairing is given by $e : \mathbb{G}_1 \times \mathbb{G}_1 \to \mathbb{G}_2$. These system parameter can be obtained using a GDH parameter generator $\mathcal{IG}$ [5].

# 3 The Model

In this section, we provide a model for a separable threshold decryption scheme, together with some security requirements for such a scheme.

There is a polynomial sender $\mathcal{S}$ in the system. There are $n$ mobile users (which could be any subset of users in an ad hoc network), denoted by $\mathcal{P} = \{\mathcal{P}_1, \ldots, \mathcal{P}_n\}$. The sender $\mathcal{S}$ would like to send a message to $\mathcal{P}$, such that a subset $\hat{\mathcal{P}} \subseteq \mathcal{P}$ can always decrypt the message, iff $|\hat{\mathcal{P}}| \geq t$. Each $\mathcal{P}_i$ has published her public key $\mathcal{PK}_i$, that is derived from any cryptosystem of her choice. Each $\mathcal{P}_i$ keeps her associated secret key $\mathcal{SK}_i$ secret.

Intuitively, a separable threshold decryption scheme consists of an encryption algorithm $E$, a decryption share generation algorithm $D$, and a share combining algorithm $C$. We require that all public keys associated with $\mathcal{P}$ be selected independently by $\mathcal{P}_i$, and there is no uniformity required. Let $k \in \mathbb{N}$ be a security parameter and $m \in \{0,1\}^*$ be a message.

**Definition 1 (Separable Threshold Decryption Scheme).**
*A* separable threshold decryption scheme *is a triple* $(E_{t,n}(\cdot), D(\cdot), C(\cdot))$, *where*

- $\{\eta_1, \ldots, \eta_n\} \leftarrow E_{t,n}(k, m)$ *is an algorithm that accepts as inputs a security parameter* $k$*, a message* $m$*, together with all public keys* $\mathcal{PK}_i$*,* $i = 1, \ldots, n$*, and outputs encrypted shares* $\{\eta_1, \ldots, \eta_n\}$*.*
- $\hat{\eta}_i \leftarrow D_{\mathcal{SK}_i}(\eta_i)$ *is a decryption share algorithm that accepts as inputs an encrypted share* $\eta_i$ *to produce a decrypted share* $\hat{\eta}_i$*.*
- $\{m, \perp\} \leftarrow C(\hat{\eta}_i, \ldots)$ *is a share-combining algorithm that accepts at least* $t$ *decrypted shares to produce the original message* $m$*. If the decrypted shares are less than* $t$*, it outputs* $\perp$*.*

**Correctness.** *We require*

$$\Pr\left[m \leftarrow C(\hat{\eta}_i \in \Gamma); \hat{\eta}_i \leftarrow D_{\mathcal{SK}_i}(\eta_i); \{\eta_1, \ldots, \eta_n\} \leftarrow E_{t,n}(k, m); |\Gamma| \geq t\right] = 1$$

*and*

$$\Pr\left[m \leftarrow C(\hat{\eta}_i \in \Gamma); \hat{\eta}_i \leftarrow D_{\mathcal{SK}_i}(\eta_i); \{\eta_1, \ldots, \eta_n\} \leftarrow E_{t,n}(k, m); |\Gamma| < t\right] \leq \epsilon.$$

**Definition 2 (Security Against Chosen Ciphertext Attack).**
*We require a separable threshold decryption scheme to provide* security

against chosen ciphertext attack *[18]. Security against chosen ciphertext attack means that any polynomial time adversary has a negligible advantage in the following game.*
Phase 1. *The adversary corrupts a fixed set of* $t-1$ *servers.*
Phase 2. *The key generation algorithm is performed. The private keys of the corrupted servers are sent to the adversary, and the other private keys of uncorrupted servers are sent to the uncorrupted servers, and kept secret from the adversary. The public key is known to the adversary.*
Phase 3. *The adversary interacts with the uncorrupted decryption servers in an arbitrary fashion, sending them ciphertext* $\eta$, *and obtaining decryption shares.*
Phase 4. *The adversary selects two plaintexts* $m_0$ *and* $m_1$ *that have the same length. These are sent to an "encryption oracle" that selects* $b \in \{0,1\}$ *at random and sends back the target ciphertext* $\eta'$ *to the adversary.*
Phase 5. *The adversary continues to interact with the uncorrupted servers, sending them ciphertext* $\eta \neq \eta'$.
Phase 6. *At the end of the game, the adversary outputs* $b' \in \{0,1\}$.
*The adversary's advantage is defined to be the*

$$\mathtt{Succ}^{IND-CCA}(k) = 2 \cdot \Pr[b = b'] - 1.$$

*We denote by* $\mathtt{Succ}^{IND-CCA}(t_{CCA}, q_D)$ *the maximum of the attacker's success over all attackers with running time* $t_{CCA}$ *and making at most* $q_D$ *decryption share generation queries. We say that the scheme is secure in the sense of IND-CCA if* $\mathtt{Succ}^{IND-CCA}(t_{CCA}, q_D)$ *is negligible in* $k$.

# 4 A Secure Scheme

In this section, we present our secure scheme that satisfies the requirements mentioned in Section 3. We divide our scheme into two main phases, namely the encryption phase (by $\mathcal{S}$) and decryption phase (by $\mathcal{P}$).

We assume each $\mathcal{P}_i \in \mathcal{P}$ has published his public key $\mathcal{PK}_i$. He can employ the RSA scheme, DLP-based scheme, ECDLP-based scheme, or ID-based scheme. The setups for these schemes are as follows (for a mobile host $\mathcal{P}_i$).

- RSA-based: Let $p_i, q_i$ be two large prime numbers and $N_i$ their product. Let $(e_i, d_i)$ be a pair of encryption and decryption keys. Let $H_1 : Z^*_{N_i} \rightarrow Z^*_q$ be a strong one-way hash function. The public tuple is denoted by $\mathcal{PK}_i = (e_i, N_i, H_1)$.

- DLP-based: Let $p$ be a safe prime and $g$ be a generator of $Z_{p_i}^*$ whose order is $q|p-1$. Let $x_i \in Z_q$ be the secret key of user $\mathcal{P}_i$ and $y_i = g^{x_i} \bmod p$. Let $H_2 : Z_{p_i}^* \rightarrow Z_q^*$ be a strong one-way hash function. The public tuple is $\mathcal{PK} = (g, p, y_i, H_2)$.

- EDLP-based: Let $E(n)$ be an additive group of prime order and $Q, P$ be generators of $E(n)$. Let $x_i \in Z_q$ be the secret key of $\mathcal{P}_i$ and $Y_i = x_i P_i$ be the corresponding public key. Let $H_3 : E(n) \rightarrow Z_q^*$ be a strong one-way hash function. The public tuple is $\mathcal{PK} = (P, Q, Y_i, H_3)$.

- ID-based: Let $\mathbb{G}_1$ be an additive group of prime order $q$ and $\mathbb{G}_2$ be an multiplicative group of prime order $q$. Let $P$ be a generator of $\mathbb{G}_1$. Let $s \in Z_q$ be a master secret key and $Y_i \in sP$ be the public key for $\mathcal{P}_i$. The secret key of $\mathcal{P}_i$ is computed from $S_{ID_i} = sH_0(ID_i)$, where $H_0$ is a strong hash function $H_0 : \{0,1\}^* \rightarrow \mathbb{G}_1$. Also assume the existence of a hash function $H_4 : \mathbb{G}_2 \rightarrow Z_q^*$.

### 4.1 Encryption Phase

A sender $\mathcal{S}$ performs the following steps to send a message to an ad hoc network. The encryption is hybrid; that is, the message is encrypted with a session (symmetric) key that is encrypted with a public key. The symmetric key algorithm can be selected by the users. The encryption protocol is described as follows.

1. Select a polynomial over $Z_q^*$ as follows,

$$f(x) = f_0 + \sum_{k=1}^{t-1} a_k x^k$$

   for random $a_k \in Z_q$.

2. Select a random integer $r \in Z_q$ (or $r_i \in Z_{N_i}$ in the case of RSA), and compute

$$k_i = \begin{cases} H_1(r_i), r_i \in Z_{N_i} & \text{if user } \mathcal{P}_i \text{ has RSA-based key pair} \\ H_2(y_i^r \bmod p) & \text{if user } \mathcal{P}_i \text{ has DLP-based key pair} \\ H_3(rQ_i) & \text{if user } \mathcal{P}_i \text{ has EDLP-based key pair} \\ H_4(e(H_0(\mathsf{ID}_i), rP_{pub})) & \text{if user } \mathcal{P}_i \text{ has ID-based key pair.} \end{cases}$$

3. Compute

$$\begin{cases} \begin{cases} C_i = r_i^{e_i} \bmod N_i \\ \text{By CRT} \\ a \bmod N_1 N_2 \ldots N_{n_1} \end{cases} & \text{if user } \mathcal{P}_i \text{ has RSA-based key pair} \\ R = g^r \bmod p & \text{if user } \mathcal{P}_i \text{ has DLP-based key pair} \\ R' = rP & \text{if user } \mathcal{P}_i \text{ has EDLP-based key pair} \\ R'' = rP' & \text{if user } \mathcal{P}_i \text{ has ID-based key pair.} \end{cases}$$

We have denoted by CRT the Chinese Remainder Theorem. $a$ is computed from $n$ equations, $C_i = r_i^{e_i} \bmod N_i$, by CRT. This method was initially introduced by Hwang [12].

4. Compute $n$ shadows
$$w_i = f(k_i) \bmod q$$
for $1 \le i \le n$.

5. Encrypt message $m$ as
$$c = E_{f_0}(m),$$
where $E_s(\cdot)$ denote a symmetric encryption scheme (such as AES) with key $s$.

6. Broadcast the ciphertext:
$$C = (c, a, R, R', R'', (\mathcal{P}_1, w_1), \ldots, (\mathcal{P}_n, w_n)).$$

## 4.2 Decryption Phase

To decrypt a ciphertext $C$, at least $t$ out of $n$ receivers in the network must work together and perform the following.

1. Compute $k_i$ as

$$k_i = \begin{cases} \begin{cases} a \bmod N_i = C_i \text{ (via CRT)} \\ r_i = C_i^{d_i} \bmod N_i \\ k_i = H_1(r_i) \end{cases} & \text{if } \mathcal{P}_i \text{ has RSA-based key pair} \\ H_2(R^{x_i} \bmod p) & \text{if } \mathcal{P}_i \text{ has DLP-based key pair} \\ H_3(x_i R') & \text{if } \mathcal{P}_i \text{ has EDLP-based key pair} \\ H_4(e(\mathsf{S}_{\mathsf{ID}_i}, R'')) & \text{if } \mathcal{P}_i \text{ has ID-based key pair.} \end{cases}$$

2. Compute the Lagrange interpolation to reveal $f_0$ as follows.

$$f_0 = \sum_{i=1}^{t} d_i w_i,$$

where

$$d_i = \sum_{j=1, j \neq i}^{t} \frac{k_i}{k_j - k_i}.$$

3. Reveal the message $m$ as follows.

$$m = D_{f_0}(c),$$

where $D_s(\cdot)$ denote a symmetric decryption scheme (such as AES) with key $s$.

# 5 Security Consideration

We now consider the security for each scheme given in the previous section. As defined in Section 3, our schemes should provide adaptively chosen ciphertext security. Observe that these schemes are based on hybrid encryption methods. That is, a session or symmetric key is encrypted by the public key, and the message is encrypted with the session key. The security analysis refers to the work by Shoup [17].

The security of the RSA-based scheme relies on the hardness of factorization of the RSA modulus and the security of the symmetric key algorithm used. To make our system secure against adaptively chosen ciphertext attacks, we can utilize Bellare and Rogaway's RSA encryption scheme using padding called OAEP [2]. Shoup [17] gives an improved version of OAEP called OAEP+ which allows long message encryption. These schemes are applicable for our RSA-based scheme.

Our DLP-based scheme can be associated with the more general security assumption: the Computational Diffie–Hellman Problem (CDHP) is hard. That is, given $g^r$ and $g^x$, find $g^{rx}$. Similar to the DL scheme, the security of our EDLP-based scheme is based on the elliptic curve CDHP; namely, given $rP$ and $xP$, find $rxP$. The security proofs in the random oracle model for such (hybrid) systems have been given by Shoup [17]. Actually, his scheme has some variation from our schemes due to consideration of the length of the messages, but his scheme is applicable to our schemes.

The ID-based scheme is based on the security assumption of the Bilinear Diffie–Hellman Problem (BDHP). That is, given $rP'$, $xP'$, and $bP' \leftarrow H_1(ID)$, find $e(P', P')^{rxb}$. This problem is believed as hard as the CDHP. The full security proof in the random oracle model was given by Boneh and Franklin [4]. It is noted that our scheme is not secure under an adaptively chosen ciphertext attack. However, a slight modification given by Boneh and Franklin [4] can make the scheme secure against such attacks; that is, the parameter $r$ used in our scheme is computed from $r = H_4(f_0, m)$.

# 6 Conclusion

We provided a model for a separable threshold decryption scheme. We note that this model is very appropriate for use in an ad hoc wireless network. We showed the security requirements of such a scheme, and finally proposed a secure scheme that satisfies this model. Our scheme will work for any public-key cryptosystems used by the receiver, including the one based on factorization (such as RSA), the discrete logarithm problem, elliptic curve discrete logarithm problem, and ID-based system that is based on bilinear pairings. We also note that our scheme can be easily extended to include new public-key cryptosystems.

# References

[1] M. Abe, M. Ohkubo, and K. Suzuki. 1-out-of-n signatures from a variety of keys. *Advances in Cryptology—Asiacrypt 2002, Lecture Notes in Computer Science 2501*, pages 415–432, 2002.

[2] M. Bellare and P. Rogaway. Optimal asymmetric encryption. In A. D. Santis, editor, *Advances in Cryptology, Proceedings of EUROCRYPT 94,* LNCS 950, pages 92–111. Springer, 1995.

[3] G. Blakley. Safeguarding cryptographic keys. *In Proceedings of AFIPS 1979 National Computer Conference*, 48:313–317, 1979.

[4] D. Boneh and M. Franklin. Identity-based encryption from the Weil pairing. In J. Kilian, editor, *Advances in Cryptology, Proceedings of CRYPTO 2001,* LNCS 2139, pages 213–229. Springer Verlag, 2001.

[5] D. Boneh and M. Franklin. Identity-based encryption from the Weil pairing. *Lecture Notes in Computer Science*, 2139:213+, 2001.

[6] C. Boyd. Digital multisignatures. In *Cryptography and Coding*, ed. H. Beker and F. Piper, Clarendon Press, Oxford, pages 241–246, 1989.

[7] J. Camenisch and M. Michels. Separability and efficiency for generic group signature schemes. *Advances in Cryptology—Crypto '99*, pages 413–430, 1999.

[8] R.A. Croft and S.P. Harris. Public-key cryptography and reusable shared secrets. *Cryptography and Coding*, H.J. Beker and F.C. Piper(eds), Clarendon Press, Oxford University, UK, pages 189–201, 1989.

[9] Y. Desmedt. Society and group oriented cryptography: A new concept. *Advances in Cryptology—Crypto '87, Lecture Notes in Computer Science 293*, pages 120–127, 1987.

[10] Y. Desmedt and Y. Frankel. Homomorphic zero-knowledge threshold schemes over any finite abelian group. *SIAM Journal on Discrete Mathematics*, 7(4), pages 667–679, 1994.

[11] H. Ghodosi, J. Pieprzyk, and R. Safavi-Naini. Dynamic threshold cryptosystems: A new scheme in group oriented cryptography. *In PRAGOCRYPT '96 (International Conference on the Theory and Applications of Cryptology)*, pages 370–379, 1996.

[12] T. Hwang. Cryptosystem for group oriented cryptography. In *Advances in Cryptology, Proceedings of CRYPTO 90,* LNCS 537, pages 353–360, 1991.

[13] J. Kilian and E. Petrank. Identity escrow. *Advances in Cryptology—Crypto 98, Lecture Notes in Computer Science 1462*, pages 169–185, 1998.

[14] J. K. Liu, V. K. Wei, and D. S. Wong. A separable threshold ring signature scheme. *In the Sixth International Conference on Information Security and Cryptology (ICISC 2003)*, pages 7–21, 2003.

[15] J. Pieprzyk, T. Hardjono, and J. Seberry. *Fundamentals of Computer Security*. Springer-Verlag, Berlin, 2003.

[16] A. Shamir. How to share a secret. *Communications of the ACM*, 22:612–613, November 1979.

[17] V. Shoup, 2001. (IACR eprint, `http://eprint.iacr.org/2001/112`).

[18] V. Shoup and R. Gennaro. Securing threshold cryptosystems against chosen ciphertext attack. *Journal of Cryptology*, 15: 75–96, 2002.

# Chapter 5

# A Secure Group Communication Protocol for Ad Hoc Wireless Networks

Yuh-Min Tseng
*Department of Mathematics, National Changhua University of Education, Jin-De Campus, Chang-Hua, Taiwan 500, R.O.C.*
E-mail: ymtseng@cc.ncue.edu.tw

Chou-Chen Yang
*Department of Information Management, National Chung-Hsing University, Taichung County, Taiwan 402, R.O.C.*
E-mail: cc.yang@dragon.nchu.edu.tw

Der-Ren Liao
*Graduate Institute of Computer Science and Information Engineering, Chaoyang University of Technology, Taichung County, Taiwan 413, R.O.C.*
E-mail: s9127614@csie.cyut.edu.tw

## 1 Introduction

### 1.1 Background

Recently, group communication has been an efficient mechanism for group-oriented applications such as video-conferencing, collaborative work, networking games, and online video. In particular, these applications require packet delivery from one or more senders to a large number of authorized receivers. Group communication and multicasting allow simultaneous delivery of messages to multiple users and thus they can reduce network bandwidth.

M.X. Cheng, D. Li (eds.) *Advances in Wireless Ad Hoc and Sensor Networks.* Signals and Communication Technology, doi: 10.1007/978-0-387-68567-0_5. 

However, internet and wireless networks are open communication channels; eavesdroppers or unauthorized users can easily obtain the transmitted message over open communication channels. A secure group communication protocol ensures that the multicast or broadcast messages are encrypted to prevent eavesdroppers or unauthorized users from obtaining the transmitted messages.

The ad hoc wireless network is defined in 802.11 standard series [1]. An ad hoc wireless network is different from other wireless networks such as access points of Wireless Local Area Networks (WLAN, 802.11g)[1, 2], or base stations of cellular mobile networks [3, 4], because it does not require a fixed infrastructure to construct communications among nodes. In an ad hoc wireless network, devices (called nodes) cooperate to forward packets over possible multihop paths.

Because ad hoc wireless networks require no fixed hardware infrastructure, communications among nodes can be quickly constructed. This property ensures that ad hoc wireless networks are mainly applied for communications in critical applications such as military, emergency, and rescue missions. But this property brings critical security requirements too. Furthermore, nodes in an ad hoc wireless network may move, leave, or join at any time, thus it causes topological changes of ad hoc wireless networks. Due to the particular characteristics, the existent secure protocols for various kinds of applications do not meet security or efficient requirements of ad hoc wireless networks. Several researchers [5–8] have focused on the design of secure routing protocols for ad hoc wireless networks. In fact, the design of an efficient and secure group communication protocol is an important issue for any network environment, and it is an especially significant challenge for an ad hoc wireless network owing to its particular characteristics.

Generally, using a common encryption key only known by authorized members to encrypt transmitted data is a practical approach for achieving secure group communications. The common encryption key can be generated by using multiparty key establishment protocols [9–12]. However, these multiparty key establishment protocols are not suitable for the scalable and dynamically changing requirements of group communications, because a secure group communication has to satisfy the following security requirements [13].

- Forward secrecy: when a new user joins the group, the joining user should not compute any old common key to decrypt the previously encrypted messages.

- Backward secrecy: when a group member leaves the group, the leaving member should not obtain any future common key to decrypt future encrypted messages.

That is, whenever there is a change in group membership, the common encryption key has to be updated to protect the past and future secure communications. Thus, efficient key management is a challenging problem for secure group communications. And the key management problem is mainly concerned with minimizing the number of key updates and the requirement of key storage. Because nodes in an ad hoc wireless network may move, leave, or join at any time, causing topological changes, it is thus a nontrivial problem to design a secure group communication protocol for an ad hoc wireless network.

## 1.2 Related Works

In the past, several secure group communication protocols [14–17] have been proposed to solve the key management problem. Most secure group communication protocols adopt a hierarchical tree to solve the key management problem. The main goal of the hierarchical tree is to decrease the cost of key update and offer easy management for frequent changes of group membership. Recently, Liu and Zhou. [17] proposed a new key management protocol for a large dynamic multicast group. The framework of their scheme is that a group controller manages several subgroup controllers. Each subgroup controller manages one subgroup using the key tree architecture. This kind of management offers efficiency for a large and dynamic multicast group but also increases the key storage of a group controller. In 2003, Tseng [15] proposed a new protocol to remedy this drawback, in which the key storage of the group controller is reduced to a constant size. Although the previously proposed protocols can work efficiently in the wired network environment, when the environment has been changed into wireless networks, these protocols [14–17]are unable to efficiently solve the key management problem.

In wireless networks, the network topology changes frequently. For example, when a user moves from one fixed access point or base station to another one, the user does not really leave the group. Thus, communication should not be broken off whether the user is using the network service or not. The action for holding network communication between access points is called "handoff." Therefore, the management requirement of group communications in wireless networks is different from that in the wired one. In secure group communication for wireless networks, the handoff problem caused by the node's dynamic move should be dealt with by an additional process. To promise secure group communication in wireless networks, several protocols [18, 19, 20] have discussed the handoff problem and given solutions for secure group communication in wireless networks. In 2002, DeCleene et al. [19] proposed a secure group communication protocol for wireless networks, in which the network nodes are separated into several unique areas. Each area contains a subgroup and a controller. Nodes can move from one area to another one randomly and the controller has to control the re-key action to hold the communication confidentiality. DeCleene et al. analyzes the re-key situation in three cases: baseline, immediate, and delayed. Baseline is the basic re-key action as the traditional key management. The immediate and delayed cases are concerned with the handoff problem.

Although the secure group communication protocols [18, 19, 20] may solve the handoff problem for wireless networks, we find that these protocols are designed with fixed hardware infrastructures such as access points of wireless local area networks (802.11g) [1, 2], or base stations of cellular mobile networks [3, 4]. In ad hoc wireless networks, there is no access point or base station for data routing. Thus, these protocols are also unsuitable for ad hoc wireless networks.

In 2002, Li et al. [21] tried to propose a secure group communication for ad hoc wireless networks by using multiparty key establishment protocols [9, 10]. However, in their protocol the maintenance problem is not discussed in detail. In their protocol, the network topology is also separated into several areas. Each area contains a controller and many nodes. However, they do not solve the moving problem of subgroup controllers. Meanwhile, the performance analysis about key update is also lacking, especially for time complexity.

## 1.3 Design Concept

Because an ad hoc wireless network has a different data routing process, critical resource requirement, and different topology architecture, the previously proposed group communication protocols cannot work well for these networks. In our opinion, the subgroup concept is a recommendable solution. The reason is presented as follows. In order to meet the requirements of forward secrecy and backward secrecy, a key update process should be performed after each joining/leaving action. The key update process includes generating a new common encryption key and distributing it to other members. This is called the <1 affects $n$> phenomenon. However, this process will cause a critical problem owing to frequent moves in ad hoc wireless networks. Therefore,in our protocol, wireless nodes are divided into several subgroups and each subgroup is managed by a controller. Each subgroup owns a different local common encryption key. This will reduce the impact of the <1 affects $n$> phenomenon. Meanwhile, how to divide nodes of a group into several subgroups is another issue. In the following, we adopt the cluster-partition method according to some previously proposed routing protocols for ad hoc wireless networks.

Some research [22–25] deals with multicast routing protocols for ad hoc wireless networks. An et al. [24] proposed a Geo-multicast protocol that is a specialized location-dependent multicasting technique. Nodes in an ad hoc wireless network are divided into several specific zones and each zone has a cluster-head to control data transmission. Sajama et al. [25] proposed an Independent-Tree Ad hoc Multicast Routing protocol (ITAMAR). The ITAMAR uniformly distributes nodes over the area of a network. Then a multicast backup tree has been generated to decide an efficient data routing path. The protocols [22, 23] have the same idea as that in [24, 25], in which nodes are divided into several small zones and each zone becomes a small subgroup. There is a difference from [24, 25]; Kozat et al. [22] and Jaikaeo and Shen. [23] proposed routing protocols based on a multicast backbone that divides nodes into several small subgroups, and each subgroup has a subgroup-controller that is chosen according to the nodes' computation ability, traffic location, and remaining power. The chosen nodes become multicast backbone nodes for transmitting data and their neighbor nodes can obtain multicast data from these multicast backbone nodes.

However, the multicast backbone-based routing protocols [22, 23] do not consider security requirements for secure group communication. In this chapter, based on multicast backbone routing protocols [22, 23] and the subgroup concept, we propose a new and secure group communication protocol for ad hoc wireless networks. In the proposed protocol, wireless nodes are also divided into several subgroups and each subgroup is managed by a controller; then these controllers consist of the set of secure multicast backbones. Meanwhile, we are also concerned with the maintenance processes for frequent topological changes of ad hoc wireless networks.

The remainder of this chapter is organized as follows. In the next section, the system model and notations are presented. Section 3 describes our secure group communication protocol. Section 4 presents the maintenance processes of our protocol. Security analysis and performance analysis are given in Sections 5 and 6, respectively. Finally, Section 7 gives conclusions and future work.

# 2 System Assumptions

## 2.1 System Model

In an ad hoc wireless network, there are many nodes located in a scope as shown in Figure 1. Each dot denotes one mobile device (or node). Nodes may cooperate to forward packets over possible multihop paths. A network route from one sending node to a receiving node requires a number of intermediate nodes to forward a packets, and each node may forward a packet to its neighborhood node. Certainly, all wireless transmission links in this network are bidirectional.

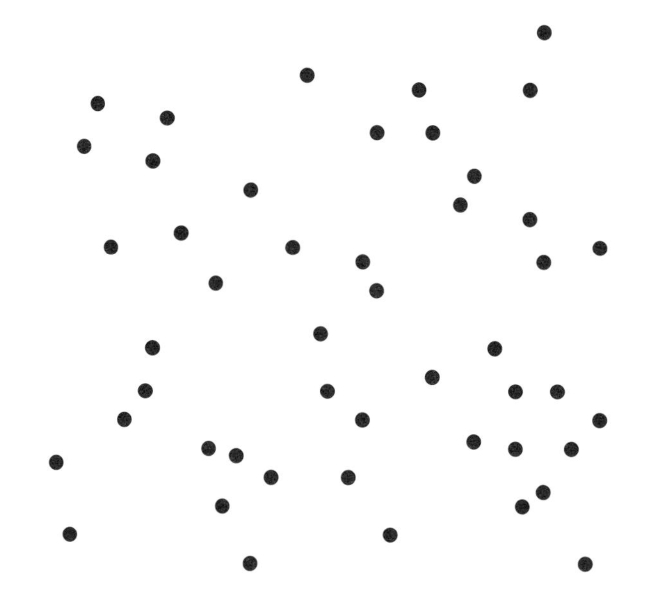

Figure 1: Example of nodes in an ad hoc wireless network.

As mentioned in Section 1.3, Kozat et al. [22] and Jaikaeo and Shen [23] proposed multicast backbone-based protocols that divide nodes into several small subgroups and each subgroup has a subgroup-controller. The chosen nodes become multicast backbone nodes for transmitting data and their neighbor nodes obtain multicasting data from these multicast backbone nodes. For example, after the multicast backbone is constructed, transmission links among nodes are as depicted in Figure 2. The transmission links denote connection paths.

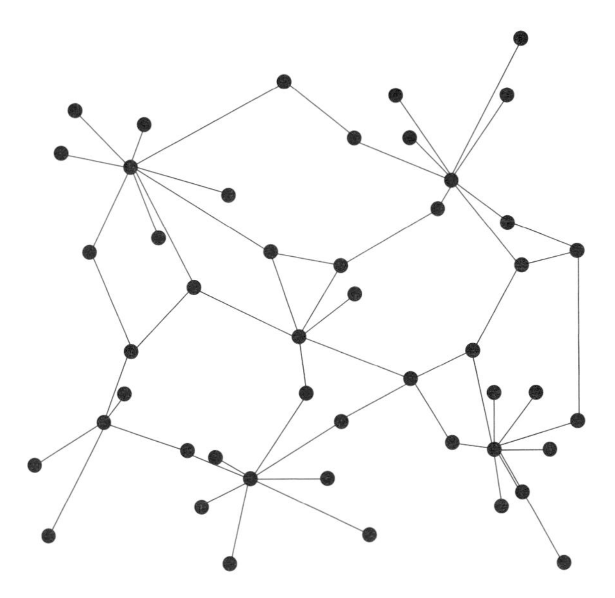

Figure 2: Architecture of multicast backbone in an ad hoc wireless network.

Based on the concept of multicast backbone-based routing protocols proposed in [22, 23], the nodes in our secure group communication protocol are classified into three modes: multicast router (MR), group member, and connector. The constructed network topology is illustrated in Figure 3. The multicast router here manages a subgroup and acts as a data gateway. All nodes are divided into several subgroups, in which each subgroup is managed by one multicast router. Each subgroup contains at least one member. A connector constructs a data path from one subgroup to another one. A member only belongs to one subgroup and is managed by the multicast router of the subgroup. Initially, there are several fundamental properties and security assumptions in our protocol which are presented as follows.

1. All nodes have to broadcast a hello message to one-hop neighborhoods periodically. This message gives nodes surrounding information about the network for communication.

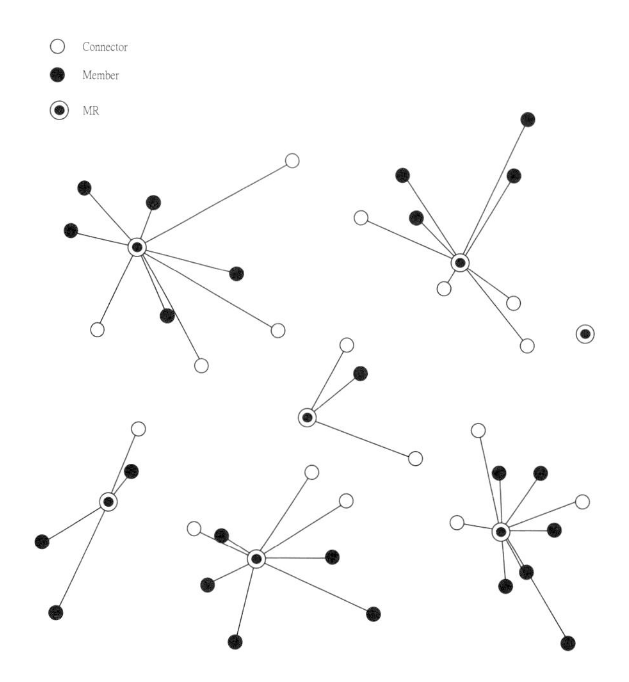


Figure 3: Three kinds of node state in our protocol.

2. One multicast router can manage only one-hop distance members.
3. The number of members managed by one multicast router must be limited in a constant.
4. An authenticated encryption scheme such as [26] is provided for the key distribution and mutual authentication services between wireless nodes.
5. Assume that all nodes have owned public-key certificates before joining the network. That is, each node has a pair of secret and related public keys in a public-key system such as RSA [27] (see ElGamal [28]).
6. We assume that when two nodes want to communicate with each other first, or a node wants to join, leave, go to, or from a subgroup, both nodes must have authenticated mutually. Thus, the related processes in the proposed protocol omit the steps for authenticating mutually.

For secure group communication, each multicast router must generate a subgroup key, and distribute it to the members. The subgroup key will be encrypted using the members' public key and be sent to members, respectively. Thus, each member of a subgroup may decrypt and get the subgroup key using his secret key. Considering the efficiency, the cryptosystem using a secure broadcast transmission should be a symmetric encryption system such as DES [29] or AES [30]. Any two adjoining multicast routers should agree on a communication key for their link path. The communication key between two adjoining multicast routers can be constructed using the existent key agreement protocols such as [31, 32, 33] under the public-key systems. Every message passing through two multicast routers will be encrypted by the communication key. The details are described in our proposed protocol.

## 2.2 Notations

Here we define notations used in our protocols. Without loss of generality, assume that the ad hoc wireless network includes $n$ nodes. In our protocol, there are $m$ nodes that are chosen as multicast routers and each multicast router manages at most $k$ members. Let $N$ be the set of nodes, $N = \{N_1, N_2, N_3, \ldots, N_n\}$, $SMR$ be a set of multicast routers, $SMR = \{MR_1, MR_2, MR_3, \ldots, MR_m\}$, and $S_i$ be a set of members where $S_i = \{S_{i1}, S_{i2}, S_{i3}, \ldots, S_{ik}\}$ and who are managed by $MR_i$, $1 \leq i \leq m$, $1 \leq k \leq (n - m)$. The other notations are defined as follows.

- $MR$: Multicast router.
- $Nei(N_i)$: One-hop neighbors of $N_i$, $1 \leq i \leq n$.
- $JOIN\_MESSAGE$: Joining request message.
- $LEAVE\_MESSAGE$: Leaving request message.
- $MR\_DISSOLVE\_MESSAGE$: This message is used to notify members that the subgroup will be dissolved.
- $W(N_i)$: Weight of $N_i$ is evaluated according to nodes' computation ability, traffic location, and remaining power.
- $ID_i$: Identity of node $N_i$.
- $MID_i$: $MR_i$'s subgroup identity.
- $SGK_i$: Subgroup key of $MR_i$.
- $MSK_{ij}$: Session key constructed by $MR_i$ and $MR_j$.

- $PK_i$, $SK_i$: Public key and secret key of node $N_i$.
- $E_{k_i}()$: Symmetric encryption with key $K_i$.
- $D_{k_i}()$: Symmetric decryption with key $K_i$.
- $PE_{PK_i}()$: Public-key encryption with public-key $PK_i$.
- $PD_{SK_i}()$: Public-key decryption with secret-key $SK_i$.
- $\{MID_i : S_{jk} \rightarrow S_{il}\}$: The connection path represents that $S_{jk}$ is a connector of $MR_j$ and it is responsible for connecting to subgroup $MR_i$ through the connector $S_{il}$ of $MR_i$.
- $MTAB_i$ : Connection path table of $MR_i$, which records the connection paths to all neighbor subgroups.
- $CTAB_{ij}$ Connection path table of connector $S_{ij}$, which records the connection path to other subgroups.
- $t_{wait}$: A time interval for waiting event.

# 3 Proposed Protocol

The proposed protocol consists of two phases: the initial phase and the data communication phase. In the initial phase, there are three processes that include how to choose multicast routers, subgroup creation and distributing the subgroup key to members securely, and constructing the communication key between two adjoining multicast routers. In the data communication phase, we present the message multicasting process of secure group communication.

## 3.1 Initial Phase

In this phase, the subgroup creation and the multicast backbone construction are proposed. Then each $MR$ generates one subgroup key and sends it to its members. Besides, any two adjoining multicast routers also agree upon a communication key.

### 3.1.1 Choosing a Multicast Router

Because our protocol is based on the methods [22, 23] of constructing a multicast backbone, we first briefly present the methods used. Kozat et al. [22] proposed a Virtual Dynamic Backbone Protocol (VDBP) to deal with the whole backbone network creation and maintenance in an ad hoc wireless network. In the VDBP protocol, nodes are marked as virtual access points, normal nodes, or unregistered nodes. Data transmission goes through virtual access points to normal users. To determine a virtual access point, all nodes in VDBP create a neighborhood information table. This table contains the records of link failure frequency, degree, and identity. Jaikaeo and Shen [23] improved the VDBP protocol to increase information of the remaining power into the neighborhood information table. An illustration of the constructed multicast backbone is shown in Figure 2.

At this phase, we use the same method to determine the neighborhood information table. We set the neighborhood information table for node $N_i$ as the weight $W(N_i)$. Here, a virtual access point is called a multicast router. The weight is used as comparison information. To ensure the weight information of nodes is correct in real-time for other nodes, all nodes have to update their weight in every period time and broadcast them with the hello message to one-hop neighborhoods periodically. Nodes may join the surrounding $MR$ and become a member of the subgroup. The whole ad hoc wireless network is divided into many subgroups.

If a node has a best weight in its communication area, it means that the node may have good power support, important traffic location, much more stability, or stronger computational ability. The way to find a local optimal node is the same as the BSP (Backbone Selection Process) proposed in VDBP [22]. In addition, we add the remaining power and computation ability into the weight. We use the process based on the BSP [22] to determine optimal nodes and let them be multicast routers. The brief process is presented as follows.

*Choosing the MR Process*

Step 1: Each node $N_i$ generates its weight $W(N_i)$ and broadcasts the hello message and $W(N_i)$ to $Nei(N_i)$.

Step 2: Each node $N_i$ begins to update the best weight based on other received weights $W(N_j)$.

Step 3: When time $t_{wait}$ expires, each node $N_i$ checks if it is the local optimal node with the best $W(N_i)$. If it is the best, node $N_i$ changes its state to a multicast router $MR$.

#### 3.1.2 Subgroup Creation and Distributing Subgroup Communication Key

After all $MR$s have been chosen, we create subgroups. The other nodes in the ad hoc network will join the chosen $MR$s. At first, the chosen $MR$s have to broadcast their weight to their neighborhoods. Other non-$MR$ nodes receive one or more weights sent by some $MR$s. Then non-$MR$ nodes choose the best weight and send a joining message to the corresponding $MR$. Thus, $MR$s receive many joining messages from many non-$MR$ nodes. Then each $MR$ randomly generates a subgroup identity and subgroup keys, then encrypts them with members' public keys and distributes them to members. The details are presented as follows.

*Subgroup Creation Process*

Step 1: Each $MR_i$,$1 \leq i \leq m$, broadcasts $W(MR_i)$ to $Nei(MR_i)$.

Step 2: Upon receiving more than a $W(MR_i)$, each non-$MR$ node $N_j$ chooses the best $W(MR_i)$ as its multicast router.

Step 3: Each non-$MR$ node $N_j$ sends $\{JOIN\_MESSAGE, ID_j, PK_j\}$ to its multicast router.

Step 4: Upon receiving the joining message sent by $N_j$, $MR_i$ adds $N_j$ to $S_i$.

Step 5: $MR_i$ generates an identity $MID_i$ and a subgroup key $SGK_i$ for its subgroup.

Step 6: Each $MR_i$ sends $PE_{PK_{S_{il}}}(SGK_i, MID_i)$ to each member $S_{il}$ in the set $S_i$, where $1 \leq l \leq k$.

Step 7: Each $S_{il}$ executes $PD_{SK_{S_{il}}}(PE_{PK_{S_{il}}}(SGK_i, MID_i))$ using his/her secret key and gets $SGK_i$ and $MID_i$.

After the subgroup creation process, the established subgroups are depicted in Figure 4. The next process is to establish the connection paths between subgroups. All members have to run the finding connector process to determine connectors. In this process, all nodes broadcast their subgroup

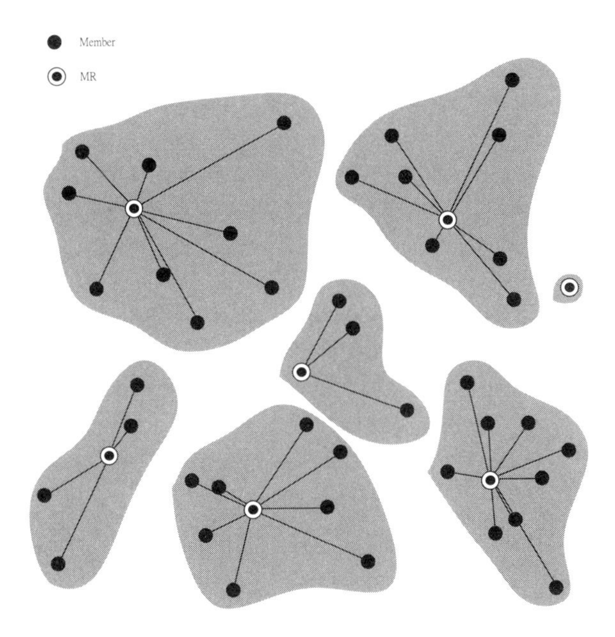


Figure 4: The established subgroup example after the subgroup creation process.

identities to their neighborhoods. The details are presented as follows.

*Finding Connector Process*

Step 1: Each $S_{ij}$ in $S_i$, $1 \leq i \leq m$, broadcasts $MID_i$ to $Nei(S_{ij})$.

Step 2: When $S_{ij}$ receives $MID_k$ from other subgroup member $S_{kl}$, $S_{ij}$ adds $(MID_k : S_{ij} \rightarrow S_{kl})$ to $CTAB_{ij}$ and sends it to $MR_i$. In this case, connector $S_{ij}$ may connect to more than one subgroup.

Step 3: Upon receiving $(MID_k : S_{ij} \rightarrow S_{kl})$ from $S_{ij}$, $MR_i$ adds it to $MTAB_i$.

In order to ensure the reliability of the connection between $MR$s, any two adjoining subgroups may own more than one connection path and each connector may connect more than two subgroups. Note that connectors are also subgroup members. After finishing this step, all backbones have been constructed as illustrated in Figure 5. Note that all nodes have to run the finding connector process in each period time to ensure the connection path

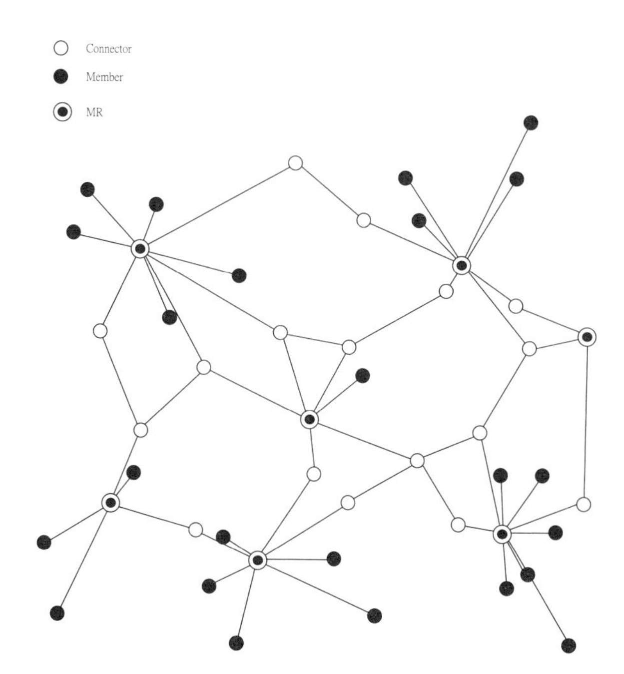


Figure 5: The network architecture after completing the initial phase.

is correct because connection paths could be changed when the network topology changes.

### 3.1.3 Creating a Communication Key Between $MR$s

Here, we use the existent public-key-based key agreement protocols [30, 31, 32] to construct the communication key between two adjoining $MR$s. For example, as in Figure 6, all nodes are assigned some serial numbers that make it easy to explain the next steps. First, all $MR$s find their neighbor subgroups through their connectors. Then all $MR$s construct the communication keys with neighbor $MR$s. We illustrate the constructed communication keys in Figure 7, where $MR_i$ and $MR_j$ construct their communication key $MSK_{ij}$.

## 3.2 Data Communication Phase

In this phase, we present the secure group transmission in our protocol. Assume that an ad hoc wireless network has constructed the multicast

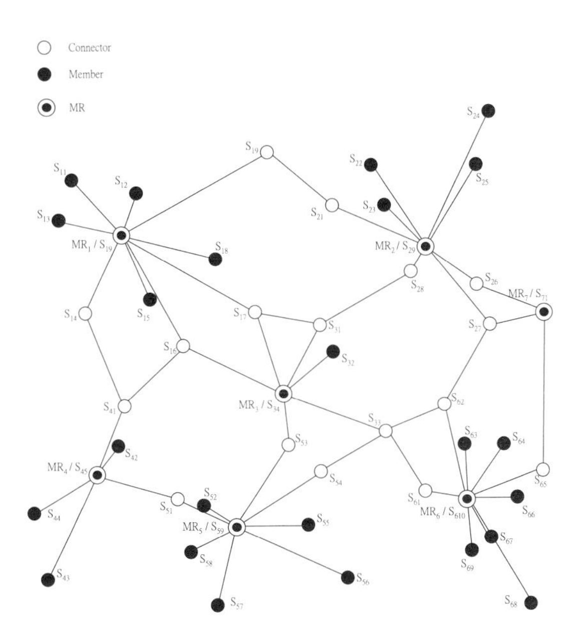


Figure 6: An example of nodes with serial numbers.

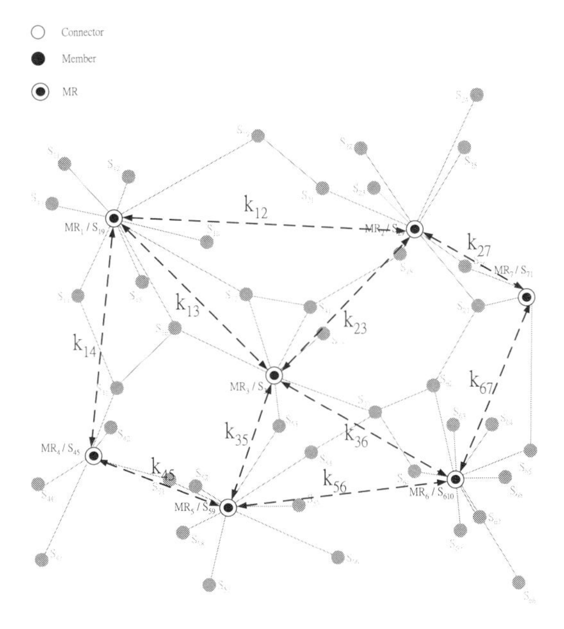


Figure 7: The constructed communication key by key agreement protocol.

backbone. When a member $S_{ij}$ is going to broadcast a message to all legal group members, the communication steps are presented as follows.

Step 1: $S_{ij}$ sends $E_{SGK_i}(message)$ to $MR_i$.

Step 2: $MR_i$ executes $D_{SGK_i}(E_{SGK_i}(message))$ to obtain the message and then sends $E_{SGK_i}(message)$ to members in $S_i$.

Step 3: All members in $S_i$ receive the encryption message and execute $D_{SGK_i}$ ($E_{SGK_i}$ $(message)$) to obtain the message.

Step 4: $MR_i$ sends $E_{MSK_{ik}}(message, MID_i)$ to its adjoining $MR_k$ according to the table $MTAB_i$.

Step 5: Upon receiving the message $E_{MSK_{ik}}(message, MID_i)$, each $MR_k$ executes $D_{MSK_{ik}}(E_{MSK_{ik}}(message, MID_i))$ to obtain the message and reencrypts the message with $SGK_k$ and distributes the encrypted message to members in $S_k$. Each $S_{ki}$ in $S_k$ may decrypt the message.

Step 6: Each $MR_k$ performs Step 4 and Step 5 until the message has been transmitted to all subgroups.

In ad hoc wireless networks, receivers may receive more than one identical message during message multicasting. This situation is a basic routing problem named the "cycle" problem for ad hoc wireless networks [22, 23]. The flexible solution of this problem is to add a sequence number to the transmitted packets. In our protocol, this problem still occurs. It could increase the overhead. To solve this problem, the message should have its sequence number.

# 4 Maintenance Processes

## 4.1 Member Joining, Leaving, or Moving

In this section, we discuss the maintenance processes for the topological changes caused by nodes in the ad hoc wireless network. We know that nodes in the ad hoc wireless network may move randomly and dynamically. According to the assumption of our protocol, nodes must contact each other periodically. Once any node changes its mode or location, the topology could be changed and some related maintenance processes need to be performed.

### 4.1.1 Member Joining

In our protocol, there are two events that will cause the "member joining" case. One is that a new and legal node joins the ad hoc wireless network. Another is that a node moves from its original subgroup to another one.

When a node wants to join a subgroup, in order to keep the forward secrecy, the subgroup must update the subgroup key. In the process, the joining node first has to find whether there is any $MR$ that is located nearby it. Then the node sends a joining message to the $MR$. The $MR$ will accept the new node if the number of members is under the limited number. On the other hand, if the node does receive any information sent by any $MR$ in an interval time, the node claims itself as a $MR$. Note that all nodes have to run the finding connector process in each period time to ensure that the connection path is correct because connection paths could be changed when the network topology changes. Thus, the new $MR$ will create the connection paths with the other subgroups after running the finding connector process in the period time. Assume that a new node $N_i$ is going to join a subgroup; the member joining process is presented as follows:

*Member Joining Process*

Step 1: $N_i$ broadcasts a joining message to $Nei(N_i)$ to seek $MR$s and waits for the response of any $MR_j$. If $N_i$ does not receive any response, it sends a message to $Nei(N_i)$ to claim itself as a $MR$ and stop this process.

Step 2: If $N_i$ receives some responses, $N_i$ chooses the best weight $W_{MR_j}$.

Step 3: $N_i$ sends $PE_{PK_j}(JOIN\_MESSAGE, ID_i, PK_i)$ to $MR_j$.

Step 4: Upon receiving $PE_{PK_j}(JOIN\_MESSAGE, ID_i, PK_i)$, $MR_j$ obtains the joining request by executing the public decryption function $PD_{SK_j}(PE_{PK_j}(JOIN\_MESSAGE, ID_i, PK_i))$.

Step 5: If $|S_j|$ is under the limited number, $MR_j$ generates a new subgroup key $SGK'_j$ and distributes $D_{SGK_j}(E_{SGK_j}(SGK'_j))$ to each $S_{jk}$ in $S_j$, where $|S_j|$ is the number of members in $S_j$. Then $MR_j$ adds $N_i$ into $S_j$ as $S'_j$ and sends $PE_{PK_i}(SGK'_j, MID_j)$ to $N_i$.

Step 6: Upon receiving the rekey message $E_{SGK_j}(SGK'_j)$, each $S_{jk}$ in $S_j$ may obtain the new subgroup key $SGK'_j$ by executing $D_{SGK_j}$ $(E_{SGK_j}$

$(SGK'_j)$). And $N_i$ also obtains the subgroup key by executing $PD_{SK_i}$ $(PE_{PK_i}\ (SGK'_i))$.

### 4.1.2 Member Leaving

Assume that a node $S_{ij}$ is a member of the subgroup controlled by $MR_i$. There are three events that will cause the "member leaving" case.

- Event A: $S_{ij}$ leaves the subgroup voluntarily.
- Event B: $MR_i$ lays off $S_{ij}$.
- Event C: $S_{ij}$ loses the connection with its $MR_i$.

In Event A, if a subgroup member actively wants to quit the group communication service, the member should notify its $MR$. Then the $MR$ can update the subgroup key for backward secrecy. In Event B, the subgroup controller $MR$ may lay off members actively. In Event C, we know that wireless nodes may move dynamically, and thus the subgroup member could lose the connection with its $MR$. The process dealing with these events is presented as follows.

*Member Leaving Process*

Step 1: If Event A occurs, $S_{ij}$ has to send $E_{SGK_i}$ $(LEAVE\_MESSAGE, ID_j)$ to $MR_i$. $MR_i$ gets the message by executing $D_{SGK_i}$ $(E_{SGK_i}$ $(LEAVE\_MESSAGE, ID_j))$ and removes $S_{ij}$ from $S_i$, and then goes to step 3.

Step 2: If Event B or Event C occurs, $MR_i$ removes $S_{ij}$ from $S_i$ and runs step 3. Note that all nodes have to broadcast the hello message to surrounding nodes periodically, so $MR_i$ or $S_{ij}$ can know whether the connection is alive.

Step 3: $MR_i$ generates a new subgroup key $SGK'_i$ and distributes $PE_{PK_k}$ $(SGK'_i)$ to each $S_{ik}$ in $S_i$, $k \neq j$.

Step 4: Upon receiving $PE_{PK_k}(SGK'_i)$, $S_{ik}$ updates the subgroup key by $PD_{SK_k}(PE_{PK_k}(SGK'_i))$.

### 4.1.3 Member Moving

Assume that $S_{ij}$ is a member of the subgroup controlled by $MR_i$. There are two events that will cause the "member moving" case.

- Event A: $S_{ij}$ is moving out of $MR_i$'s communication area and misses the connection.
- Event B: $MR_i$ is moving and lets $S_{ij}$ miss $MR_i$.

When a member disconnects with its $MR$, the $MR$ also disconnects with the member. However, if the $MR$ changes its subgroup key immediately and the member again moves into the communication area of the $MR$ in a short time, the $MR$ will again change its subgroup key. This situation will cause the "handoff" problem. Solving the handoff problem is an important design issue in secure group communication protocols for wireless networks [18, 19, 20].

Here we use the retransmission method to solve the handoff problem. Assume that a member is moving out of the $MR$'s communication area. The $MR$ will notify the neighbor $MR$ to find this member. If some neighbor $MR$ acknowledges the original $MR$ in the time $t_{wait}$ and a broadcast message needs to be sent, then the original $MR$ will send the broadcast message to the member by passing through the acknowledged $MR$ during $t_{wait}$. If $t_{wait}$ expires, the $MR$ still disconnects the member, and the $MR$ updates its subgroup key immediately.

To deal with the above events, the member has to run the member moving process and the $MR$ has to run the MR moving process, described in detail in Section 4.2.1. The member moving process is described as follows.

*Member Moving Process*

Step 1: If $S_{ij}$ reconnects with $MR_i$ in $t_{wait}$, then $S_{ij}$ stops this process. Otherwise, $S_{ij}$ leaves the subgroup and runs the member joining process.

Step 2: If $S_{ij}$ receives $PE_{PK_i}(MR\_DISSOLVE\_MESSAGE)$ from $MR_i$, then $S_{ij}$ runs the member joining process to find another subgroup.

## 4.2 MR Moving or Dissolving

In this subsection, we discuss the topological changes caused by $MR$'s moving and dissolving actions.

### 4.2.1 MR Moving

In this case, $MR$ moving could cause some members to lose their multicast router. And the multicast router will disconnect its members too. In this situation, the multicast router will wait for the disconnecting members. Once the member does not come back after the time $t_{wait}$, the multicast router treats it as the member leaving. The details are presented as follows.

*MR Moving Process*

Step 1: If $MR_i$ disconnects its member $S_{ij}$, then $MR_i$ sends a message to its neighbor $MR_k$, $1 \leq k \leq m$, $k \neq i$ to find $S_{ij}$. If $MR_k$ receives the message from $MR_i$ and $S_{ij}$ is located in its communication area. $MR_k$ acknowledge $MR_i$. If $MR_i$ receives the acknowledgment from $MR_k$ and a broadcast message needs to be sent during the time $t_{wait}$, then the original $MR$ will send the broadcast message to the member by passing through $MR_k$. That is, $MR_i$ sends $E_{SGK_i}(data)$ to $S_{ij}$ passing through $MR_k$.

Step 2: If $S_{ij}$ does not come back in the time $t_{wait}$, $MR_i$ removes $S_{ij}$ from the set $S_i$. Thus, $MR_i$ generates a new $SGK_i'$ and distributes $PE_{PK_k}(SGK_i')$ to $S_{ik}$, $1 \leq k \leq m$, $k \neq j$. Meanwhile, $S_{ij}$ should run the member joining process.

Step 3: Upon receiving $PE_{PK_k}(SGK_i')$, each $S_{ik}$ obtains the new subgroup key $SGK_i'$ by executing $PD_{SK_K}(PE_{PK_k}(SGK_i'))$.

### 4.2.2 MR Dissolving

The $MR$ dissolving case occurs in the situation where the subgroup has lost all connections to other subgroups. In this case, the $MR$ must send a $MR\_DISSOLVE\_MESSAGE$ to its members. In another situation, if the $MR$ has crashed, members will lose the connection with the $MR$ and then members will wait for the time $t_{wait}$. When time $t_{wait}$ expires, members run the member joining process automatically to find a new subgroup.

*MR Dissolving Process*

Step 1: If $MR_i$ has no subgroup members and no connection paths connect to other multicast routers $MR_j$, $MR_i$ changes its state into the member mode and runs the member joining process.

Step 2: If $MR_i$ has subgroup members and no connection paths connect to other multicast routers $MR_j$, $MR_i$ sends $E_{SGK'_i}(MR\_DISSOLVE\_MESSAGE)$ to $S_{ik}$ in $S_i$ changes its state into the member mode runs the member joining process.

## 4.3 Connector Moving

Assume that $S_{ij}$ is a connector of the subgroup controlled by $MR_i$. There are two events that will cause the "connector moving" case. The steps for the two events are different and are respectively described as follows.

*Connector Moving Process*

- Event A: Connector $S_{ij}$ of $MR_i$ disconnects from the connector $S_{kl}$ of $MR_k$.

  Step 1: $S_{ij}$ deletes the connection path of the table $CTAB_{ij}$.

  Step 2: $S_{ij}$ sends a message to notify $MR_i$ to drop the connection path.

  Step 3: If the table $CTAB_{ij}$ is null, $S_{ij}$ changes its state to the member mode.

  Step 4: $MR_i$ deletes the connection path in the table $MTAB_i$.

- Event B: Connector $S_{ij}$ disconnects from its $MR_i$ and it has some links to connect to the connectors $S_{kl}$ of $MR_k$, $1 \leq k \leq m$, $k \neq i$.

  Step 1: $S_{ij}$ sends a message to notify $S_{kl}$ to drop the connection paths and $S_{ij}$ deletes all connection paths in the table $CTAB_{ij}$. Then, $S_{ij}$ changes its state into the member mode and runs the member moving process.

  Step 2: Upon receiving the drop message from $S_{ij}$, each $S_{kl}$ drops this connection path and deletes the connection path of the table $CTAB_{kl}$.

  Step 3: $S_{kl}$ sends a message to notify $MR_k$ to drop the connection path.

  Step 4: $MR_k$ deletes the corresponding connection path in the table $MTAB_k$.

## 5 Security Analysis

In this section, we analyze the security of the proposed group communication protocol. The security of our proposed protocol is based on three security assumptions: secure symmetric cryptosystems [29, 30], secure public-key-based cryptosystems [26, 27, 28], and secure key agreement protocols [31, 32, 33] based on public-key systems. Note that here we do not want to discuss in detail the security of the used secure key agreement protocols and cryptosystems. To get further understanding about used secure protocols and cryptosystems, we suggest readers refer to some related papers [26–33]. We mainly focus on the fact that the proposed protocol can provide both forward secrecy and backward secrecy of secure group communication.

In our protocol, each $MR$ generates a subgroup key and distributes it to its members using the secure public-key-based cryptosystem. That is, the subgroup key is encrypted using the member's public key, and each member may decrypt the subgroup key by her secret key. According to the system model mentioned in Section 2, a legal and authorized node must own a legal public-key pair and other nodes may authenticate the node. Therefore, if an attacker wants to enter one subgroup, the attacker has no legal public-key pair, so he could not obtain the subgroup key. Furthermore, any two adjoining multicast routers may construct a communication key using the secure key agreement protocols [31, 32, 33] based on public-key systems. If an attacker tries to impersonate a multicast router and sends the key update message to other members, and we have assumed that any two nodes can authenticate each other, the attackers cannot masquerade as multicast routers or members because an attacker is unable to get the subgroup keys and the communication keys between any two adjoining multicast routers. Thus, our protocol can achieve secure group communication.

In the following, we show that both forward secrecy and backward secrecy are kept in our protocol.

- Forward secrecy: To achieve forward secrecy, the new joining member cannot decrypt the past group communication message. In our maintenance processes, the member joining process has defined that once a new member is going to join a subgroup, the $MR$ has to update the subgroup key first and then send the new subgroup key to the new member. Thus, the new joining member has no old subgroup key and cannot decrypt the past group communication message.
- Backward secrecy: Backward secrecy is such that if a group member

leaves the subgroup, he cannot decrypt the future group communication messages. For the leaving member, the $MR$ will immediately send the new subgroup key through a secure channel to the remaining members. In the maintain phase, if a $MR$ disconnects a member after the time $t_{wait}$, the $MR$ will treat it as if the member has left the subgroup and update the subgroup key immediately. These actions promise backward secrecy in our protocol.

# 6 Performance Analysis

The performance of the key update process is a critical issue for the design of secure group communication protocols. As mentioned in Section 1.2, Li et al. [21] proposed a secure group communication protocol for ad hoc wireless networks. In their protocol, nodes in the ad hoc wireless network are also divided into several subgroups and each subgroup has a subgroup controller. These subgroup-controllers construct a group key by using multiparty key establishment protocols [9, 10]. Their protocol has a maintenance problem: because a subgroup controller could dissolve or move frequently, all subgroup controllers must reconstruct a new group key by multiparty key establishment protocols. In their protocol, Li et al. do not consider this situation and are only concerned with the member joining and leaving processes. As we know, multiparty key establishment protocols require expensive computation time. Thus, the <1 affect $n$> phenomenon will decrease the whole performance.

Therefore, in our protocol, wireless nodes are divided into several subgroups, each subgroup is managed by a controller, and then controllers consist of a set of secure multicast backbones. Each subgroup owns a different local common encryption key. This will reduce the impact of the <1 affects $n$> phenomenon. Meanwhile, although the moving of the $MR$ will cause topological changes, the impact of the <1 affects $n$> phenomenon is small because we use a two-party key agreement protocol to construct their communication keys. Only the surrounding $MR$s of the moving $MR$ need to agree again upon the new communication key. Our protocol can offer better performance when the number of $MR$s is large and $MR$ moves are highly dynamic.

In the following, we analyze the proposed protocol in terms of time complexity, the size of key storage, and the number of key updates. For convenience, the following notations are used to analyze the performance of the protocol. Assume that an ad hoc wireless network contains $n$ nodes and $m$ $MR$s. Let $k$ denote $|S_i|$ such that $\sum |S_i| = n$ and $1 \leq |S_i| \leq n - m + 1$. $T_{PE}$ denotes the time of a public-key encryption. $T_{PD}$ denotes the time of a public-key decryption. $T_{EK}$ denotes the time of a symmetric encryption and $T_{DK}$ denotes the time of a symmetric decryption. $T_{agree}$ denotes the time required for two $MR$s to agree upon a communication key. The performance evaluation about the related processes in our protocol is summarized in Table 1. Note that the Member Moving Process and MR moving process could include one member joining process and one member leaving process, or nothing because our protocol may manage the handoff problem. For Event B in the member moving process, it requires the time of the member moving process.

Table 1: Performance evaluation.

| | | Time complexity | S_MR | S_M | NU |
|---|---|---|---|---|---|
| Subgroup creation process | | $mT_{PE} + (n-m)T_{PD}$ | 1 | 1 | 0 |
| Finding connector process | | 0 | 1 | 1 | 0 |
| $MR$ key agreement $(min, max)$ | | $((m-1)T_{agree}, (\frac{m(m-1)}{2})T_{agree})$ | $(2, m)$ | 1 | 0 |
| Member joining process | | $2(T_{PE} + T_{PD}) + T_{EK} + kT_{DK}$ | $(2, m)$ | 1 | 1 |
| Member leaving process | | $T_{EK} + T_{DK} + (k-1)(T_{PE} + T_{PD})$ | $(2, m)$ | 1 | 1 |
| $MR$ dissolving process | | $T_{PE} + kT_{PD}$ | $(2, m)$ | 1 | 0 |
| Data communication phase | Communication inside subgroup | $T_{PE} + kT_{DK}$ | $(2, m)$ | 1 | 0 |
| | Communication between $MR$s (min, max) | $\left((m-1)(T_{EK} + T_{DK}), (\frac{m(m-1)}{2})(T_{EK} + T_{DK})\right)$ | $(2, m)$ | 1 | 0 |

S_MR = the size of stored keys in each MR.
S_M = the size of stored keys in each member.
NU = the number of key update.

In Table 1, there is a situation that needs to be discussed here. Because the topology of an ad hoc wireless network changes randomly, we give minimal and maximal time complexity here. Figure 8 illustrates the best and worst topology for $MR$ agreement time. In Figure 8a, each $MR$ has only one neighbor $MR$. So each $MR$ only does a key-agreement action. In Figure 8b,

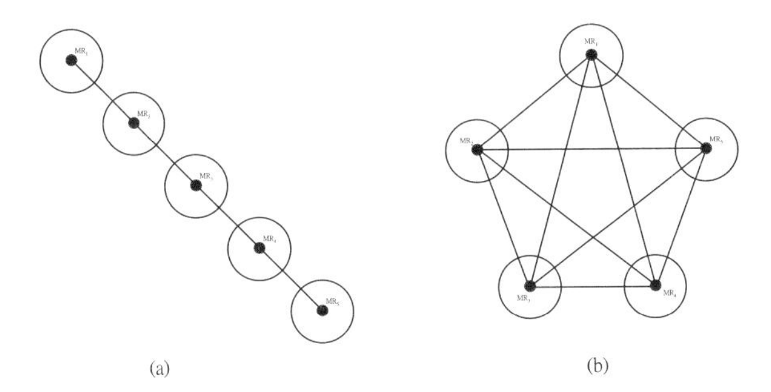


Figure 8: Topologies of subgroups. (a) Linear graph; (b) complete graph.

all $MR$s are connected as a complete graph. It is the worse topology for key agreement because each $MR$ has to do $(m-1)$ key agreement actions and will keep $(m-1)$ communication keys. So the time complexity by adding the total cost of $MR$s will require ((m(m-1))/2)$T_{agree}$. However, note that in a complete graph there is minimal communication delay. On the other hand, there is maximal communication delay in a linear graph.

# 7 Conclusions and Future Works

A new and secure group communication protocol for ad hoc wireless networks has been proposed. Our protocol considers the real network topology and combines the subgroup concept with a multicast backbone-based routing method to provide efficient maintenance processes for topological changes. Considering frequent topological changes of ad hoc wireless networks such as joining, leaving, and dissolving of wireless nodes ($MR$s, connectors, and members), we have also proposed the corresponding maintenance processes. For security, we have shown that the proposed protocol possesses both forward and backward secrecy of a secure group communication. Meanwhile, performance analysis regarding time complexity, key storage, and key update is given to validate our protocol. Although the decryption/re-encryption of data transmission between $MR$s will delay the communication and increase the computation cost, our management framework is more suitable for ad hoc wireless networks than previously proposed protocols.

Mutual authentication services for wireless nodes are not discussed in our protocol in detail. We assume that each node has a pair of secret key and related public keys in the public-key system to achieve key distribution and the mutual authentication services between wireless nodes. Generally, there is a certification authority that is responsible for issuing each node's public-key certificate. But, how to mutually authenticate efficiently with each other in ad hoc wireless networks will be an interesting and critical issue in the future.

## References


[1] IEEE Std 802.11. *Wireless LAN media access control (MAC) and physical layer (PHY) specifications* (ANSI/IEEE Std. 802.11: 1999 (E) Part 11, ISO/IEC 8802-11, 1999).

[2] J. Ala-Laurila, J. Mikkonen, and J. Rinnemaa, Wireless LAN access network architecture for mobile operators, *IEEE Communications Magazine* 39, 11 (2001) pp. 82–89.

[3] *General Packet Radio Services (GPRS) Service Description Stage 2* (TS 122 060, ETSI, 2002).

[4] *Mobile Execution Environment (MExE) service description Stage 1* (3GPP TS 22.057 Version 5.4.0, Jul, 2002).

[5] S. Gupte and M. Singhal, Secure routing in mobile wireless ad hoc networks, *Ad Hoc Networks*, 1 (2003), pp. 151–174.

[6] K. Sanzgiri, B. Dahill, B.N. Levine, C. Shields, and E.M. Belding-Royer, A secure routing protocol for ad hoc networks. *In Proceedings of 2002 IEEE InternationalConference on Network Protocols (ICNP)* (November 2002) pp. 1–10.

[7] Y.C. Hu, D.B. Johnson, and A. Perrig, SEAD: Secure efficient distance vector routing for mobile wireless ad hoc networks, *Ad Hoc Networks* 1 (2003), pp. 175–192.

[8] M.G. Zapata, Secure ad hoc on-demand distance vector (SAODV) routing, *Internet-Draft draft-guerrero-manetsaodv-00.txt, August 2002* (first published in the IETF MANET Mailing List October 8, 2001).

[9] G. Ateniese, M. Steiner, and G. Tsudik, New multiparty authentication services and key agreement protocols, *IEEE J. Sel. Areas Comm.* 18(4), (2000) pp. 628–639.

[10] M. Steiner, G. Tsudik, and M. Waidner, CLIQUES: A new approach to group key agreement. *In Proceedings of the Eighteenth International Conference Distributed Computing. Systems. (ICDCS'98)* (1998) pp. 380–387.

[11] Y.M. Tseng and J.K. Jan, Anonymous conference key distribution systems based on the discrete logarithm problem, *Comput. Communi,* 22(8), (1999) pp. 749–754.

[12] Y.M. Tseng, Multi-party key agreement protocols with cheater identification, *Appli. Math. and Comput.,* 145(2–3) (2003) pp. 551–559.

[13] M.J. Moyer, J.R. Rao, Jr, and P. Rohatg, A survey of security issues in multicast communications, *IEEE Network,* 13(6), (1999) pp. 12–23.

[14] C.K. Wong, M. Gouda, and S.S. Lam, Secure group communications using key graphs, *IEEE/ACM Transa. Netw.* 8(1), (2000) pp. 16–30.

[15] Y.M. Tseng, A Scalable Key Management Scheme with Minimizing Key Storage for Secure Group Communications, *Int. J. Netw. Manage.* 13(6), (2003) pp. 419–425.

[16] M. Li, R. Poovendran, and C. Berenstein, Design of secure multicast key management schemes with communication budget constraint, *IEEE Communi. Lett.,* 6(3), (2002) pp. 108–110.

[17] J. Liu and M. Zhou, Key management and access control for large dynamic multicast group, *WECWIS 2002* (2002) pp. 121–128.

[18] Y. Sun, W. Trappe, and K.J. Liu, An efficient key management scheme for secure wireless multicast. *In Proceedings of IEEE International Conference on Communications* (2002) pp. 1236–1240.

[19] B. Decleene, L. Dondeti, S. Griffin, T. Hardjono, D. Kiwior, J. Kurose, D. Towsley, S. Vasudevan, and C. Zhang, Secure group communication for wireless networks, *Military Communications Conference. IEEEE MILCOM* (2001) pp. 113–117.

[20] B.S. Kim and K.J. Han, Multicast handoff agent scheme for micro-mobility in all-IP wireless network, *IEE Electronics Letters,* 38(12) (2002) pp. 596–597.

[21] X.Y. Li, Y. Wang, and O. Frieder, Efficient hybrid key agreement protocol for wireless ad hoc networks. *In Proceedings. Eleventh International Conference on Computer Communications and Networks* (2002), pp. 404–409.

[22] U.C. Kozat, G. Kondylis, B. Ryu, and M.K. Marina, Virtual dynamic backbone for mobile ad hoc networks, *IEEE International Conference on Communications (ICC) (Helsinki, Finland)* (June 2001).

[23] C. Jaikaeo and C.C. Shen, Adaptive backbone-based multicast for ad hoc networks. *In ICC 2002. IEEE International Conference on Communications* (2002) pp.3149–3155.

[24] B. An and S. Papavassiliou, Geomulticast: Architectures and protocols for mobile ad hoc wireless networks, *Journal of Parallel Distributed Computing,* 63 (2003) pp. 182–195.

[25] Sajama and Z.J. Haas, Independent-tree ad hoc multicast routing (ITAMAR), *Mobile Networks and Applications,* 8 (2003) pp. 551–566.

[26] Y.M. Tseng, J.K. Jan, and H.Y. Chien, Authenticated encryption schemes with message linkages for message flows, *International Journal of Computers & Electrical Engineering,* 29(1), (2003), pp. 101–109.

[27] R. Rivest, A. Shamir, and L. Adleman, A method for obtaining digital signatures and public-key cryptosystem, *Communications of the ACM* 21(2), (1978), pp. 120–126.

[28] T. ElGamal, A public-key cryptosystem and a signature scheme based on discrete logarithms, *IEEE Transactions on Information Theory,* 31(4), (1985), pp. 469–472.

[29] R.M. Davies, The data encryption standard in perspective, *Computer Security and the Data Encryption Standard, National Bureau of Standards Special Publication* (Feb. 1978).

[30] NIST, Announcing the Advanced Encryption Standard (AES), *Federal Information Processing Standards Publication* n. 197 (November 26, 2001).

[31] W. Diffie, P.C. Van Oorschot, and M.J. Wiener, Authentication and authenticated key exchanges, *Designs, Codes and Cryptography,* 2 (1992), pp. 107–125.

[32] S. Blake-Wilson, D. Johnson, and A. Menezes, Key agreement protocols and their security analysis. *In Proceedings of the Sixth IMA International Conference on Cryptography and Coding, Lecture Notes in Computer Science 1355* (1997) pp. 30–45.

[33] Y.M. Tseng, Robust generalized MQV key agreement protocol without using one-way hash functions, *Computer Standards and Interfaces,* 24(3), (2002), pp. 241–246.

Chapter 6

# Reliable Routing in Ad Hoc Networks Using Direct Trust Mechanisms

Asad Amir Pirzada
*School of Computer Science and Software Engineering*
*The University of Western Australia, Crawley, WA 6009, Australia.*
E-mail: pirzada@csse.uwa.edu.au

Chris McDonald
*School of Computer Science and Software Engineering*
*The University of Western Australia, Crawley, WA 6009, Australia.*
E-mail: chris@csse.uwa.edu.au

## 1 Introduction

The miniaturization of computing devices and the need for ubiquitous communication has augmented the demand for improvised networks that can be created on the fly. As a consequence, ad hoc networks have become a focus of active research, primarily due to their independence from external hardware and to their self-organized nature. These networks, which were mainly meant for military, law enforcement, and rescue operations, became a vital candidate for pervasive computing. Ad hoc networks are formed and operated by their constituent wireless nodes. As the network only appears when these nodes start cooperating with each other, it is also called a spontaneous network. These networks are based upon a trust-your-neighbor paradigm with the immediate neighbors making up the primary communication links.

M.X. Cheng, D. Li (eds.) *Advances in Wireless Ad Hoc and Sensor Networks.*
Signals and Communication Technology, doi: 10.1007/978-0-387-68567-0_6.

These implicit trust relationships require expression of benevolent behavior by all nodes. However, this is not always the case and due to a number of monetary, social, or selfish reasons some of the nodes deviate from the agreed set of principles.

Routing protocols act as the building blocks in ad hoc networks by finding and maintaining virtual connections between the nodes. Several types of routing protocols have been specially developed for ad hoc networks and have been classified into two categories as reactive and proactive [13]. In reactive routing protocols, in order to conserve a node's battery, routes are only discovered when required, whereas in proactive routing protocols routes are established prior to use and hence avoid the latency that incurs in new route discoveries.

Routing in ad hoc networks encompasses some major security problems. One problem is to ensure that data are routed securely through trusted nodes and the second is the security of the routing protocol itself. In view of the fact that both data and control messages employ the same wireless transmission medium, routing protocol messages can be modified to vary routing behavior. This means that if a routing protocol message is altered to generate a false route, then no amount of security on data packets can correct this routing misbehavior. The accurate execution of the routing protocols is hence imperative for establishing correct routes through the ad hoc network.

A number of protocols were thereby developed to secure the routing process against Byzantine [9] and malicious behavior. The comparison of these protocols [3] revealed that all the secure routing protocols were reliant upon a central trust authority for implementing conventional cryptographic algorithms. These protocols gave either the assurance of existence of 100% security or its nonexistence, but none of these presented a transitional level of security protection. Authentication, being one of the initial requirements of any secure communication, necessitates the need for preshared keys or digital certificates by participating nodes [2]. The requirement of a central trust authority or preconfiguration of nodes does resolve many of the core security issues in wired networks but it is neither practical nor feasible for wireless ad hoc networks. To differentiate this environment the term "managed ad hoc network" was introduced in which the nodes could be configured before the network was actually established. However, this contradicts the very aim of ad hoc networks, which endeavor to spontaneously establish an improvised network. We distinguish between the two types of network and call the latter a "pure ad hoc network," which has no infrastructure requirements and is created on the fly.

In this chapter, we present a unique way of establishing trust in pure ad hoc networks without using any cryptographic add-ons. We accentuate that trust can be established between unfamiliar nodes based upon a give-and-take mechanism, where the necessity of getting a thing done is more vital than its security. Using the intrinsic knowledge existent in the network, the participating nodes compute trustworthy routes. These routes although not secure in terms of cryptography, still carry an accurate measure of reliability with them. In Section 2, we explain trust and security issues related to ad hoc networks. In Section 3, we discuss specific attacks that are carried out against these networks. Some relevant previous work is then discussed in Section 4. In Section 5, we describe in detail our proposed trust model and show its application to the DSR protocol in Section 6. Simulation results are presented in Section 7. The rest of the chapter consists of an analysis of the proposed model in Section 8 with concluding remarks in Section 9.

## 2 Trust and Security Issues

According to Jøsang [6], trust and security represent two sides of the same thing. Both these terms are so highly interconnected that they cannot be evaluated independently. To secure any information system, cryptographic mechanisms are generally employed. These mechanisms necessitate trusted key exchange prior to their application. Similarly, for trusted key exchange to take place, a secure channel is required for key transport. Due to this interdependence between trust and security, we need to consider both these issues when defining a secure system. In wired networks, trust is usually realized using indirect trust mechanisms that are comprised of trusted certification agencies and authentication servers. However, these mechanisms still necessitate some out-of-band mechanism for initial authentication and are frequently dealt with by physical or location-based verification schemes.

In comparison, establishing trust in ad hoc wireless networks is an extremely challenging task. These networks are primarily based upon naive "trust-your-neighbor" relationships with relatively short life spans. This simplistic approach, although workable, makes the trust relationships prone to attacks by malicious nodes present in the network. In addition, the lack of permanent trust infrastructure, restricted resources, ephemeral connectivity and availability, shared wireless medium, and physical vulnerability make trust establishment virtually impossible. In order to overcome these problems, a number of impractical assumptions have to be made including the creation of an omnipresent trust authority and preconfiguration of nodes.

Mayer et al. [18] define trust as "the willingness of a party to be vulnerable to the actions of another party based on the expectation that the other party will perform a particular action important to the trustor, irrespective of the ability to monitor or control the party." Jøsang [6] defines trust in a passionate entity (human) as the belief that it will behave without malicious intent and trust in a rational entity (system) as the belief that it will resist malicious manipulation. Trust in entities is based on one's belief that the trusted entity will not act maliciously in a particular situation. Trust may be achieved directly based on previous similar experiences with the same party, or indirectly based on recommendations from other trusted parties. In addition, trust is also time dependent; it grows and decays over time.

A pure ad hoc network strongly resembles the human behavior model in which people are able to establish trust relationships based upon their individual experiences. According to Denning [12], trust cannot be treated as a property of trusted systems but rather it is an assessment based on experience that is shared through networks of people. We believe that as in real life, trust levels are determined by the particular actions that the trusted party can carry out for the trustee. In the same way trust levels in ad hoc networks can be computed based upon the effort that one node is prepared to expend for another node. This effort can be in terms of packets forwarded or dropped, battery consumption, or any other such parameter that helps to establish a reciprocal trust level. A trust model that is based on experience alone may not be sheltered from attacks in an ad hoc network but it can definitely discover routes with an explicit confidence level.

## 3 Attacks on Wireless Networks

Two kinds of attacks can be launched against ad hoc networks [23], passive and active. In passive attacks the attacker does not disturb the routing protocol. It only eavesdrops on the routing traffic and endeavors to extract valuable information such as node hierarchy and network topology from it. For example, if a route to a particular node is requested more often than to other nodes, the attacker might infer that the node is important for the functioning of the network, and that disabling it could bring the entire network down. Similarly, even when it might not be possible to isolate the exact position of a node, one may be able to find out information about the network topology by analyzing the contents of routing packets. A less vivid but understated malicious behavior is node selfishness in which nodes, in order to preserve their batteries, may be tempted not to relay packets. These types of attacks are virtually impossible to detect in a wireless environment and are hence also extremely difficult to prevent.

In active attacks, the aggressor node has to consume some of its energy in order to perform the attack. Nodes that perform active attacks, with the aim of disrupting other nodes and the network, are considered to be malicious. In active attacks, malicious nodes can disturb the correct functioning of a routing protocol by modifying routing information, by fabricating false routing information, and by impersonating other nodes [10]. Attacks using modification are normally targeted against the integrity of routing computation. By modifying routing information an attacker can cause network traffic to be dropped, redirected to a different destination, or to take an extended route to the destination. A more subtle type of modification attack is the creation of a tunnel or wormhole [8] in the network between two colluding malicious nodes linked through a private network connection. This exploit allows a node to short-circuit the normal flow of routing messages creating a virtual vertex cut in the network that is controlled by the two colluding attackers. Fabrication attacks are performed by generating deceptive routing messages. The rushing attack [24] is a typical example of a fabrication attack, in which an attacker rapidly spreads routing messages throughout the network so that nodes drop legitimate routing messages by evaluating them as duplicates. These attacks are difficult to identify as they are received as legitimate routing packets. During impersonation attacks a malicious node can launch many attacks in a network by masquerading as another node through spoofing. This occurs when a malicious node fakes its identity by altering its MAC or IP address in order to change the perspective of a benevolent node regarding the network.

# 4 Previous Work

## 4.1 ARIADNE

ARIADNE [23] is an on-demand secure ad hoc routing protocol based on the Dynamic Source Routing (DSR) protocol that protects against node compromise and relies only on extremely efficient symmetric cryptography. The security of ARIADNE is based upon the secrecy and authenticity of keys that are kept at the nodes. ARIADNE prevents a large number of denial-of-service attacks from malicious or compromised nodes. ARIADNE provides assurance that the target node of a route discovery process can verify the initiator, that the initiator can verify each transitional node that is on the path to the destination present in the `ROUTE REPLY` message, and that no intermediate node can reduce the node list in the `ROUTE REQUEST` or `ROUTE REPLY` messages. Route discovery is performed in two stages: the initiator floods the network with a `ROUTE REQUEST` that solicits a `ROUTE REPLY` from the target. During route discovery the target authenticates each node in the node list of the `ROUTE REQUEST` and the initiator authenticates each individual node in the node list of the `ROUTE REPLY`.

For node authentication, ARIADNE has three alternative techniques: TESLA (Timed Efficient Stream Loss tolerant Authentication) [7], digital signatures, or pairwise shared secret keys. The authentication mechanisms used by ARIADNE require excessive preconfiguration of nodes during the network establishment phase. ARIADNE prefers using the TESLA broadcast authentication scheme with delayed key disclosure both for its speed and efficiency. However, TESLA in turn requires clock synchronization between communicating nodes, which is considered to be an unrealistic constraint upon ad hoc wireless networks.

## 4.2 ARAN

The Authenticated Routing for Ad hoc Networks (ARAN) [10] secure routing protocol is an on-demand routing protocol that identifies and shields against malevolent actions by malicious nodes in the ad hoc network environment. ARAN relies on the use of digital certificates and can successfully operate in the managed-open scenario where no network infrastructure is predeployed, but a small amount of prior security coordination is expected. ARAN provides authentication, message integrity, and nonrepudiation in ad hoc networks by using a preliminary certification process that is followed by a route instantiation process that guarantees end-to-end provisioning of security services.

All nodes are supposed to keep fresh certificates with a trusted server and should know the server's public key. Prior to entering the ad hoc network, each node has to apply for a certificate that is signed by the certificate server. The certificate contains the IP address of the node, its public key, a timestamp of when the certificate was generated and a time at which the certificate expires, along with the signature by the certificate server. ARAN accomplishes the discovery of routes by a broadcast route discovery message from a source node, which is replied to in a unicast manner by the destination node. All the routing messages are authenticated at every hop from the source to the destination, as well as on the reverse path from destination to source.

ARAN requires the use of a trusted certificate server in the network. This requirement imposes a number of presetup restrictions that must be catered to before establishing the ad hoc network. Also, having a centralized certificate repository in a physically insecure environment creates a single point of compromise and capture.

## 4.3 Distributed Trust Model

The distributed trust model [5] makes use of a protocol to exchange, revoke, and refresh recommendations about other entities. By using a recommendation protocol each entity maintains its own trust database. This ensures that the trust computed is neither absolute nor transitive. The model uses a decentralized approach to trust management and uses trust categories and values for computing different levels of trust. The integral trust values vary from −1 to 4 signifying discrete levels of trust from complete distrust (−1) to complete trust (4). Each entity executes the recommendation protocol either as a recommender or a requestor and the trust levels are computed using the recommended trust value of the target and its recommenders. The model has provision for multiple recommendations for a single target and adopts an averaging mechanism to yield a single recommendation value. The model is most suitable for less formal, provisional, and temporary trust relationships and does not specifically target ad hoc networks. Moreover, as it requires that recommendations about other entities be passed, the handling of false or malicious recommendations has to be supported via some out-of-band mechanism.

## 4.4 Distributed Public-Key Model

The distributed public-key model [15] makes use of threshold cryptography to distribute the private key of the certification authority over a number of servers. An (`n`, `t+1`) scheme allows any `t+1` servers out of total of `n` servers to combine their partial keys to create the complete secret key. Similarly, it requires that at least `t+1` servers must be compromised to acquire the secret key. The scheme is quite robust but has a number of factors that limit its application to pure ad hoc networks. Primarily it requires an extensive preconfiguration of servers and a distributed central authority, secondly the `t+1` servers may not be accessible to any node desiring authentication, and lastly asymmetric cryptographic operations are known to drain precious node batteries.

# 5 The Trust Model

Our trust model [1] is influenced by Marsh's trust model [21] and has been optimized for application to pure ad hoc networks. Marsh's model works out the situational trust in agents using the general trust in the trustor. Each agent also assigns importance and utility to the situation in which it finds itself. General trust is defined as the trust that an entity places in another entity based upon all prior transactions in all situations. Importance and utility are considered similar to knowledge so that an agent can weigh the costs and benefits that a particular situation holds at a particular time. In our model, we use a single variable called weight to represent the utility and importance of a situation. This weight increases or decreases with time and is set according to the hostility of the environment. The trust model is executed by means of agents that reside on network nodes. Each agent operates autonomously and upholds its own point of view with regard to the trust hierarchy. Each agent accumulates data from events that are being experienced by a node in the current environment. These events are filtered and assigned weights so as to compute the trust in other nodes. Every trust agent essentially performs three functions: trust derivation, quantification, and computation. The estimated division of these functions with respect to the OSI reference model and the TCP/IP protocol suite is shown in Figure 1.

<table>
<tr><th>OSI</th><th>PROPOSED</th><th>TCP/IP</th></tr>
<tr><td>Application</td><td rowspan="3">Computation<br>&<br>Quantification</td><td rowspan="3">Application</td></tr>
<tr><td>Presentation</td></tr>
<tr><td>Session</td></tr>
<tr><td>Transport</td><td rowspan="4">Derivation</td><td>Transport</td></tr>
<tr><td>Network</td><td>Internet</td></tr>
<tr><td>Data link</td><td rowspan="2">Host to Network</td></tr>
<tr><td>Physical</td></tr>
</table>

Figure 1: Structure of the trust agent.

## 5.1 Trust Derivation

Trust derivation is carried out in passive mode, that is, without entailing the use of special interrogation packets. Essential information regarding other nodes can be gathered by examining the local network traffic. Taps are inserted at different points in the protocol stack so as to analyze events that are currently in progress. Likely events that can be recorded in passive mode are the measure and precision of:

1. Frames received
2. Data packets forwarded
3. Control packets forwarded
4. Data packets received
5. Control packets received
6. Streams established
7. Data forwarded
8. Data received

The information obtained from these events is split into one or more trust categories. These categories specify the explicit feature of trust that is pertinent to a particular relationship. This categorization of events facilitates the computation of situational trust in other nodes.

## 5.2 Trust Quantification

Trust generally portrays a continuous trend and hence discrete representation is insufficient to clearly represent trust. Secure routing protocols represent trust in a binary form by either the presence of security or its absence. As discussed earlier, the distributed trust model [5] symbolizes trust using six values ranging from distrust to complete trust. Similarly, Pretty Good Privacy (PGP) [19] uses four values ranging from unknown to fully trusted to signify trust levels. Discrete values, although straightforward to represent and categorize trust, are not suitable for application to ad hoc networks. Trust in an ad hoc network is never static because the trust relationships are consistently varying due to the dynamic topology. The period of interaction with other nodes being significantly succinct necessitates that trust be represented as an incessant range to differentiate between nodes with comparable trust levels. In our model we signify trust from −1 to +1, representing an unremitting range from complete distrust to absolute trust.

### 5.3 Trust Computation

During trust computation, weights are assigned to the previously monitored and quantified events. All nodes dynamically allocate these weights depending upon their own criteria and situation. These weights are represented in a continuous range from 0 to +1, where 0 signifies unimportant and +1 signifies most important. The aggregate trust level is then computed by combining the situational trusts according to their importance and utility. We represent the aggregate trust in node $y$, by node $x$ as $T_x(y)$ and as given by the following equation,

$$T_x(y) = \sum_{i=1}^{n} [W_x(i) \times T_x(i)],$$

where $W_x(i)$ is the weight of the $i$th trust category to $x$ and $T_x(i)$ is the situational trust of $x$ in the $i$th trust category. The total number of trust categories $n$ is dependent on the protocol and scenario to which the trust model is being applied.

## 6 Extension to DSR

### 6.1 Dynamic Source Routing Protocol

The Dynamic Source Routing (DSR) protocol [11] is a reactive routing protocol. As the name suggests it makes use of the strict source routing feature of the Internet protocol. All data packets that are sent using the DSR protocol contain the complete list of nodes that the packet has to traverse. During route discovery, the source node broadcasts a `ROUTE REQUEST` packet with a unique identification number. The `ROUTE REQUEST` packet contains the address of the target node to which a route is desired. All nodes that have no information regarding the target node, or have not previously seen the same `ROUTE REQUEST` packet, append their IP addresses to the `ROUTE REQUEST` packet and rebroadcast it. In order to control the spread of the `ROUTE REQUEST` packets, the broadcast is done in a nonpropagating manner with the IP TTL field being incremented in each route discovery.

The ROUTE REQUEST packets keep spreading in the network until the time they reach the target node or any other node that has a route to the target node. The recipient node creates a ROUTE REPLY packet, which contains the complete list of nodes that the ROUTE REQUEST packet had traversed. Depending upon the implementation, the target node may respond to one or more incoming ROUTE REQUEST packets. Similarly, the source node may accept one or more ROUTE REPLY packets for a single target node.

For optimization reasons, nodes maintain a PATH CACHE or a LINK CACHE scheme [22]. The former stores complete paths to a particular destination, whereas the latter only caches information related to individual links. The advantage of the LINK CACHE scheme is that it allows alternate paths to a destination even when some of the intermediate links have failed. Each node either forwarding or overhearing data and control packets, adds all useful information to its respective route cache. This information is used to limit the spread of control packets for subsequent route discoveries.

## 6.2 Vulnerabilities in DSR

The major vulnerabilities [4] present in the DSR protocol are as follows.

### 6.2.1 Deceptive Alteration of IP Addresses

During propagation of the ROUTE REQUEST packet, intermediate nodes append their IP addressees to it for route buildup. However, any malicious node may modify, delete, or add IP addressees to create routes as per its own requirement as shown in Figure 2. Doing so enables malicious nodes to launch a variety of attacks in the network including creation of wormholes and blackholes.

### 6.2.2 Deceptive Alteration of Hop Count

The hop count field of the IP packet usually informs the recipient of the total number of hops that the packet has traversed so far. Malicious nodes may increase this count so as to portray longer routes or decrease it for shorter routes. By doing so a malicious node is able to degrade or upgrade routes, thereby creating a topology that is most favorable to it.

| Option Type | Option Data Length | Identification |
|---|---|---|
| Target Address | | |
| Address [1] | | |
| Address [2] | | |
| Malicious Node IP Address | | |
| Address [n] | | |

Figure 2: Deceptive alteration of DSR `ROUTE REQUEST` packet.

## 6.3 Trust Derivation

In DSR, we use the following inherent features to build up trust categories for our model.

### 6.3.1 Acknowledgments

Any node can obtain information about the successful transmission of a packet through the following three methods.

**Link-Layer Acknowledgments.** By means of link-layer acknowledgments the core MAC protocol provides feedback of the successful delivery of the transmitted data packets.

**Passive Acknowledgments.** When passive acknowledgments are used, the sending node places itself into promiscuous mode after the transmission of a packet so as to eavesdrop upon the subsequent retransmission by the next node.

**Network Layer Acknowledgments.** This method permits the sender to explicitly request a network layer acknowledgment from the next hop using the DSR options header.

All three methods inform the sender regarding the successful receipt of a packet. However, the passive acknowledgment method also provides the following information about the next hop, including: it is acting like a black hole if the packet is dumped and not retransmitted; it is carrying out a modification attack if the contents have been fallaciously modified; it is carrying out a fabrication attack if a self-generated fallacious packet is transmitted; it is carrying out an impersonation attack if the MAC or IP addresses have been spoofed; it is showing selfish behavior by not retransmitting a packet; and it is inducing latency by delaying the retransmission of the packet.

The method of passive acknowledgment can be further classified into acknowledgments for data packets and acknowledgments for control packets. The number of these acknowledgments occurring with respect to every node is maintained and tabulated as shown in Table 1. For every transmitted packet, the appropriate counter for success or failure is incremented, depending if the neighboring node has correctly forwarded it or not.

Table 1: Trust table based on passive acknowledgments.

| Node Acknowledgement | Route Request ($R_q$) | | Route Reply ($R_p$) | | Route Error ($R_e$) | | Data (D) | |
|---|---|---|---|---|---|---|---|---|
| | Success $R_{qs}$ | Fail $R_{qf}$ | Success $R_{ps}$ | Fail $R_{pf}$ | Success $R_{es}$ | Fail $R_{ef}$ | Success $D_s$ | Fail $D_f$ |

### 6.3.2 Packet Precision

The precision of received data and routing packets permits computation of trust levels. For example, if routing packets are received that are found to be accurate and effective, then the originator of the packets can be assigned an elevated trust level. The above technique can be classified into data and control packet types as shown in Table 2, which are respectively allotted trust values. Success and failure counters are maintained for every received packet, which are incremented based upon the accuracy or inaccuracy of the packet.

### 6.3.3 Gratuitous Route Replies

The DSR protocol provides the facility of "route shortening" to avoid unnecessary intermediate nodes. For example, if a node overhears a data packet that is supposed to traverse a number of nodes before passing through it, then this node creates a shorter route known as the `GRATUITOUS ROUTE REPLY` and sends it to the original sender. The `GRATUITOUS ROUTE REPLY`

Table 2: Trust table based on packet precision.

| Node Packet Precision | Route Request ($R_q$) | | Route Reply ($R_p$) | | Route Error ($R_e$) | | Data (D) | |
|---|---|---|---|---|---|---|---|---|
| | Success $R_{qs}$ | Fail $R_{qf}$ | Success $R_{ps}$ | Fail $R_{pf}$ | Success $R_{es}$ | Fail $R_{ef}$ | Success $D_s$ | Fail $D_f$ |

can be considered as a trust category as it provides the following information about the sender: it is displaying either malicious or benevolent behavior, and it is not showing selfish behavior.

If the GRATUITOUS ROUTE REPLY is found to be accurate, then the originator can be allotted a higher trust value along with the set of nodes provided in that route. The above method can be used to allocate different trust values to different nodes, as shown in Table 3. All GRATUITOUS ROUTE REPLY packets that are found to be correct or incorrect are recorded using appropriate counters.

Table 3: Trust table based on gratuitous route replies.

| Node | Gratuitous Route Replies (G) | |
|---|---|---|
| | Success $G_s$ | Fail $G_f$ |

### 6.3.4 Blacklists

DSR maintains blacklists for nodes displaying unidirectional behavior, that is, if a neighboring node has received a packet and either due to a unidirectional link or selfish behavior the sender cannot hear it retransmitting. If the MAC protocol is expected to provide feedback (like IEEE 802.11) then this implies that the links must be bidirectional and the neighboring node is acting selfishly. The blacklists can be used to provide trust values for nodes while computing route confidence levels. The format of the trust table based on blacklists is shown in Table 4.

Table 4: Trust table based on blacklists.

| Node | Present in Blacklist (B) |
|---|---|

### 6.3.5 Salvaging

If an intermediate node receives a packet for which its next hop is not available, it may drop the packet and inform the sender. However, if it has a route to the final recipient it can salvage that route from its cache, send the packet on the new route, and inform the sender about the failed link through a `ROUTE ERROR` packet. If the salvaged route is found to be correct then it reveals that the sender of the `ROUTE ERROR` is displaying benevolent and altruistic behavior. Hence, this information can be used to build up trust levels and be considered as a trust category. All salvaged `ROUTE ERROR` packets found to be correct or incorrect are recorded using counters, as shown in Table 5.

Table 5: Trust table based on salvaging.

| Node | Salvage Route Error (S) | |
|---|---|---|
| | Success $S_s$ | Fail $S_f$ |

## 6.4 Trust Quantification

During trust quantification the recorded events from the trust derivation phase are measured and allotted weights so as to calculate the situational trust values for other nodes.

### 6.4.1 Trust Category $P_A$

The trust category derived using passive acknowledgments is denoted $P_A$. The events recorded in Table 1 are quantized as per the following equations to provide trust levels.

$$\begin{aligned} R_p &= \frac{R_{ps} - R_{pf}}{R_{ps} + R_{pf}} \quad \text{for} \quad R_{ps} + R_{pf} \neq 0 \quad \text{else} \quad R_p = 0 \\ R_q &= \frac{R_{qs} - R_{qf}}{R_{qs} + R_{qf}} \quad \text{for} \quad R_{qs} + R_{qf} \neq 0 \quad \text{else} \quad R_q = 0 \end{aligned}$$

$$\begin{aligned} R_e &= \frac{R_{es} - R_{ef}}{R_{es} + R_{ef}} \quad \text{for} \quad R_{es} + R_{ef} \neq 0 \quad \text{else} \quad R_e = 0 \\ D &= \frac{D_s - D_f}{D_s + D_f} \quad \text{for} \quad D_s + D_f \neq 0 \quad \text{else} \quad D = 0. \end{aligned}$$

In order to limit the trust values between –1 to +1, we normalize the values of $R_p$, $R_q$, $R_e$, and $D$. Complete distrust is represented by –1, zero stands for a noncontributing event, and a value of +1 implies absolute trust in a particular event. If the number of failures is more than the number of successes then a negative trust value can also occur.

These normalized trust levels are then assigned weights in a static or dynamic manner depending upon their utility and importance. The situational trust $T_n(P_A)$ in node $n$ for trust category $P_A$ is computed using the following equation,

$$T_n(P_A) = W(R_p) \times R_p + W(R_q) \times R_q + W(R_e) \times R_e + W(D) \times D,$$

where W is the weight assigned to the specific type of events that took place with node $n$.

### 6.4.2 Trust Category $P_P$

Trust category $P_P$, which is based on packet precision, is derived from the events recorded in Table 2. The following equations are used to quantize the events.

$$\begin{aligned} R_p &= \frac{R_{ps} - R_{pf}}{R_{ps} + R_{pf}} \quad \text{for} \quad R_{ps} + R_{pf} \neq 0 \quad \text{else} \quad R_p = 0 \\ R_q &= \frac{R_{qs} - R_{qf}}{R_{qs} + R_{qf}} \quad \text{for} \quad R_{qs} + R_{qf} \neq 0 \quad \text{else} \quad R_q = 0 \\ R_e &= \frac{R_{es} - R_{ef}}{R_{es} + R_{ef}} \quad \text{for} \quad R_{es} + R_{ef} \neq 0 \quad \text{else} \quad R_e = 0 \\ D &= \frac{D_s - D_f}{D_s + D_f} \quad \text{for} \quad D_s + D_f \neq 0 \quad \text{else} \quad D = 0. \end{aligned}$$

All these events are then assigned weights in a similar manner and the situational trust in category $P_P$ for node $n$ is computed using the following equation.

$$T_n(P_P) = W(R_p) \times R_p + W(R_q) \times R_q + W(R_e) \times R_e + W(D) \times D.$$

### 6.4.3 Trust Category $G_R$

The trust category based upon `GRATUITOUS ROUTE REPLY` packets is derived using Table 3 and is denoted $G_R$. The trust levels are quantized using the following equation,

$$G \quad = \quad \frac{G_s - G_f}{G_s + G_f} \qquad \text{for} \qquad G_s + G_f \neq 0 \qquad \text{else} \qquad G = 0.$$

The situational trust in category $G_R$ for a node $n$ is computed using the equation,

$$T_n(G_R) = W(G) \times G.$$

### 6.4.4 Trust Category $B_L$

Table 4 is used to derive trust category $B_L$, which is based on blacklists. The value of $B$ is Boolean reflecting the presence or absence of a node in the trust table. The situational trust for category $B_L$ in node $n$ is computed using the following equation:

$$T_n(B_L) = W(B) \times B.$$

### 6.4.5 Trust Category $S_G$

The trust category derived using the salvaging information from Table 5 is denoted $S_G$. The trust levels are quantized using the following equation:

$$S \quad = \quad \frac{S_s - S_f}{S_s + S_f} \qquad \text{for} \qquad S_s + S_f \neq 0 \qquad \text{else} \qquad S = 0.$$

The situational trust in category $S_G$ for a node $n$ is computed using the equation

$$T_n(S_G) = W(S) \times S.$$

## 6.5 Trust Computation

To compute the aggregate trust level for a particular node, the situational trust values from all trust categories ($P_A$, $P_P$, $G_R$, $B_L$, $S_G$) are then assigned weights and combined. Aggregate trust $T$ in node $y$ by node $x$ is represented as $T(y)$ and given by the following equation,

$$T(y) = W(P_A) \times T_y(P_A) + W(P_P) \times T_y(P_P) + W(G_R) \times T_y(G_R) + W(B_L) \times T_y(B_L) + W(S_G) \times T_y(S_G),$$

where $W$ represents the weight assigned to a trust category by node $x$ and $T_y$ is the situational trust in node $y$ by node $x$ for that trust category. The aggregate trust table for node $x$ with respect to node $y$ is shown in Table 6.

Table 6: Aggregate trust table.

| Node | Passive Acknowledgment | Packet Precision | Gratuitous Route Replies | Black Lists | Salvage Route Errors | Aggregate Trust Level |
|---|---|---|---|---|---|---|
| $y$ | $T_y(P_A)$ | $T_y(P_P)$ | $T_y(G_R)$ | $T_y(B_L)$ | $T_y(S_G)$ | $T(y)$ |

Each agent dynamically updates the aggregate and situational trust values depending upon the occurrence of events and severity of the situation. These computed trust levels are then associated as weights with the routes present in the `LINK CACHE` in real-time. Nodes that subsequently access their `LINK CACHE` seeking a particular route use the Dijkstra algorithm [14] to find the most trustworthy path to a particular destination. The routes discovered in this way may not be the shortest in terms of hop count between any two communicating entities. However, these routes carry along an associated level of trustworthiness with them that provides the requisite belief factor in an unfamiliar environment.

# 7 Simulation

## 7.1 Setup

The simulation of the trust model was performed using NS2 [16]. The simulation parameters are listed in Table 7.

During the simulation, 50 nodes randomly move in a 1000 × 1000 meter area as shown in Figure 3. Each node starts at a random position, waits for the pause time, and then moves to another random position with a velocity chosen between 0 m/s to the maximum simulation speed. Twenty nodes imitate malicious or selfish behavior and launch attacks against the network.

Table 7: Simulation parameters.

| | |
|---|---|
| Examined Protocol | DSR |
| Simulation time | 900 seconds |
| Simulation area | 1000 × 1000 m |
| Number of nodes | 50 |
| Transmission range | 250 m |
| Movement model | Random waypoint |
| Maximum speed | 20 m/s |
| Pause time | 100–900 seconds |
| Traffic type | CBR (UDP) |
| Maximum connections | 30 |
| Payload size | 512 bytes |
| Packet rate | 4 pkt/s |
| Malicious nodes | 20 |
| Types of attacks | Modification, black and grey hole |

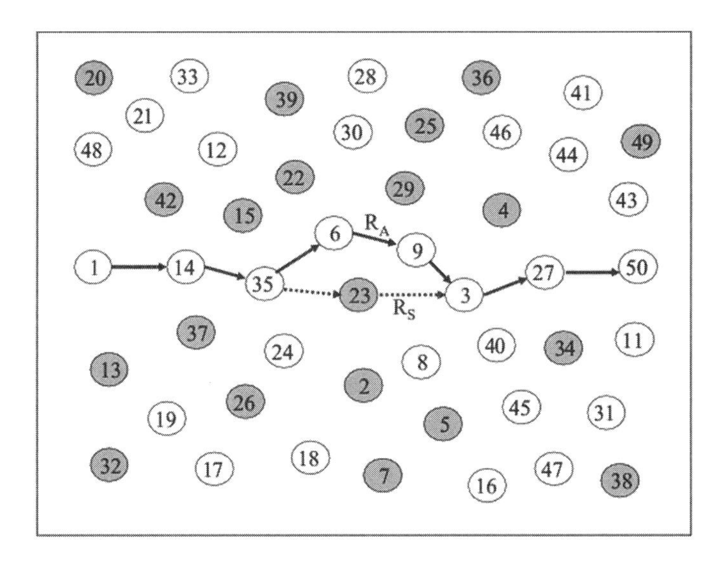


Figure 3: Simulation environment.

All nodes except the malicious nodes execute the trust model. As the model matures with the passage of time, the aggregate trust levels become more accurate. Each node associates these trust levels with the routes present in the LINK CACHE. By doing so, a node is able to find trusted routes ($R_A$) instead of the standard shortest routes ($R_S$) from its respective LINK CACHE.

### 7.2 Attack Pattern

Twenty malicious nodes were used to corrupt and suppress the routing and data packets. These nodes worked in a noncolluding manner with each one having its own attack profile.The following types of attacks were simulated.

*Modification Attack.* These attacks are carried out by adding, altering, or deleting IP addresses from the ROUTE REQUEST, ROUTE REPLY, ROUTE ERROR, and data packets, which pass through the malicious nodes.

*Black Hole Attack.* In this attack the malicious node drops all packets, which it is supposed to forward.

*Grey Hole Attack.* In the grey hole attack the malicious node selectively dumps data and control packets at random intervals.

### 7.3 Trust Weight Computation

The assignment of weights to different trust categories was carried out using a trial-and-error mechanism. The weights for each category were initially adjusted according to the average number of packets received per node as shown in Figure 4. These weights were then varied in both directions so as to increase the overall throughput of the network with minimal packet loss. The optimal weights and mean trust levels thus computed for each trust category are shown in Table 8.

### 7.4 Results and Discussion

To evaluate the performance of the trust model with optimal weights, we have used the following metrics.

*Packet Loss.* It is the count of packets that were dumped by malicious nodes without any notification.

*Throughput.* It is the ratio between the number of packets received by the application layer of destination nodes to the number of packets sent by the application layer of source nodes.

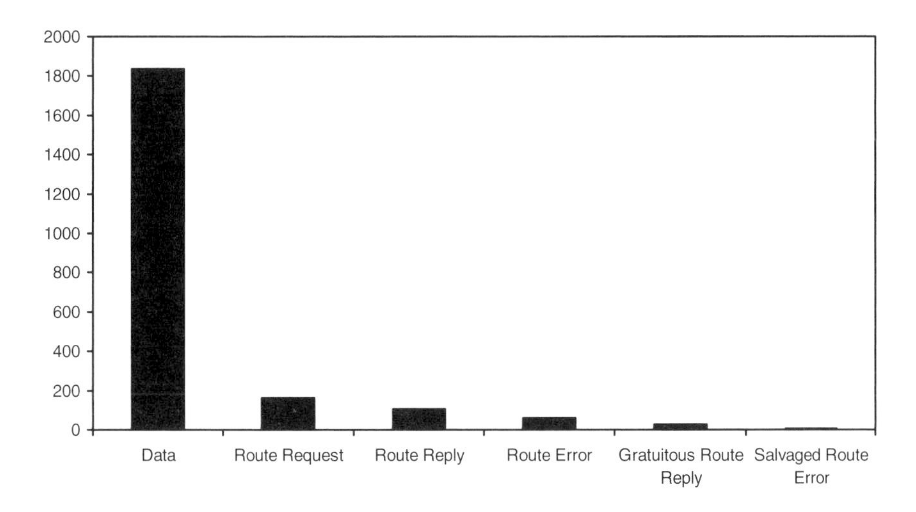


Figure 4: Average number of packets received per node.

Table 8: Optimal weights and mean trust levels.

| Situational Trust Category | Optimal Subsituational Weights | | | | Mean Situational Trust Level $T_n$ | Optimal Situational Weights W | Aggregate Situational Trust Level $T_n$ x W | Aggregate Trust Level per Node $\Sigma\{T_n \text{x} W\}$ |
|---|---|---|---|---|---|---|---|---|
| Passive Acknowledgements $P_A$ | $W(R_q)$ 0.92 | $W(R_p)$ 0.86 | $W(R_e)$ 0.95 | W(D) 0.28 | $T_n(P_A)$ 2.6 | $W(P_A)$ 0.2225 | 0.5785 | +2.1359 |
| Packet Precision $P_P$ | $W(R_q)$ 0.82 | $W(R_p)$ 0.94 | $W(R_e)$ 0.91 | W(D) 0.42 | $T_n(P_P)$ 2.2 | $W(P_P)$ 0.6675 | 1.4685 | |
| Gratuitous Route Replies $G_R$ | W(G) 0.75 | | | | $T_n(G_R)$ 0.75 | $W(G_R)$ 0.05 | 0.0375 | |
| Blacklists $B_L$ | W(B) 0.67 | | | | $T_n(B_L)$ 0.67 | $W(B_L)$ 0.02 | 0.0134 | |
| Salvaging $S_G$ | W(S) 0.95 | | | | $T_n(S_G)$ 0.95 | $W(S_G)$ 0.04 | 0.038 | |

*Packet Overhead.* This is the ratio between the total number of control packets generated to the total number of data packets received during the simulation time.

*Byte Overhead.* This is the ratio between the total number of control bytes generated to the total number of data bytes received during the simulation time.

*Average Latency.* Gives the mean time (in seconds) taken by the packets to reach their respective destinations.

*Path Optimality.* It is the ratio between the number of hops in the optimal path to the number of hops in the path taken by the packets.

Figure 5 depicts the performance results for the trust-based DSR protocol compared with that of the standard DSR protocol in the presence of malicious nodes. The results indicate that the performance of the trust model improves with the increase in node pause time. This is due to the fact that the nodes maintain their positions for longer durations and so the information in the `LINK CACHE` remains consistent over time. The average packet loss in nodes using the trusted DSR protocol was up to 5% lower than those executing the standard DSR protocol. This also results in an increase in the throughput of the network by a similar amount.

As the trust model matures with the passage of time, the nodes become more knowledgeable about the network and are hence able to bypass malicious nodes in subsequent route discoveries. However, the routes being retrieved from the `LINK CACHE` are based upon a minimal trust threshold, so we see a control packet overhead when no currently available route in the `LINK CACHE` suffices the threshold. An increase in the packet latency and deviation from the optimal paths has also been observed. This can be attributed to the fact that the routes obtained from the `LINK CACHE` are not optimal in terms of hops but instead consist of nodes that have been found to be more trustworthy than the others. This indirectly leads to a decrease in the path optimality, as the packets now traverse a longer route, which in turn increases the latency of the network.

## 8 Analysis

The amount of trust that can be established in the network routes was simulated and the trust model was found to be simple, flexible, and pragmatic for use in pure ad hoc networks. Any node that can place its interface into promiscuous mode can passively receive a lot of information about the network. This information can be further used to build trust levels for different nodes. However, this method has certain drawbacks that have been highlighted by Marti et al. [20]. The foremost is the ambiguous collision problem in which a node A cannot hear the broadcast from neighboring node B to node C, due to a local collision at A. In the receiver collision problem node

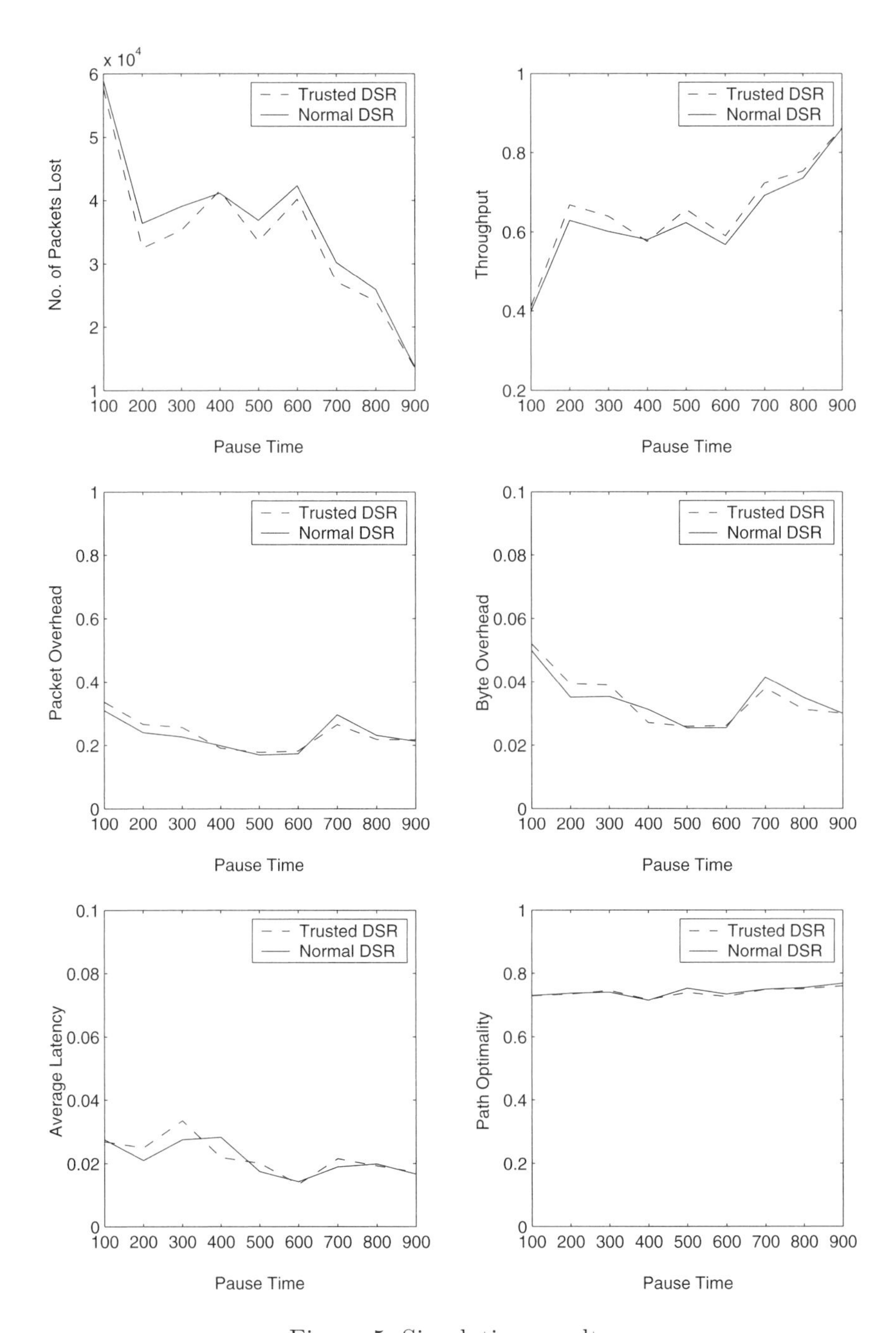


Figure 5: Simulation results.

A overhears node B broadcast a packet to C but cannot hear the collision that occurs at node C. Similarly, if nodes have varying transmission power ranges, the mechanism of passive acknowledgments might not work properly.

The trust model maintains the level of trust for nodes based upon a node's IP address. This technique is susceptible to a number of impersonation attacks where a node may frequently change its IP address after launching an attack in order to attain a higher trust level. Likewise, in order to cheat the trust model, any node can initially behave in a benevolent manner so as to carry out malicious activity later on. However, once it starts the malicious activity, its corresponding trust levels that are being maintained by other nodes start decaying. This results in circumventing the malicious node in subsequent route discoveries. The precise impact of such attacks is very difficult to gauge, especially if the malicious node operates at the threshold between benevolent and malevolent behavior. We suggest using intrusion detection systems such as those proposed by Zhang and Lee [25] and Kachirski and Guha [17] for identifying such nodes in ad hoc networks.

## 9 Conclusion

Security in ad hoc networks is currently being achieved by means of different cryptographic techniques. Key distribution and revocation being minimal requirements are frequently accomplished using a number of impractical assumptions. In this chapter, we have presented a pragmatic approach for establishing and managing trust in ad hoc networks using an effort-return trust model. This model is not another type of hard-security cryptographic or certification mechanism that is typically based upon superfluous assumptions and requirements. Instead it aims at building confidence measures regarding route trustworthiness in nodes by dynamically computing and modifying the effort expended by nodes. In an ad hoc network where doubt and uncertainty are inherent, our trust model creates and maintains trust levels using the knowledge that a node can gather from its experience and surroundings. The routes thus created using our model may not be cryptographically secure but they do establish relative levels of trustworthiness with them. The trust model is applicable to both pure and managed ad hoc networks as it provides confidence measures regarding the reliability of routes, which are computed using direct trust mechanisms. We believe that our model will be most suited to ad hoc networks that can be created on the fly without any trust infrastructure or preconfiguration of nodes.

## References

[1] A. A. Pirzada and C. McDonald, Establishing trust in pure ad hoc networks. *In Proceedings of the 27th Australasian Computer Science Conference (ACSC)*, Vol. 26 (2004), pp. 47–54.

[2] A. A. Pirzada and C. McDonald, Kerberos assisted authentication in mobile ad hoc networks. *In Proceedings of the 27th Australasian Computer Science Conference (ACSC)*, Vol. 26 (2004), pp. 41–46.

[3] A. A. Pirzada and C. McDonald, Secure routing protocols for mobile ad hoc wireless networks. In T. A. Wysocki, A. Dadej, and B. J. Wysocki (Eds.), *Advanced Wired and Wireless Networks* (Springer, Newyouk, 2004).

[4] A. A. Pirzada and C. McDonald, Secure routing with the DSR protocol. *In Proceedings of the Third International Workshop on Wireless Information Systems (WIS)* (2004), pp. 24–33.

[5] A. A. Rahman and S. Hailes, A distributed trust model. *In Proceedings of the ACM New Security Paradigms Workshop* (1997), pp. 48–60.

[6] A. Jøsang, The right type of trust for distributed systems. *In Proceedings of the ACM New Security Paradigms Workshop* (1996), pp. 119–131.

[7] A. Perrig, R. Canetti, D. Tygar, and D. Song, The TESLA broadcast authentication protocol, RSA CryptoBytes, Vol. 5 (2002).

[8] A. Perrig, Y. C. Hu, and D. B. Johnson, Wormhole protection in wireless ad hoc networks. *In Proceedings of the Eighth Annual International Conference on Mobile Computing and Networking* (2001).

[9] B. Awerbuch, D. Holmer, C. Nita-Rotaru, and H. Rubens, An on-demand secure routing protocol resilient to byzantine failures. *In Proceedings of the ACM Workshop on Wireless Security (WiSe)* (2002).

[10] B. Dahill, B. N. Levine, E. Royer, and C. Shields, A secure routing protocol for ad hoc networks. *In Proceedings of the International Conference on Network Protocols (ICNP)* (2002), pp. 78–87.

[11] D. B. Johnson, D. A. Maltz, and Y. Hu, The dynamic source routing protocol for mobile ad hoc networks (DSR), *IETF MANET*, Internet Draft (work in progress) (2003).

[12] D. Denning, A new paradigm for trusted systems. *In Proceedings of the ACM New Security Paradigms Workshop* (1993), pp. 36–41.

[13] E. M. Royer and C. K. Toh, A review of current routing protocols for ad hoc mobile wireless networks, *IEEE Personal Communications Magazine,* 6 (1999), pp. 46–55.

[14] E. W. Dijkstra, A note on two problems in connection with graphs, *Numerische Mathematik* (1959), pp. 83–89.

[15] L. Zhou and Z. J. Haas, Securing ad hoc networks, *IEEE Network Magazine,* 13 (1999), pp. 24–30.

[16] NS, The Network Simulator, http://www.isi.edu/nsnam/ns/ (1989).

[17] O. Kachirski and R. Guha, Intrusion detection using mobile agents in wireless ad hoc networks. *In Proceedings of the IEEE Workshop on Knowledge Media Networking (KMN)* (2002).

[18] R. C. Mayer, J. H. Davis, and F. D. Schoorman, An integrative model of organizational trust. *In Proceedings of the Sixth Annual ACM/IEEE International Conference on Mobile Computing and Networking*, Vol. 20 (1995), pp. 709–734.

[19] S. Garfinkel, *PGP: Pretty Good Privacy,* (O'Reilly & Associates, Inc., 1995).

[20] S. Marti, T. Giuli, K. Lai, and M. Baker, Mitigating routing misbehavior in mobile ad hoc networks. *In Proceedings of the Sixth Annual ACM/IEEE International Conference on Mobile Computing and Networking* (2000), pp. 255–265.

[21] S. P. Marsh, Formalizing trust as a computational concept, Ph.D. Thesis in Department of Mathematics and Computer Science: University of Stirling, 1994.

[22] Y. C. Hu and D. B. Johnson, Caching strategies in on-demand routing protocols for wireless ad hoc networks. *In Proceedings of the Sixth Annual International Conference on Mobile Computing and Networking* (2000), pp. 231–242.

[23] Y. C. Hu, A. Perrig, and D. B. Johnson, Ariadne: A secure on-demand routing protocol for ad hoc networks. *In Proceedings of the Eighth Annual International Conference on Mobile Computing and Networking* (2002), pp. 12–23.

[24] Y. C. Hu, A. Perrig, and D. B. Johnson, Rushing attacks and defense in wireless ad hoc network routing protocols. *In Proceedings of the 2003 ACM Workshop on Wireless Security* (2003), pp. 30–40.

[25] Y. Zhang and W. Lee, Intrusion Detection in Wireless Ad Hoc Networks. *In Proceedings of the Sixth International Conference on Mobile Computing and Networking* (2000).

Chapter 7

# On the Power Optimization and Throughput Performance of Multihop Wireless Network Architectures

G. Bhaya
*Department of Computer Science and Engineering*
*University of Washington, WA 98195, USA*
E-mail: gbhaya@cs.washington.edu

B. S. Manoj
*Department of Computer Science and Engineering*
*Indian Institute of Technology Madras, Chennai 600036, INDIA*
E-mail: bsmanoj@cs.iitm.ernet.in

C. Siva Ram Murthy[1]
*Department of Computer Science and Engineering*
*Indian Institute of Technology Madras, Chennai 600036, INDIA*
E-mail: murthy@iitm.ac.in

## 1 Introduction

The mobile terminals in the wireless domain lack significant advantages that their counterparts in the wired domain have. These include the computing resources, power storage, potentially large bandwidth, and of course

[1]Author for correspondence.

M.X. Cheng, D. Li (eds.) *Advances in Wireless Ad Hoc and Sensor Networks.*
Signals and Communication Technology, doi: 10.1007/978-0-387-68567-0_7.

the threats posed by the wireless domain such as high bit error rates and security issues. With fast-developing computing technology, the problem of limited computing resources is alleviated to some extent, but applies greater pressure on other scarce resources provided by the network. The increasing number of applications mobile devices places an additional demand on the limited bandwidth and available battery power. Limited battery power imposes many constraints in the deployment of large-scale wireless networks. A significant amount of power is consumed at the mobile nodes during transmission and reception, display, and I/O access. Some algorithms for power saving during disk access are suggested in [4]. Earlier attempts for design of power-saving protocols for Mobile Ad hoc NETworks (MANETs) that allow mobile hosts to switch off to a low-power sleep mode are suggested in [5]. The authors of[5] propose three power management protocols based on the above scheme. PAMAS, discussed in [6] is one such solution. In PAMAS, mobile nodes conserve power by intelligently powering off when not actively transmitting or receiving. PAMAS uses a separate signaling channel for RTS-CTS that enables other nodes not involved in transmitting or receiving to turn off for the time of the transmission.

An attempt is made in [7] to solve the problem of adjusting transmit power in an ad hoc network to create the desired network topology. It discusses algorithms for optimum power usage in the case of static networks and some heuristics in the case of networks in which nodes are mobile. A Power Control Multiple Access protocol (PCMA) is proposed in [8] which uses variable transmission power. It uses Request-Power-To-Send/ Accept-Power-To-Send to determine the power usage for the transmission. The power for transmitting data at the sender is determined by the noise level at the receiver. PCMA uses a separate channel to indicate channel busy over which the receiver transmits a busy signal during data reception.

Although this scheme gives a significant power improvement, it does not account for mobility of the sender and the receiver. Furthermore it uses a separate channel for the busy signal to indicate that the channel is busy. The presence of mobility might cause fluctuations in the strength of the busy signal that is crucial for determining transmission power. Also the sender decides on the transmitting power based on the noise level at the receiver which is time varying. In addition, the receiver needs to be enabled for the above protocol mechanism. We attempt to solve the above problems for the MCN and the MuPAC architectures such that the protocol works in a hybrid environment where some nodes may not have power optimization enabled.

Jung and Vaidya [9] show that using variable power for transmission in fact leads to a reduction in throughput. However, their study is based on using the maximum possible power for RTS/CTS and the minimum possible power for the data and ACK transmission. However, by doing so they prevent some nodes that could have otherwise been permitted to send data from transmission. In our scheme we do not use the maximum possible power thus preventing the barest minimum number of nodes from transmission. We prevent the nodes from transmission only because of the possibility of collision due to their transmission.

The MCN architecture although primarily proposed with the aim of improving the throughput and bandwidth utilization show a significant improvement in power utilization per successfully delivered packet as compared to the Single-hop Cellular Networks (SCN)[10]. The Multihop Cellular Networks (MCNs) proposed in [1] and[2] are examples of such schemes proposed for multihop communications, which show a significant power optimization (i.e., average power consumption per successfully delivered packet) in addition to very high throughput as compared to the SCN[10]. But, these network architectures require mobile nodes to expend power for relaying traffic generated by other nodes. Also, the use of multiple interfaces might consume higher total power as compared to SCNs. Hence we propose a power optimization scheme for these architectures. The MultiPower Architecture for wireless packet data networks (MuPAC) proposed in [3] extends the idea of MCN to use multiple transmission ranges. This provides flexibility to the mobile nodes to use a lower transmission power when the receiver is nearer. But the choice of the transmission range is restricted to a limited set of data channels decided a priori. We extend this idea further to incorporate variable power transmission ranges. This leads to better power utilization and also leads to improvement in the throughput over the existing MCN and MuPAC architectures. Hence, in the future multihopping is likely to be used extensively in architectures such as HWN[11], MCN, and MuPAC. Our proposed scheme can be used with any of the multihop architectures without being confined to the above.

The organization of the rest of this chapter is as follows. We first briefly describe the MCN and the MuPAC architectures in Sections 2 and 3, respectively. Section 4 describes our proposed extension to the above architectures and Section 5 presents the analysis of the proposed scheme. In Section 6 we present the performance analysis with the simulation results. Finally, Section 7 concludes this chapter and discusses future work.

# 2 Description of MCN Architecture

The MCN [1] was originally designed to provide spatial reuse of the channel by decreasing the transmission power of the mobile nodes without having to pay the penalty for a large number of Base Stations (BSs). The MCN architecture uses a transmission range that is a fraction $1/k$ of the cell radius, $R$. This means that in the same cell up to a maximum of $k^2$ nodes can transmit simultaneously. The analysis in [1] proved that the hop count increases linearly with $k$. Hence the throughput increase is expected to increase linearly with $k$. The extension of MCN architecture and unicast routing protocol proposed in [2] used a single data channel of transmission range $R/2$ and a control channel with the transmission range of $R$. The control channel is used for delivering the topology information to the BS. The registration and *Route Request* messages use the control channel for communication with the BS.

## 2.1 Routing Protocol

The routing protocol used in MCN is Base Assisted Ad hoc Routing (BAAR) and is briefly described here.

1. The nodes periodically exchange *Hello* beacons thus updating the neighbor information. This neighbor information is relayed to the BS which builds the connectivity graph for the network.

2. When a node has packets to send, it sends a *Route Request* packet to the BS. The BS uses the topology information to compute the path between the source and destination (using Dijkstra's shortest-path algorithm), and responds with a *Route Reply* packet. Both the *Route Request* and *Route Reply* packets are sent over the control channel.

3. A *Route Error* is generated by a node that finds the next hop unreachable. The BS responds to this with a *Route Correct* message.

Figure 1 depicts the MCN system and routing protocol. The mobile node A sends a *Route Request* (to the BS) for a route to mobile node B over the control channel. The BS replies with a *Route Reply* containing the path A-C-D-B, over the control channel. Transmissions from A to B now take place over the data channel of transmission range $R/k$ via C and D.

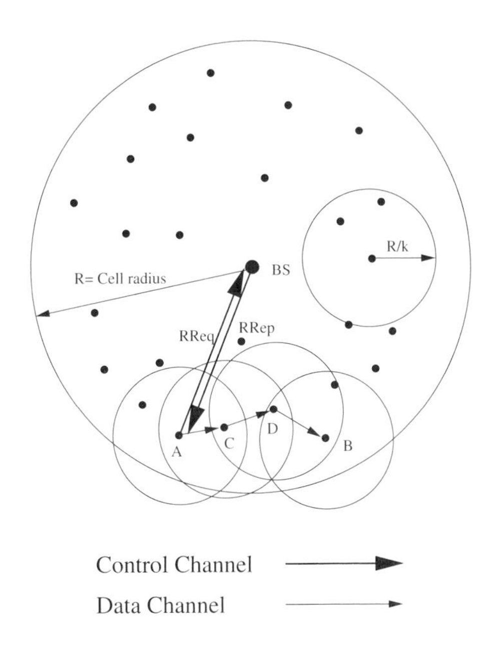


Figure 1: Routing protocol in MCN.

# 3 Description of MuPAC Architecture

The MuPAC [3] is a multipower scheme for packet data cellular networks. This scheme was primarily proposed to overcome the limitations posed by the MCN architecture and to enhance the throughput further. In the MCN architecture [2] the nodes use a transmission range of $r$ which is $1/k$ fraction of the cell radius. The parameter $k$ is called the reuse factor. The $n$-channel MuPAC uses $(n+1)$ channels with different transmission ranges. The available bandwidth is divided among $n$ data channels and a control channel, thus denoted $(n, r_1, b_1, \ldots, r_n, b_n)$, where $r_i$ is the transmission range of the $i$th data channel and $b_i$ is the bandwidth of the $i$th data channel. The transmission range of the control channel is $R$, the cell radius. The idea behind the use of multiple channels is to control the number of hops in the transmission and to lower the transmission power thus providing a greater reuse and hence a better throughput. The use of a higher-power data channel serves the following purpose.

1. Partitioning in the network is avoided by using higher transmission power and thus making secluded nodes reachable.

2. In the case of a route established using multiple hops, higher channels serve as backups to avoid link breaks. This means, if the node on the next hop that was reachable using the $i$th data channel is no longer reachable using the same, then MuPAC upgrades the transmission to the $(i+1)$th data channel.
3. They may be used to reduce the number of hops in a multihop transmission.

### 3.1 Routing Protocol

The routing protocol of MuPAC is very similar to that of MCN. As in MCN, the nodes transmit *Hello* beacons using the transmission power of $r_n$ and the nodes hearing this record their neighbors and the received power from the neighbors. This information is conveyed to the BS using the control channel which helps the BS to maintain the connectivity graph of the network. The connectivity graph is necessary to answer route requests sent over the control channel by nodes to the BS. The rest of the details remain the same as the routing protocol of MCN described in Section 2.1. Because MuPAC is a multipower scheme unlike MCN which uses a fixed transmission power to transmit data over the next hop, MuPAC chooses the transmission power as follows. For a given route the data are transmitted to the next hop using the data channel given by $r_i > \alpha * d_{approx}$; where $\alpha > 1$ is the safety factor and $d_{approx}$ is the distance estimated using the received power of the *Hello* beacons. If the next hop is not reachable over the $n$th data channel (the highest power data channel) then a *Route Error* message is sent to the BS and handled the same way as in MCN described in Section 2.1. The packet transmission in a two-channel MuPAC is shown in Figure 2. For further details on routing in MuPAC refer to [3].

### 3.2 MuPAC-2 and MuPAC-3

The MuPAC-2 refers to the specific case of the MuPAC architecture in which the number of data channels is two, and similarly MuPAC-3 has three data channels. From the simulation studies in [3] with MuPAC-2, the optimal performance is obtained when $r_1$ is $R/3$ and $r_2$ is $R/2$ with $b_1 = b_2 = 2.5$ Mbps when the bandwidth for control channel is 1 Mbps. MuPAC-3 uses $r_1 = 170$ m, $r_2 = 220$ m, and $r_3 = 250$ m for a cell radius ($R$) of 500 m. $b_1 = 0.75$ Mbps, $b_2 = 1.25$ Mbps, and $b_3 = 3$ Mbps. Henceforth MuPAC-2 and MuPAC-3 mean the division with the above characteristics and we

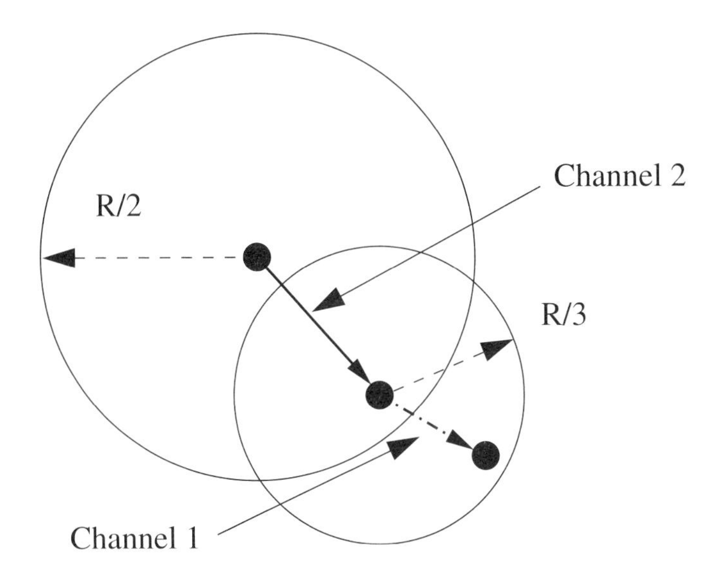


Figure 2: Packet transmission in a two-channel MuPAC.

use the same for our study. We study only the effect on two- and three-channel MuPAC because using a larger number of channels may not lead to significant benefit [3].

# 4 Our Work

The MCN and MuPAC architectures provide limited flexibility in terms of the transmission ranges. The transmission power must be chosen from a limited set of available values in the case of MuPAC and is fixed in the case of MCN. Although MuPAC does not restrict the number of data channels that can be used, increasing the number of data channels does not increase the flexibility because the available bandwidth is divided into many different channels thus increasing the transmission time. Also, the increase in number of data channels increases the number of interfaces, which may add to the complexity. Furthermore at low loads this is not advisable because most of the data channels will remain idle. MCN on the other hand forces the use of a single data channel for nodes near and far alike.

## 4.1 Issues in Using Variable Power

Apart from the problem mentioned above using variable power directly having a large number of data transmission ranges leads to an increase in the number of collisions. This can be explained as follows.

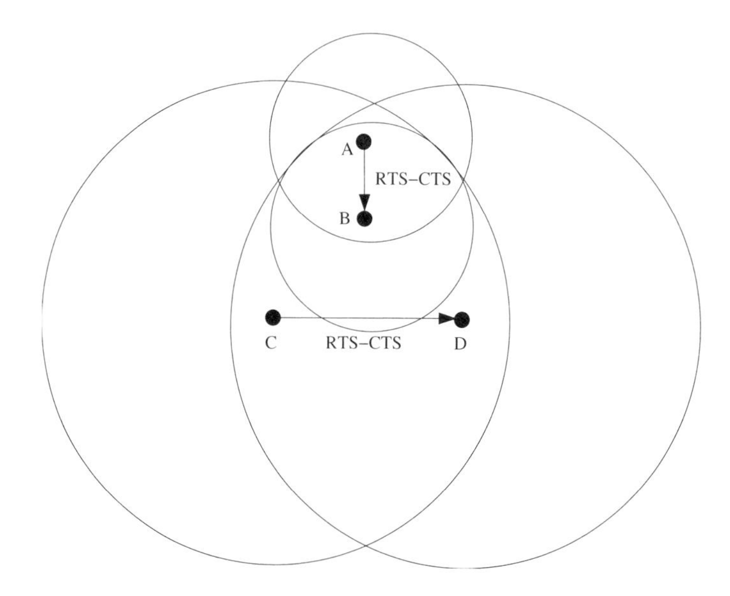


Figure 3: Problem in using variable power directly on MAC protocols such as IEEE 802.11.

The IEEE 802.11 [12] MAC protocol uses the Request-to-Send/Clear-to-Send (RTS-CTS) mechanism for gaining access to the channel. In this mechanism the sender senses the channel to be idle and sends an RTS message to the receiver. The receiver replies with a CTS message. These messages are heard by the neighbors who then avoid transmission so as to avoid collisions for a time interval specified by the Network Allocation Vector (NAV; the NAV carries the information about the time interval for which the transmission is likely to go on). Thus the effect of the hidden terminal problem [13] is reduced. However, when multiple transmission powers are used this effect will be nullified. This is illustrated with an example in Figure 3. Consider a wireless network as shown in Figure 3. Node A wishes to send data to Node B. Node A finds the channel idle and hence sends an RTS message to Node B. Because Node B is close enough to Node A, A uses a lower transmission power sufficient for Node B to hear. Node B replies with the CTS with the same power. Both the RTS and CTS are not heard by Nodes C and D who are relatively farther apart. Thus the transmission between Nodes A and B is initiated. Now, Node C wishes to send data to Node D and also finds the channel idle, but Node D being relatively farther away, Node C uses a higher transmission power. This transmission is audible at Node B and hence a collision occurs at Node B. Thus, the origination of a new packet transmission causes interference with an existing packet transmission, due to the use of variable transmission power.

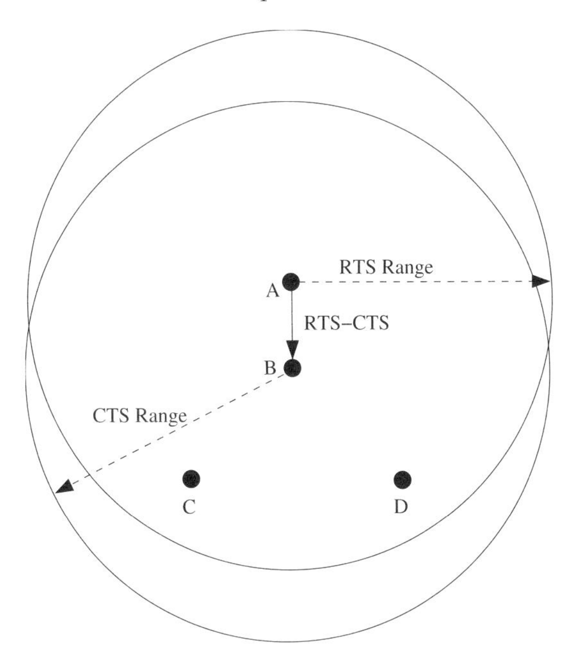


Figure 4: The collision is avoided when the same transmission range is used.

This problem would not have occurred if Nodes A and B had used the same transmission power as Node C as shown in Figure 4. In this case, the RTS or the CTS of the transmission between Nodes A and B would be heard at Nodes C or D or both and hence the collision would be avoided.

## 4.2 Application of Proposed Scheme to MCN and MuPAC Architectures

The optimization scheme proposed for these architectures is aimed at power saving. We divide the available bandwidth into $n$ different transmission range channels similar to the MuPAC architecture. We do not consider MCN separately because it can be considered as a special case of the MuPAC with one data channel. Thus we use the same representation as MuPAC to represent this division: $(n, r_1, b_1, \ldots, r_n, b_n)$, where $r_i$ is the transmission range of the $i$th data channel and $b_i$ is the bandwidth of the $i$th data channel. The $r_n$ is used to exchange the *Hello* messages as in the MuPAC architecture.

Using the *Hello* message and received power at every node, an approximate distance to the sender can be calculated as in the MuPAC. The MuPAC uses $\alpha$ as a safety factor to calculate this distance. In addition to this, we define *mobility margin* to be the mobility safety factor. This is necessary because the factor $\alpha$ only compensates for the error in the calculation of the distance. But, mobility may cause two nodes to move farther apart until the next *Hello* message is transmitted. This is accounted for by the mobility margin. The mobility margin may be transmitted by the BS to every node at the time of registration. The value of the mobility margin may be defined based on the desired need for power consumption (in which case the value needs to be less) and the mobility of the network (in which case the value needs to be more). Also, the value of the mobility margin depends on the frequency of neighbor updates. In the case of high update frequency a low value of the mobility margin may suffice. However, in the case of low update frequency the value of the mobility margin needs to be increased.

The transmission power is estimated from the sum of estimated distance and the mobility margin value. This is done as follows. Let the estimated distance between the two nodes be $x$ m. Let $s$ be the mobility margin. Thus a transmission power needed to transmit up to a distance of radius $x + s$ is calculated to be $t_x$. Let $r_i$ be the transmission range of the channel selected by MuPAC for transmission. If the transmission power $t_x$ is greater than the transmission power of $r_i$ then the transmission takes place as in MuPAC, without any change. Otherwise, as shown in Figure 5 (in function SendMessage()), the RTS-CTS are exchanged over the $r_i$th channel but the data transmission takes place with the reduced transmission power. The acknowledgment is also sent over the reduced transmission range. Although we are affecting the transmission range (power) of various data, we do not alter the transmission power of the RTS and CTS messages. Hence, we may expect a higher throughput due to increased reusability of bandwidth as opposed to the study by Jung and Vaidya [9] which increases the transmission range of RTS and CTS to the maximum possible value. Hence, we prevent only the nodes that pose a threat to the ongoing transmission from transmitting.

```
SendHelloMessage()
begin
    if the time since last broadcast > MIN_HELLO_INTERVAL
            Broadcast Hello Message
end

HelloMessageReceived()
begin
    Update the neighbour table
    Inform the changes to the BS
end

SendMessage(Message m)
begin
    if the next hop of m is reachable
    begin
            Estimate the distance to the next hop
            Apply the mobility magin correction to the estimated distance
            Calculate the required transmission power based on the above distance
            Calculate the transmission power estimated by MuPAC scheme
            if the transmission power by MuPAC is lesser
                    Send the packet using MuPAC scheme
            else
            begin
                    Send the RTS/CTS with power calculated using MuPAC
                    Send the data using reduced power
            end
    end
    else
            Send a route error to the BS
end
```

Figure 5: The power optimization scheme.

The effect of the above scheme is as follows. The number of nodes that hear the RTS or the CTS is more than the number of nodes that hear the actual transmission. In fact, this is the same number of nodes that are blocked by the nonoptimized version of the scheme. It is important that these nodes are prevented from transmitting data in order to prevent a collision with the ongoing transmission. Therefore, in the above examples (Figure 3) Nodes C and D hear the RTS sent by Node A or the CTS sent by Node B although they may not hear the actual transmission. Thus Nodes C and D are prevented from transmitting data on the same data channel as Nodes A and B for the transmission interval specified by RTS-CTS. Hence the chances of collision during the data transmission interval are reduced (see Figure 6).

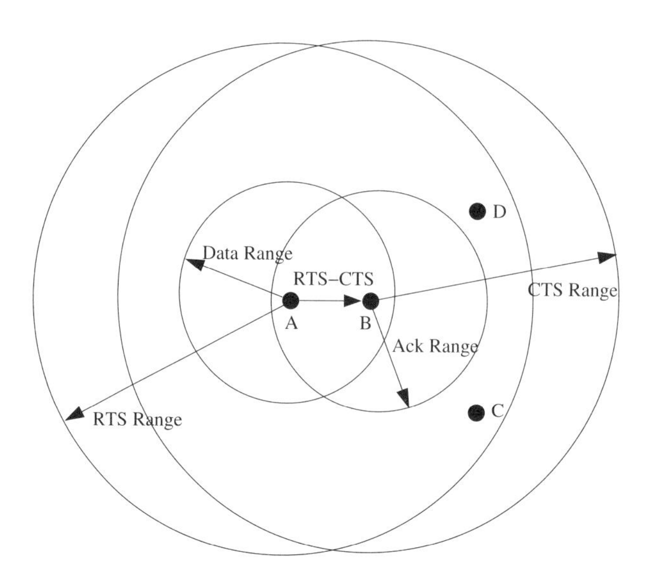


Figure 6: The collision is avoided in the proposed power optimization scheme.

The above method of preventing certain nodes from transmission even though they are not directly affected by the ongoing data transmission is necessary in the interest of the nodes that are currently involved in transmission. Besides, these nodes were prevented from transmission by the MCN and MuPAC schemes. We continue to prevent these nodes from transmission in order to avoid the probability of collision. In our proposed optimization to the MCN and MuPAC architecture we depend on the estimated distance between the nodes for the calculation of the reduced transmission range. We assume that it is possible to estimate the distance between two nodes approximately. This can be done in many ways. If the GPS information is available the receiver can convey this to the sender in the CTS. Alternatively, the sender can translate the signal strength of the *Hello* messages received from the neighbors, thus estimating the distance. In this case the transmission power information may be included in the packets by all the nodes in their transmissions. Furthermore, to do away with the variable noise levels a node may average the received power for the last few receptions from the same sender. In this work we assume that provisions exist for approximate calculation of distance between nodes.

# 5 Theoretical Analysis

We analyze the improvement in the power consumption of the above proposed architectures for MCN and MuPAC. Let $\alpha$, $\beta$, and $\gamma$ be the traffic generated on the data, neighbor, and the control channel, respectively. Let $\delta$ denote the mobility margin. By traffic here we refer to bytes of data generated and not the number of packets. We do not take into account the reception energy because it is a constant and unaffected in the case of the optimized scheme as compared to the nonoptimized MCN/MuPAC. So in a power control scheme like the one proposed in this chapter, it is more logical to study the power consumed by transmission only. Because, transmission power consumed is proportional to the square of the distance between the nodes, we have the following.

## 5.1 MCN

In the case of MCN, the mobile node uses a transmission range of $R/2$ for transmitting data to any other mobile node within a distance of $R/2$. Regions 1, 2, and 3 in Figure 7 correspond to the area covered by the transmission range of $R/2$. However, in our proposed optimization, the mobile node uses a transmission range $r$ for transmitting to the nodes in Region 2. Region 2 corresponds to a circular ring of thickness $dr$ and radius $r$. In the calculations below we integrate over $r$ to find the average transmission power per data transmission.

The average power consumed per node per transmission in the MCN architecture for transmitting data alone over one hop is proportional to

$$\int_0^{R/2} \frac{2\pi r dr (R/2)^2}{\pi R^2/4} = \frac{R^2}{4}. \tag{1}$$

Similarly, the average power consumed in the case of the power-optimized MCN architecture for transmitting data over one hop is given by

$$\int_0^{R/2} \frac{2\pi r dr (r+\delta)^2}{\pi R^2/4} = \frac{R^2}{8} (\text{ignoring } \delta). \tag{2}$$

Here $\delta$ is the mobility margin. Hence the ratio of the total power consumed by power-optimized MCN to MCN without power-optimization is given by

$$\frac{\alpha R^2/8 + \beta R^2/4 + \gamma R^2}{\alpha R^2/4 + \beta R^2/4 + \gamma R^2}. \tag{3}$$

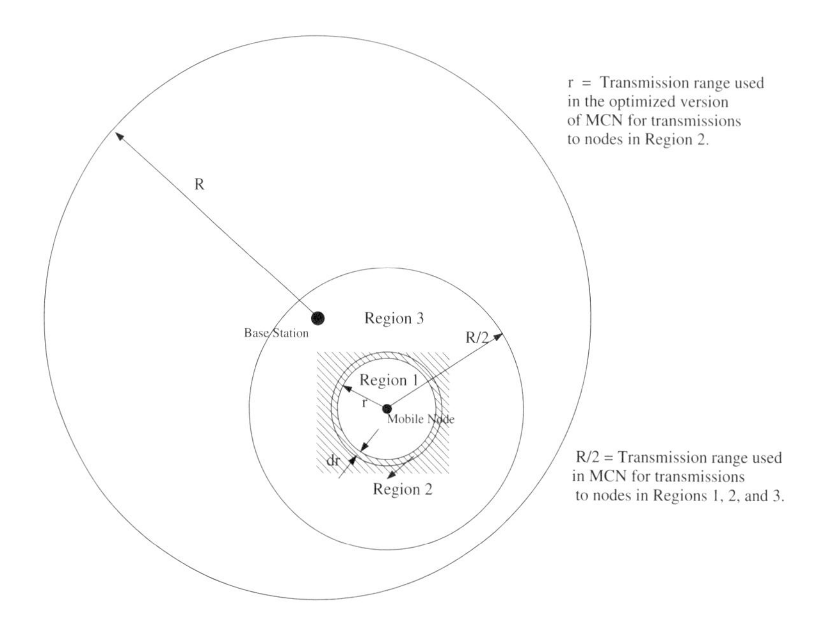


Figure 7: Analysis for power-optimized MCN.

Here, the terms from left to right (in both the numerator and the denominator) refer to the power consumed in transmissions over the data channel, neighbor exchanges, and the transmissions over the control channel, respectively. These can be compared directly owing to the fact that the routes chosen for data transmission in both schemes will be the same and no additional traffic is generated by the power-optimized version of the protocol.

## 5.2 MuPAC

In the case of MuPAC-2, the mobile node uses a transmission range of $R/3$ for transmitting data to any other mobile node within a distance of $R/3$ but a transmission range of $R/2$ for transmitting to mobile nodes beyond a distance of $R/3$ but within a distance of $R/2$. Regions 1, 2, and 3 in Figure 8 correspond to the former and Region 4 corresponds to the latter. In our proposed optimization, a transmission range of $r$ is used for the nodes in Region 2. The value of r varies up to $R/2$.

The average power consumed per node per transmission in the MuPAC

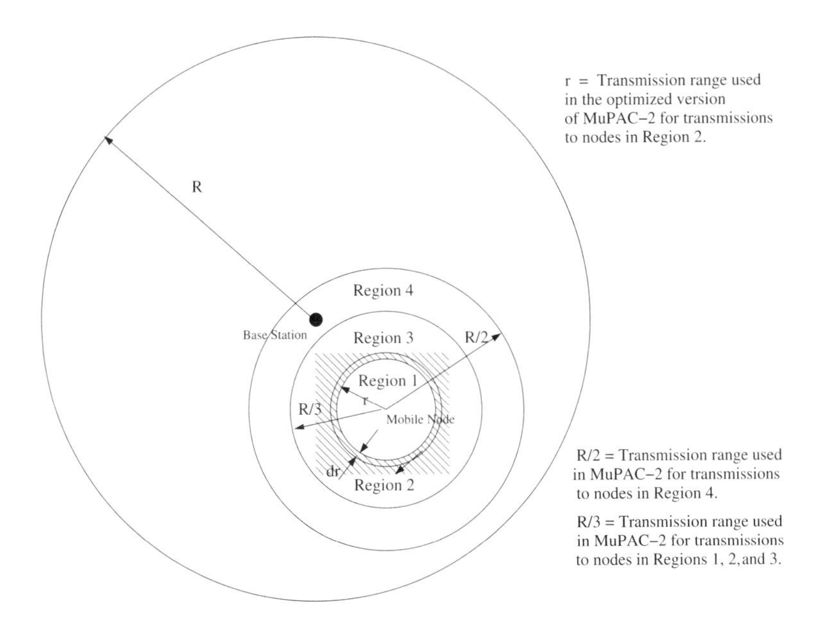


Figure 8: Analysis for power-optimized MuPAC.

architecture for transmitting data alone over one hop is proportional to

$$\int_0^{R/3} \frac{2\pi r dr (R/3)^2}{\pi R^2/9} + \int_{R/3}^{R/2} \frac{2\pi r dr (R/2)^2}{\pi R^2/4 - \pi R^2/9} = \frac{13 \times R^2}{36}. \tag{4}$$

Similarly, the average power consumed in the case of the power-optimized MuPAC architecture is given by

$$\int_0^{R/2} \frac{2\pi r dr (r+\delta)^2}{\pi R^2/4} = \frac{R^2}{8} (\text{ignoring } \delta). \tag{5}$$

Hence the ratio of the total power consumed by power-optimized MuPAC to MuPAC without power optimization is given by

$$\frac{\alpha R^2/8 + \beta R^2/4 + \gamma R^2}{\alpha 13 R^2/36 + \beta R^2/4 + \gamma R^2}. \tag{6}$$

Here, the terms from left to right (in both the numerator and the denominator) refer to the power consumed by data transmission, neighbor exchanges, and the transmissions over the control channel, respectively. In the Equations (3) and (6) the numerator is proportional to the power-consumed by the power-optimized version of MCN and MuPAC, respectively and the denominator refers to the power consumed by the MCN and MuPAC without power optimization. From the above we see that the numerator in both the MCN and the MuPAC cases remains the same but the denominator is more in the case of MCN. Hence we expect more saving in the case of MCN than in the case of MuPAC. Hence, the power consumed by the power-optimized version of MCN or MuPAC remains the same irrespective of the backbone architecture.

Comparing the above analysis for the data channel alone (i.e., the terms with $\alpha$ alone) we see up to 50% improvement in the case of power-optimized MCN and 34% in the case of power-optimized MuPAC-2. When we consider the power consumed by all the interfaces this saving percentages is reduced. However, the values of the quantities in Equations (3) and (6) determined from the simulations in Section 6 show about 10 to 15% improvement in power consumption.

## 6 Performance Analysis

In order to evaluate the performance of the power optimization scheme we simulated the MCN and MuPAC architectures using GloMoSim [14]. The region of simulation was a hexagonal area consisting of seven BSs and the statistics were collected in the center cell as shown in Figure 9. The simulation was performed for the optimization applied to various schemes including MCN, MuPAC-2, and MuPAC-3. The performance was compared at various node densities, and various values of the mobility margin at varying load and

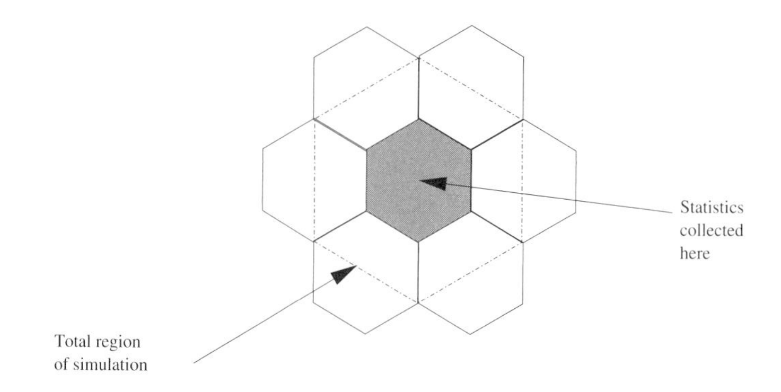


Figure 9: Region of simulation.

mobility. Each simulation was run for 60 seconds and averaged over 8 seed values. The bandwidth of the system was fixed at 6 Mbps (1 Mbps for the control channel and the rest for data channels). The cell radius $R$ was set to 500 m. For the MCN architecture we used the data channel transmission ranges as $R/2$. The results are discussed in the following sections.

## 6.1 Effect of Network Load

We evaluated the performance of MCN, MuPAC-2, and MuPAC-3 with and without power optimization on the basis of throughput obtained and the transmission power expended at various values of UDP load. The UDP load generated at each node varies from 2 packets/second/node to 10 packets/second/node. The maximum mobility of the nodes is restricted to 10 m/s and the value of mobility margin is fixed at 4 m. The results were studied under two scenarios: locality = 1 and locality = 0. (Here locality = 1 corresponds to the case when all the mobile nodes in a cell have data meant for nodes in the same cell. Locality = 0 refers to the case where all the mobile nodes have data meant for nodes in other cells.) We compare the performance at locality = 1 with MuPAC-3 and MCN using the same scheme.

### 6.1.1 Locality = 1

At a value of 4 m for the mobility margin, the power consumption in all the power-optimized versions is around 10 to 15% lower (see Figures 10–12) The power consumption at various values of mobility margin does not show any major difference in MuPAC-2 (see Figure 13). One interesting observation is that the power improvement was independent of load. As shown in Figure 14, the MCN and MuPAC-3 schemes give almost the same throughput with or without power optimization. MuPAC-2 with optimization, however, gives a slightly poorer performance at lower values of load but outperforms its counterpart without optimization at higher loads. But when the mobility margin was increased to 6 m or 8 m the performance in terms of throughput matched MuPAC-2 without power optimization (see Figure 15).

### 6.1.2 Locality = 0

We studied MuPAC-2 without power optimization with the optimized version at the mobility margin of 4 m, 6 m, and 8 m when traffic locality is zero (see Figures 16 and 17). The power consumption shows minor differences

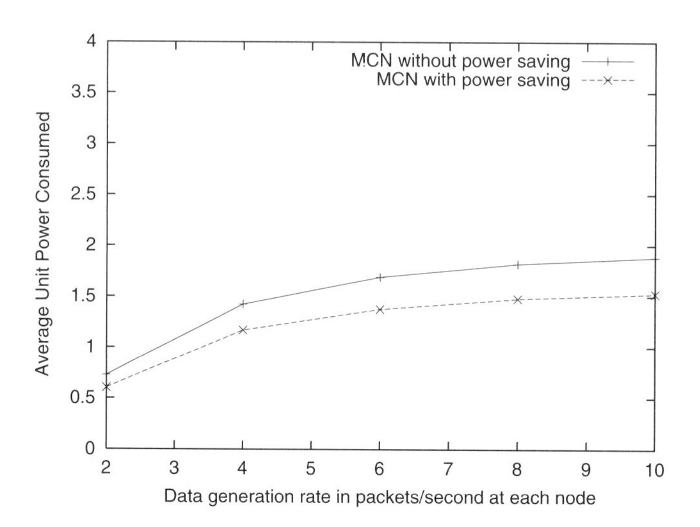


Figure 10: Power consumption versus network load for MCN (mobility = 10 m/s; mobility margin = 4 m; locality = 1 ).

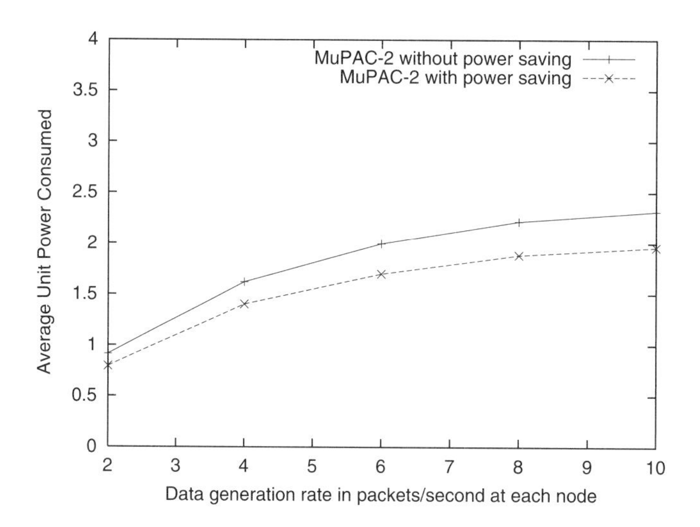


Figure 11: Power consumption versus network load for MuPAC-2 (mobility = 10 m/s; mobility margin = 4 m; locality = 1).

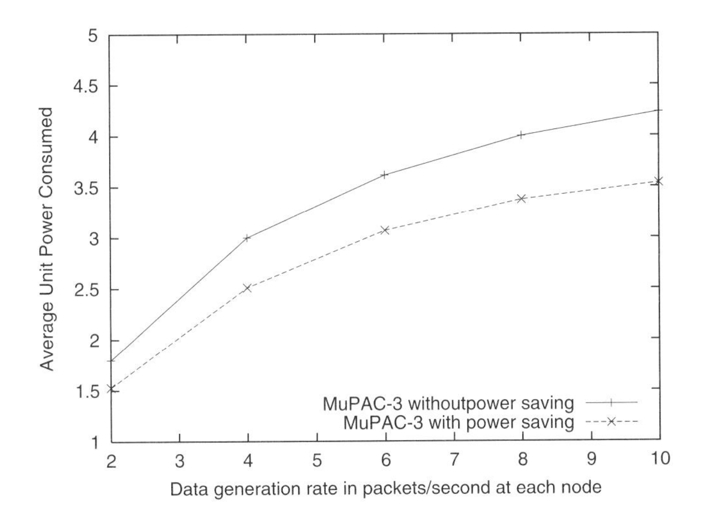


Figure 12: Power consumption versus network load for MuPAC-3 (mobility = 10 m/s; mobility margin = 4 m; locality = 1).

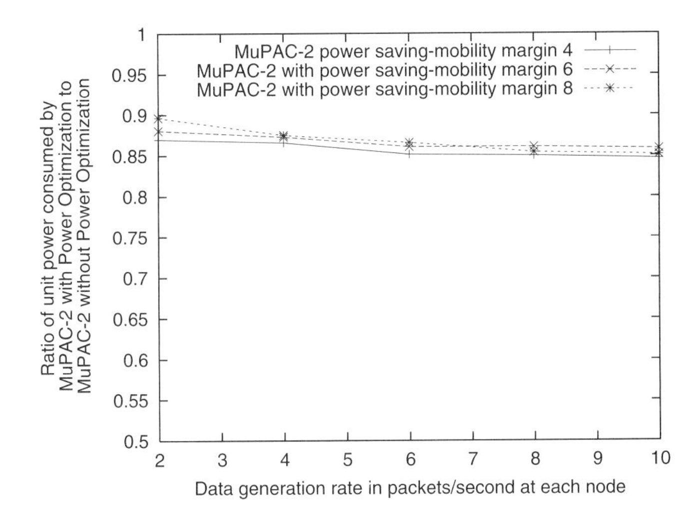


Figure 13: Power consumption versus network load for MuPAC-2 at various mobility margin values (mobility = 10 m/s; locality = 1; nodes = 160).

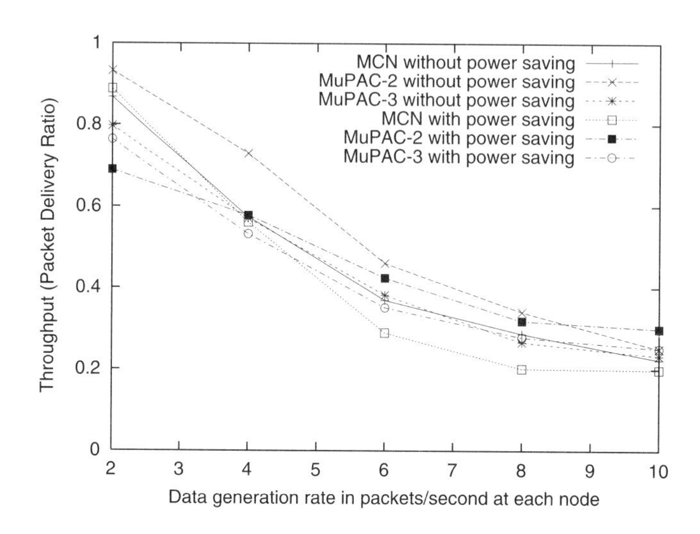


Figure 14: Packet delivery ratio versus network load for MCN, MuPAC-2, MuPAC-3 (mobility = 10 m/s; mobility margin = 4 m; locality = 1).

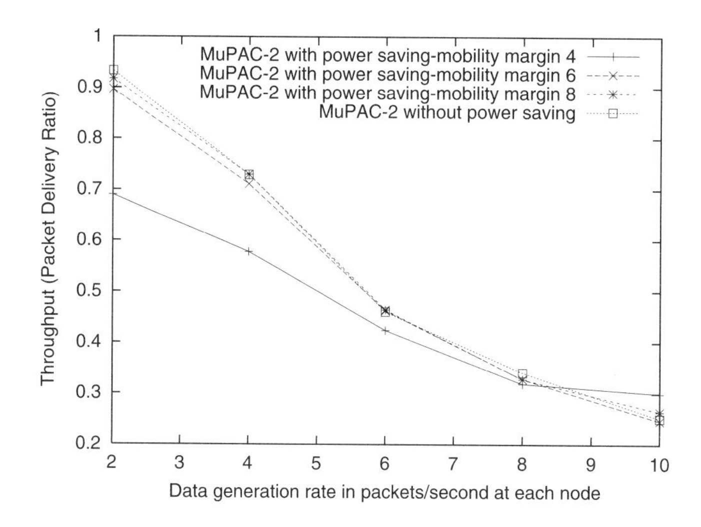


Figure 15: Packet delivery ratio versus network load for MuPAC-2 at various mobility margin values (mobility = 10 m/s; locality = 1; nodes = 160).

across various values of the mobility margin. The throughput in all the above cases matches the throughput obtained by the nonoptimized version of MuPAC-2 (see Figure 18).

## 6.2 Effect of Mobility

We simulated the system at various maximum mobility values, from 2 m/s to 18 m/s, at varying values of the mobility margin from 2 m to 10 m. This was at a fixed UDP load of 5 packets/second/node and at the locality of 1. For the power-optimized MuPAC-2 the throughput was not significantly different from its nonoptimized counterpart but the power consumed showed an improvement of 10 to 15%, as shown in Figures 18 and 19. The increase in mobility margin did not affect the power usage significantly but the throughput improved with the mobility margin, more so at high mobility values. When the same conditions were applied to MCN architecture, the power gain was significantly higher, about 20 to 25%, whereas the throughput remained slightly lower than the nonoptimized version for low mobility values (about 2 to 5% lower). But at high mobility values the throughput for the optimized version of MCN was higher than the nonoptimized version for a mobility margin of 10 m (see Figures 21 and 22).

## 6.3 Effect of Network Size

We simulated the system at varying sizes of the network ranging from 40 nodes to 240 nodes, at a fixed locality of 1, load of 5 packets/second/node and a maximum mobility of 10 m/s. In the case of low load density the power-optimized MuPAC-2 performed significantly better both in terms of throughput and power usage and the power gain was as high as 25%. Increasing the network size, the throughput gain was no longer significantly different from that of MuPAC-2 without power optimization and the power consumption also increased but seemed to stabilize at about 12% gain as compared to MuPAC-2 without power optimization (see Figures 23, 24, 25, and 26).

The MCN results were almost similar to those of MuPAC-2, except the fact that the throughput in the case of 160 nodes was lesser in the case of the optimized version than in the nonoptimized version (see Figure 27). This can be attributed to the fact that the mobility margin value of 4 m and 8 m are insufficient in this case. But at a mobility margin of 10 m, the power-optimized version outperforms the nonoptimized version in terms of

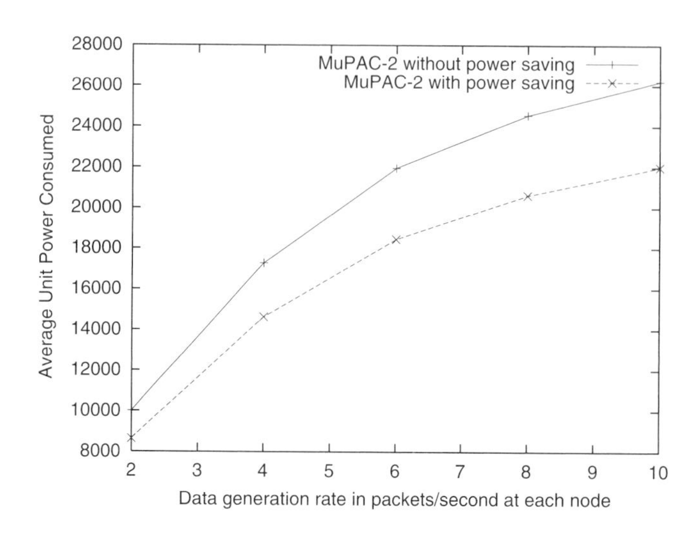


Figure 16: Power usage versus network load for MuPAC-2 (mobility = 10 m/s; mobility margin = 4 m; locality = 0).

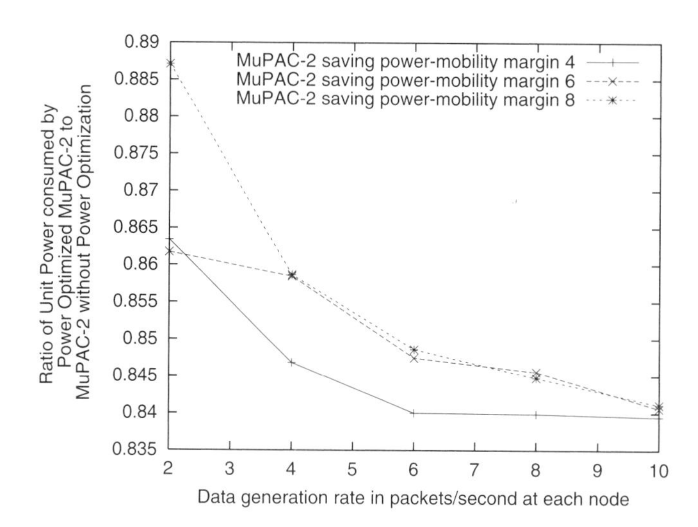


Figure 17: Power consumption versus network load for MuPAC-2 at various mobility margin values (mobility = 10 m/s; locality = 0; nodes = 160).

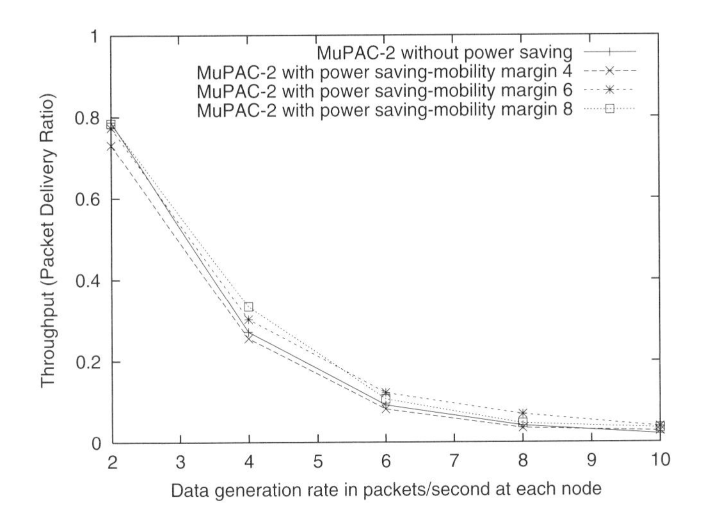


Figure 18: Packet delivery ratio versus network load for MuPAC-2 at various mobility margin values (mobility = 10 m/s; locality = 0; nodes = 160).

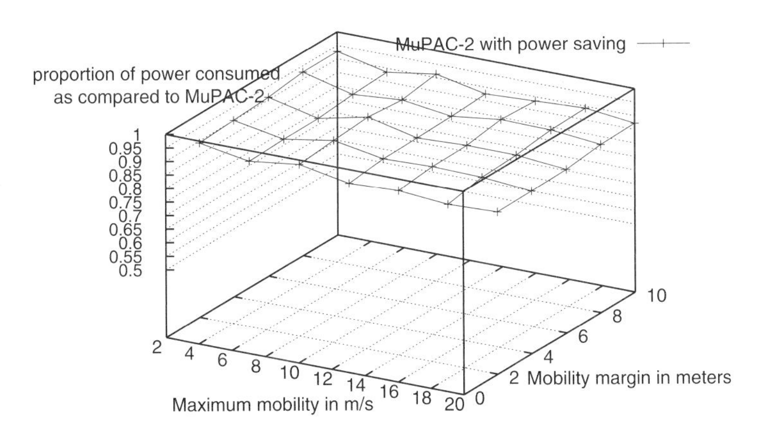


Figure 19: Power consumption versus mobility versus mobility margin for MuPAC-2 (locality = 1; load = 5 pkts/s/node; nodes = 160).

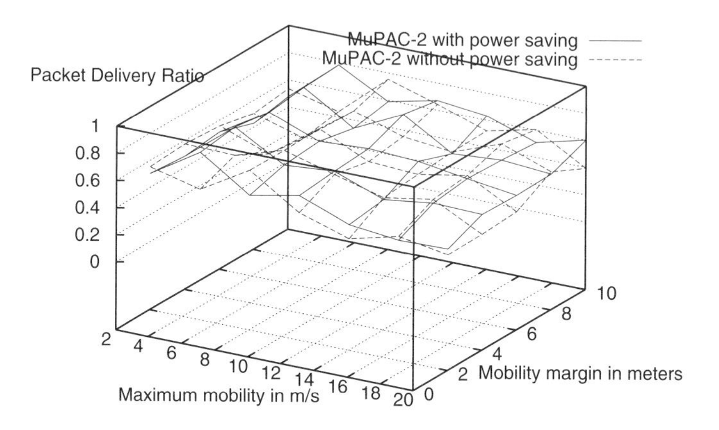


Figure 20: Packet delivery ratio versus mobility versus mobility margin for MuPAC-2 (locality = 1; load = 5 pkts/s/node; nodes = 160).

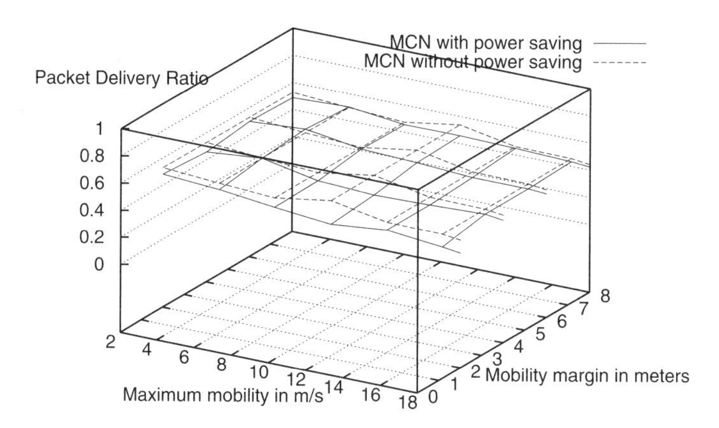


Figure 21: Packet delivery ratio versus mobility versus mobility margin (locality = 1; load = 5 pkts/s/node; 160 nodes).

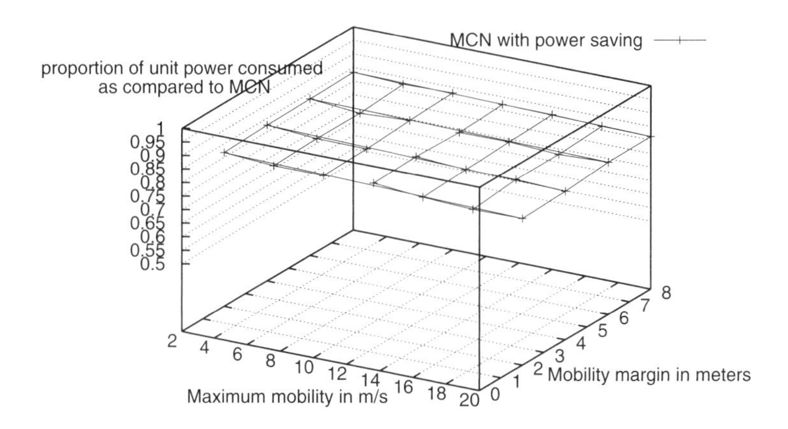


Figure 22: Power consumption versus mobility versus mobility margin (locality = 1; load = 5 pkts/s/node; 160 nodes).

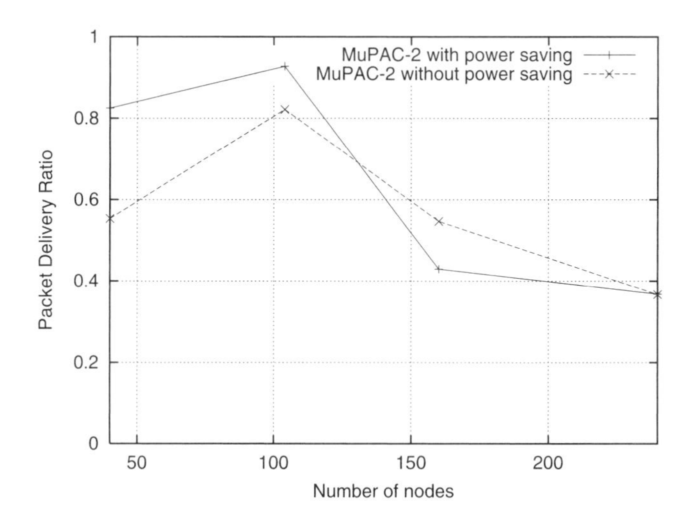


Figure 23: Packet delivery ratio versus number of nodes (locality = 1; load = 5 pkts/s/node; mobility = 10 m/s; mobility margin = 4 m).

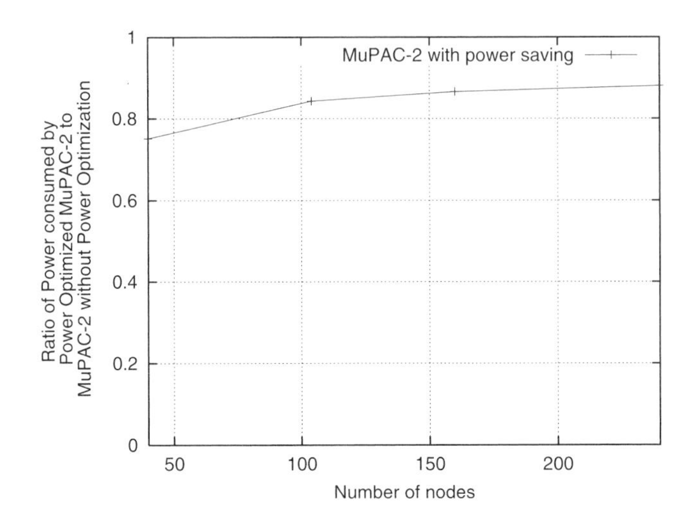


Figure 24: Power consumption versus number of nodes (locality = 1; load = 5 pkts/s/node; mobility = 10 m/s; mobility margin = 4 m).

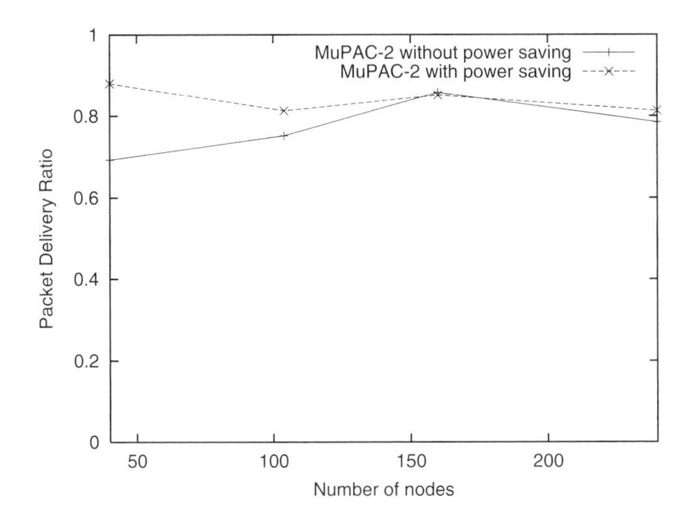


Figure 25: Packet delivery ratio versus number of nodes (locality = 1; load = 2 pkts/s/node; mobility = 10 m/s; mobility margin = 4 m).

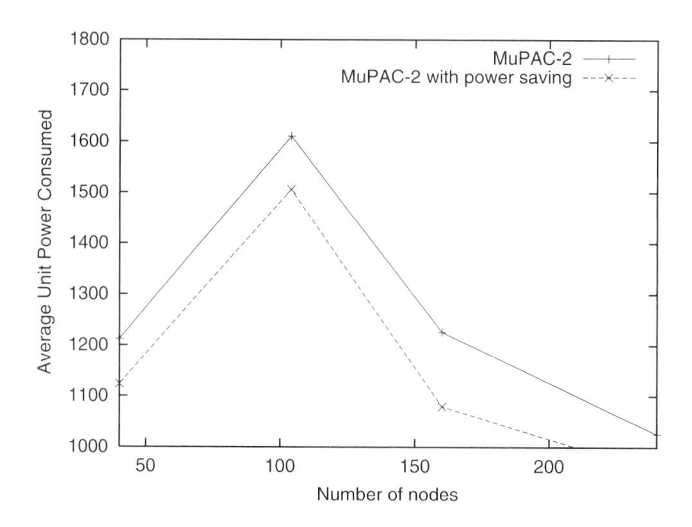


Figure 26: Power consumption versus number of nodes (locality = 1; load = 2 pkts/s/node; mobility = 10 m/s; mobility margin = 4 m).

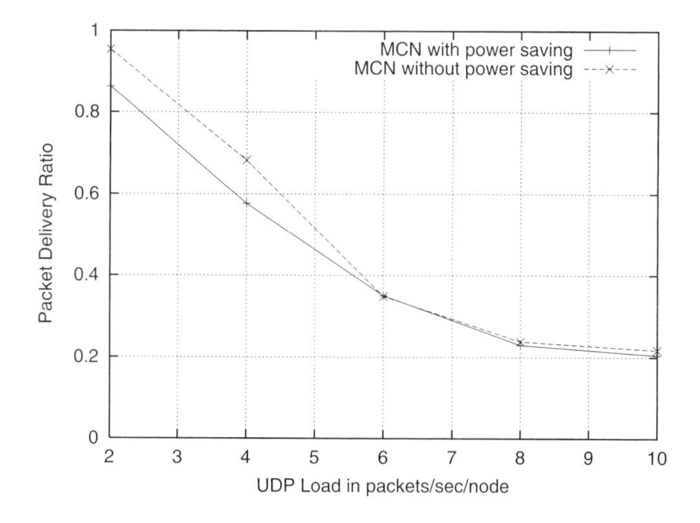


Figure 27: Packet delivery ratio versus UDP load (locality = 1; mobility = 10 m/s; mobility margin = 4 m; 160 nodes).

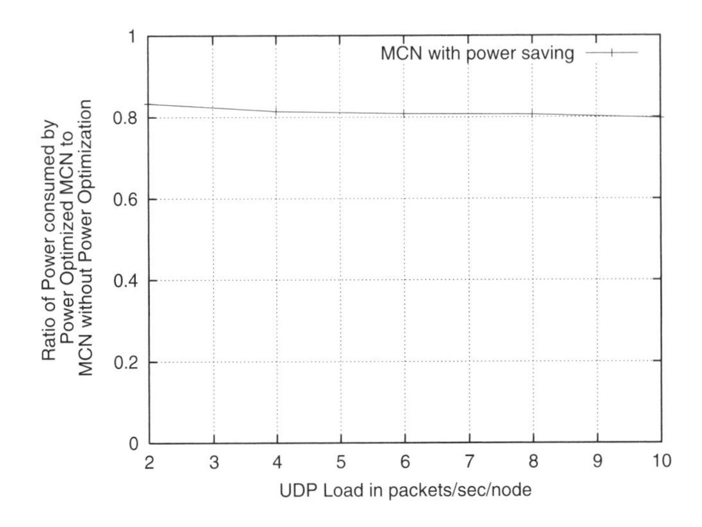


Figure 28: Power consumption versus UDP load (locality = 1; mobility = 10 m/s; mobility margin = 4 m; 160 nodes).

throughput (see Figure 20). The power gain is once again independent of the number of nodes (see Figure 28).

# 7 Conclusions

In this chapter we proposed a power optimization scheme for multihop architectures such as MCN [2] and MuPAC [3] to improve power consumption performance and throughput. Although the main aim of the scheme was to conserve power, up to 10% improvement of the throughput was achieved. The main cause of improvement of throughput was that some collisions caused as a result of mobility were avoided because the data transmissions affect a smaller portion of the network. On the other hand the factor that had a reducing effect on the throughput was the mobility margin. From our simulations we obtained about 10 to 15% improvement in power consumption which is a significant saving in the case of power-constrained systems such as laptops.

Theoretical analysis shows a significant saving in power consumed. Considering the data channels alone in MCN and MuPAC we get almost 50% and 34% reduction of transmission power. But the presence of a control channel reduces this gain considerably. Hence the application of similar mechanisms for improvement of power consumed on the control channel may lead to better performance. But care must be taken while applying this to the control channel because this will have a direct bearing on the throughput.

## References


[1] Y. D. Lin and Y. C. Hsu, "Multihop cellular: A new architecture for wireless communications." In *Proceedings of IEEE INFOCOM 2000*, March 2000, pp. 1273–1282.

[2] R. Ananthapadmanabha, B. S. Manoj, and C. Siva Ram Murthy, "Multihop cellular networks: The architecture and routing protocols." In *Proceedings of IEEE PIMRC 2001*, October 2001, Vol. 2, pp. 78–82.

[3] K. J. Kumar, B. S. Manoj, and C. Siva Ram Murthy, "MuPAC: Multi power architecture for wireless packet data network." In *Proceedings of IEEE PIMRC 2002*, September 2002, Vol. 4, pp. 1670–1674.

[4] F. Douglis, F. Kaaskoek. B. Marsh, R. Caceres, K. Lai, and J. Tauber, "Storage alternatives for mobile computers." In *Proceedings of Symposium on Operating System Design and Implementation (OSDI) 1994*, November 1994, pp. 25–37.

[5] Y. Tseng, C. Hsu, and T. Hsieh, "Power saving protocols for IEEE 802.11-based multihop ad hoc networks." In *Proceedings of IEEE INFOCOM 2002*, June 2002, Vol. 1, pp. 200–209.

[6] S. Singh and C. S. Raghavendra, "PAMAS: Power aware multi-access protocol with signaling for ad hoc networks." *ACM Computer Communication Review*, 28(3) (1998), pp. 5–26.

[7] R. Ramanathan and R. Rosales-Hain, "Topology control of multihop wireless networks using transmit power adjust." In *Proceedings of IEEE INFOCOM 2000*, March 2000, pp. 404–413.

[8] J. P. Monks, V. Bharghaven, and W. M. W. Hwu, "A Power controlled multiple access protocol for wireless packet networks." In *Proceedings of IEEE INFOCOM 2001*, April 2001, pp. 219–228.

[9] E. S. Jung and N. H. Vaidya, "A power control MAC protocol for ad hoc networks." In *Proceedings of ACM MOBICOM 2002*, June 2002, pp. 36–47.

[10] K. J. Kumar, B. S. Manoj, and C. Siva Ram Murthy, "On the use of multiple hops in the next generation wireless architectures." In *Proceedings of IEEE ICON 2002*, August 2002, pp. 283–288.

[11] H. Y. Hsieh and R. Sivakumar, "Performance comparison of cellular and multihop wireless networks: A quantitative study." In *Proceedings of ACM SIGMETRICS 2001*, June 2001, pp. 113–122.

[12] IEEE Standards Board, "Part 11: Wireless LAN medium access control (MAC) and physical layer (PHY) specification." *The Institute of Electrical and Electronics Engineers Inc.*, 1997.

[13] C. L. Fullmer and J. J. Garcia-Luna-Aceves, "Solutions to hidden terminal problems in wireless networks." In *Proceedings of ACM SIGCOMM 1997*, September 1997, pp. 39–49.

[14] X. Zeng, R. Bagrodia, and M. Gerla,"GloMoSim: A library for parallel simulation of large-scale wireless networks." In *Proceedings of PADS-98*, May 1998, pp. 154–161.

[15] C. Siva Ram Murthy and B. S. Manoj, *Ad hoc Wireless Networks: Architectures and Protocols*, Prentice Hall PTR, New Jersey, May 2004.

## Chapter 8

# Provisioning the Performance in Wireless Sensor Networks Through Computational Geometry Algorithms

Luiz Filipe Menezes Vieira
*Department of Computer Science*
*Federal University of Minas Gerais, Belo Horizonte, MG 31270-010, Brazil*
E-mail: lfvieira@dcc.ufmg.br

Marcos Augusto Menezes Vieira
*Department of Computer Science*
*University of Southern California, Los Angeles, CA 90089-0781, USA*
E-mail: mvieira@usc.edu

Linnyer Beatriz Ruiz
*Department of Eletrical Engineer*
*Federal University of Minas Gerais, Belo Horizonte, MG 31270-010, Brazil*
E-mail: linnyer@dcc.ufmg.br

Antonio Alfredo F. Loureiro
*Department of Computer Science*
*Federal University of Minas Gerais, Belo Horizonte, MG 31270-010, Brazil*
E-mail: loureiro@dcc.ufmg.br

Diógenes Cecílio da Silva, Jr.
*Department of Eletrical Engineer*
*Federal University of Minas Gerais, Belo Horizonte, MG 31270-010, Brazil*
E-mail: diogenes@eee.ufmg.br

M.X. Cheng, D. Li (eds.) *Advances in Wireless Ad Hoc and Sensor Networks.*
Signals and Communication Technology, doi: 10.1007/978-0-387-68567-0_8.

Antônio Otávio Fernandes
*Department of Computer Science*
*Federal University of Minas Gerais, Belo Horizonte, MG 31270-010, Brazil*
E-mail: otavio@dcc.ufmg.br

---

## 1 Introduction

The maturing of integrated circuitry, microelectromechanical systems, digital signal processing and low-range radio electronics on a single node has led to the design of the wireless sensor network. This network may have hundreds or thousands of sensor nodes, each one with the ability to sense its environment, perform simple computations, and communicate with its neighbors.

A large number of sensor nodes allow the sensing on a larger geographical region with a greater accuracy than previously possible. This type of network has the potential for innumerable applications [1], including weather monitoring, security and tactical surveillance, and environment monitoring.

A Wireless Sensor Network (WSN) differs from other networks, having some unique characteristics. The most important feature is the need to be energy efficient. A sensor node has a finite energy reserve supplied by a battery. It is often infeasible to recharge the node's battery. Thus, the design of a WSN should be as energy efficient as possible. In the case of a network with a high density of sensor nodes, some problems may arise such as the intersection of sensing area, redundant data, communication interference, and energy waste.

We propose a mechanism to control the network density based on a criterion to decide which nodes should be turned off or on. Then, we present a management function to solve this problem, which can take the sensor node out of service temporally. Our solution is based on the Voronoi diagram, which decomposes the space into regions around each node, to determine which sensor node should be turned off or on.

To evaluate our design, we perform a simulation comparison. We evaluate the scheduling of nodes varying the network density. We show that our design can save energy without losing sensing area. This schema could be used in management architecture for WSN [2].

We also propose an efficient algorithm for incremental deployment of nodes in a WSN. By examining the distribution of node density, its energy level, and the sensing cover area, the algorithm indicates which position should have more nodes deployed and how many new nodes are necessary to cover the desired monitoring area. Intelligent sensor deployment facilitates the unified design and operation of sensor systems, and decreases the need for excessive network communication. Most work has been done in determining the sensor deployment of the initial network [3, 4]. To our knowledge, this is the first to consider deterministic incremental deployment after the initial network has been used and considering the energy consumed. Another advantage of the proposed algorithm is its computation time. Combining well-known results in computational geometry, the worst case to compute the position to add a node is $\Theta(n \log n)$, where $n$ is the total number of nodes in the network.

To evaluate our design, we performed a simulation comparison. We compared the incremental deployment algorithm with a random insertion algorithm. To strengthen our result, we considered four different energy models. As predicted, our design can save energy without losing sensing area.

In this chapter, two designs based on computational geometry algorithms are evaluated. They can be combined to improve the WSN performance, minimizing energy consumption, maximizing the lifetime of the network, and minimizing the cost involved. This chapter is organized as follows: in the next section we give an overview of the WSN. In Section 3 we summarize the related work. Section 4 describes the scheduling schema based on the Voronoi diagram. Then, in Section 5, we present the incremental algorithm based on the largest empty circle. Section 6 contains a wide array of experimental results. Section 7 is a brief discussion of future work and the conclusion.

## 2 WSN Architecture

This section gives an overview of the WSN architecture used in this work. WSNs are networks composed of a great number of sensor nodes. The objective of these networks is to collect information. WSNs usually do not have a previous infrastructure, like cellular or local wireless networks. WSN is an ad hoc network, because its topology is dynamic, due to the fact that sensor nodes can wake up, joining the network, or go to sleep, leaving the WSN.

The major resource at WSN is energy. Each sensor node is composed of a small battery, with a limit capacity. It is almost infeasible to recharge all batteries because WSN can be composed of thousands of sensor nodes. Therefore, the WSN project focus, from hardware design to network protocols, is saving energy.

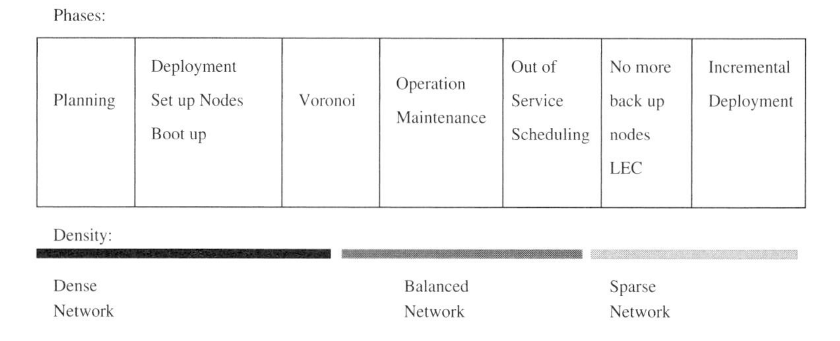


Figure 1: WSN lifetime.

Figure 1 illustrates the lifetime of a WSN. The first phase is planning. Depending on the application, the main characteristics of the WSN are decided upon, such as number of sensor nodes, protocols, data rate, frequency of requests, and the like. Then the WSN is deployed (deployment phase). The sensor nodes wake up, self-organize (setup phase), and start to operate (bootup phase). It is necessary to decide which sensor nodes will be initially turned off or on, forming the set of backup nodes and allowing the dense network to be balanced. This can be done by using our approach based on the Voronoi diagram. After that, the network will operate and be maintained (operation and maintenance phase) during the lifetime of the network, sensor nodes will die due to the power consumption (out of service). The backup sensor nodes chosen during by our approach can be awakened to keep the network operational (scheduling phase). In a given time, the backup nodes set will be empty (no more backup nodes). To keep the network operational it is required to deploy new sensor nodes. Our second design is to identify how many sensor nodes and their exact locations are required by using an approach base on LEC, Largest Empty Circle. After that, the incremental deployment can be done. Therefore, the two designs based on a computational geometry algorithm can provision the performance of WSN.

### 2.1 Assumptions

We work with a flat and homogeneous WSN distributed on a 2D plane field, but we could easily extend to $n$ dimensions. Each node is immobile, although the network topology can be dynamic, because nodes can become unavailable permanently or temporarily. Each node knows its location on the plane. The position does not need to be global, and can be relative to the base station or to a known point. The ultimate sensor nodes have support for localization mechanisms [5–7]. We also assume that we have the approximate energy level in each node. The energy map of the network can be obtained as shown in [8]. Furthermore, there are components such as DS2438 that allow monitoring battery energy [9].

## 3 Related Work

Minimizing energy consumption, maximizing the lifetime of the network, and minimizing the cost involved has been a major design goal for WSNs.

A node-scheduling scheme was developed by Tian and Georganas [10]. In their approach, nodes take turns in saving energy without affecting the service provided. The node-scheduling scheme turns some nodes on or off and certain redundancy is still guaranteed. A node decides to turn off when it discovers that its neighbors can help it to monitor its whole area. The solution does not suppose a global knowledge of the network and is performed locally at each node. Thus, it does not guarantee the optimal solution. The proposed scheme increases communication cost, requires synchronization, and involves the calculation of a geometric representation's intersection. The cost of calculating the regions defined on a plane by $n$ circles is exponential (with $n$ circles, there can be $2n$ such regions).

T. Clouquer et al. [11] realized a study in minimizing the cost of deployment sensors to achieve the desired detection performance for target detection. The minimum path exposure is used as a measure of the goodness of deployment. They did not work with deterministically placed sensors or consider the energy consumed by the network.

K. Chakrabarty et al. [3] present an Integer Linear Programming (ILP) solution for minimizing the cost of the sensor for complete coverage of the sensor field. Energy consumed by the network is not considered. The exact solution to the ILP takes an excessive amount of computation time. Our new approach reduces the computation time, taking advantage of known results in computational geometry [12].

A Voronoi diagram has already been applied to solve other problems in a WSN. Meguerdichian et al. [13] proposed an algorithm for calculating the maximal breach and maximal support paths in a sensor network based on a Voronoi diagram.

To our knowledge this is the first time that computational geometry has been combined with an algorithm to incrementally deploy sensors in a used WSN, to reduce computation time, and to maintain the network.

# 4 Scheduling Nodes and Voronoi Diagram

In this work we have a desired area that we wish to monitor, and a set of $N$ wireless sensors that together define a cover monitoring area. This area gives the fraction of the desired area that is actually being monitored. We define it as a Quality-of-Service (QoS) metric of the network.

## 4.1 Scheduling Nodes

As mentioned before, some problems may arise if the network has a high density of sensor nodes. In the following, we present a management schema to deal with this problem. The management may take a sensor node out of service temporally, scheduling the nodes that will be turned on and off.

Suppose we have a network with a topology as depicted in Figure 2 and all nodes transmit at the same frequency. Node 1 wants to transmit information to node 5 and/or node 2 wants to transmit to node 3. But, if node 4 transmits something to another node, it will cause interference at both transmissions, as illustrated in Figure 3. It could be argued that the MAC layer with carrier sense would solve it. The MAC layer will still be necessary, but if the density of the network is too high at that region, the number of collisions will increase and energy will be wasted unnecessarily. A management solution could temporally turn node 4 off in the case where the monitoring application does not need another node in the sensing area already covered by other nodes. An important point is that increasing the number of hops does not necessarily save energy. An interesting work that discusses this and other issues is [14].

The scheduling node schema discussed in this work may be used for different purposes. Here we present two ideas. When an active node leaves the network, due to energy problems for instance, the management may activate some nodes that are off. Energy is saved and coverage of the monitoring area is not affected. The second purpose is for security reasons. The nodes that

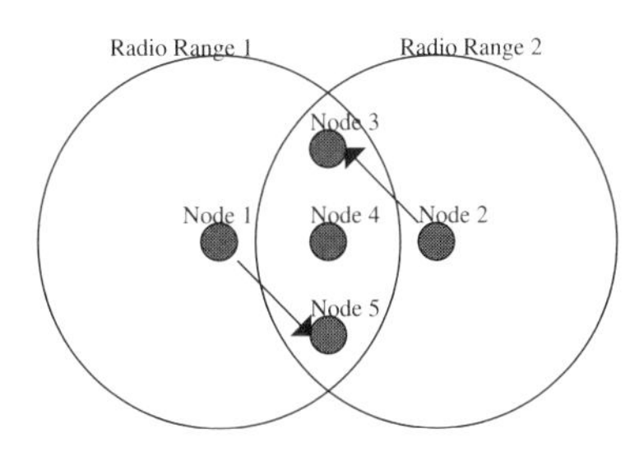


Figure 2: Example of a network topology.

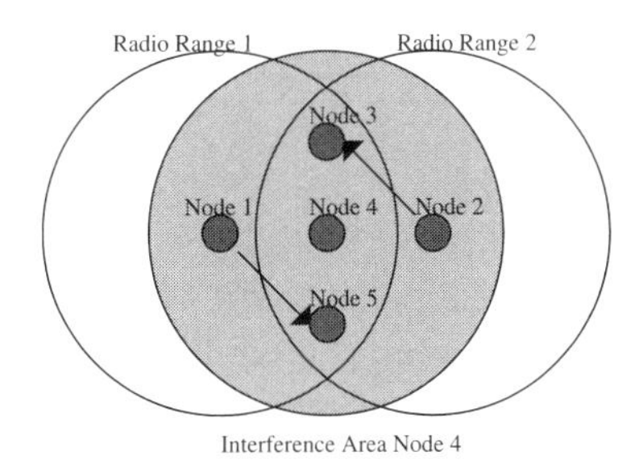


Figure 3: Interference of node 4.

are not sensing could stay in a mode listening to the network, verifying if the nodes are transmitting what they received and increasing the security of the network. When necessary, the node that is off would become an active node. To determine if a node is going to be turned off, we use a Voronoi diagram, which is explained in the following section.

## 4.2 Voronoi Diagram

Let $S = \{p_1, p_2, \ldots, p_i, \ldots, p_n\}$ be a set of points in a two-dimensional Euclidean plane. These points are called sites. A Voronoi diagram decomposes the space into regions around each site, such that all points in the region around $p_i$ are closer to $p_i$ than any other point in $S$.

Using the definition in [15], the Voronoi region $V(p_i)$ for each $p_i$ is expressed as

$$V(p_i) = \{x : |p_i - x| \leq |p_j - x|, \forall j \neq i\}.$$

$V(p_i)$ consists of all points that are closer to $p_i$ than any other site. The set of all sites forms the Voronoi diagram $V(S)$.

The following example, extracted from [15], illustrates a simple Voronoi diagram. Consider two points $p_1$ and $p_2$. Let $B(p_1, p_2) = B_{12}$ be the perpendicular bisector of the segment $\overline{p_1 p_2}$. Then every point $x$ on $B_{12}$ is equidistant from $p_1$ and $p_2$. This can be seen by drawing the triangle $(p_1, p_2, x)$ as depicted in Figure 4. By Euclid's side-angle-side theorem, $|p_1 x| = |p_2 x|$.

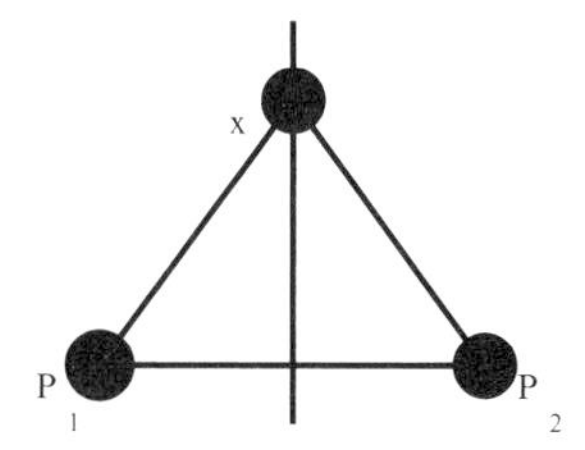


Figure 4: Two points $|p_1 x| = |p_2 x|$.

To sum up, given input points presented in Figure 5, the corresponding Voronoi diagram is depicted in Figure 6.

Figure 5: Input points.

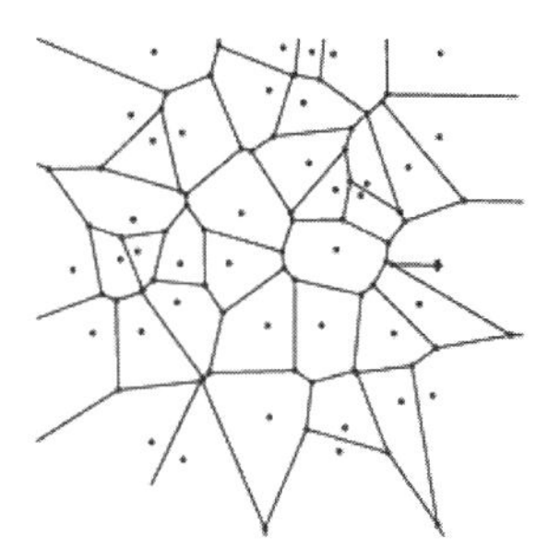

Figure 6: Voronoi diagram for the input points.

```
Input: set of points.
Output: nodes that should be turned off.
Calculate Voronoi();
do begin
      for every node begin
         calculate Vonoroi_Area();
      end
      get SmallestArea();
      if (smallest_area < threshold) then
         begin
            node_responsible ← turn off;
            update_Voronoi();
            keep_searching ← true;
         end
      else keep_searching ← false;
   end
while (keep_searching)
```

Figure 7: Algorithm for selecting backup nodes.

## 4.3 The Algorithm

In this section, we discuss the algorithm used to calculate which nodes are turned on or off. Given the location of the nodes and the area to be monitored, each node represents a point, and the desired area to be monitored is the polygon that is defined by the Voronoi diagram. The objective is to determine the area for which each node is responsible. Then, we pick up the node with the smallest area and if it is too small, the node should be turned off. The neighbors of that node become responsible for that area, updating the Voronoi diagram. This process continues until there is a node responsible for an area smaller than a given threshold. Figure 7 shows the pseudo-code of the algorithm.

The algorithm is illustrated taking the network topology of Figure 2. Figure 8 shows the Voronoi diagram. Because node 4 is responsible for a small area (smaller than a threshold), the algorithm decides that node 4 should be turned off. Figure 9 [16] illustrates the new topology. At this step, there is no node responsible for a smaller area than the threshold, and it ends.

The worst-case complexity of calculating a Voronoi diagram is $\Theta(n \log n)$.

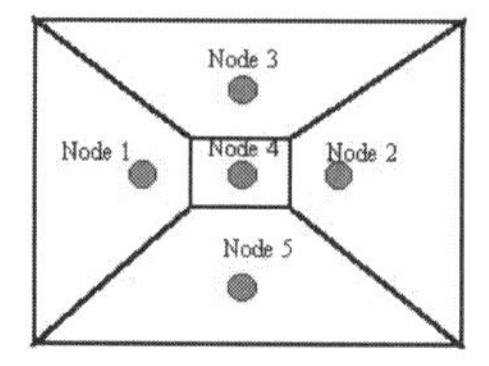


Figure 8: Initial Voronoi diagram.

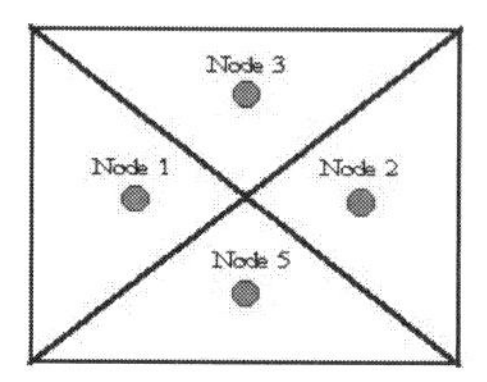


Figure 9: New Voronoi diagram.

Thus, the algorithm seems feasible to be executed in a base station. A simple naive approach, to update the Voronoi diagram, is to rebuild it, with a worst-case complexity of $\Theta(n^2 \log n)$. However, an incremental approach to update the Voronoi diagram could be used.

A local algorithm for deciding the scheduling of nodes could use the idea presented in [17]. A self-elected cluster head collects data positions from all sensor nodes in its cluster, calculates the Voronoi diagram, and transmits its decision back to the nodes in a distributed fashion. However, there are some disadvantages: it expends energy choosing a cluster head and transmitting the information to each node; the decision if a node should be turned on or off is very important and should be done in a management layer because it could affect the entire network; and it does not solve the problem in the neighborhood of each cluster head.

If the network is hierarchical, we can devise three options. If the cluster head does not sense, it can be left out of the Voronoi diagram algorithm. The second option is to treat the cluster head as a common node using the previous design. The third option is to assign a weight to each node. Some choices are the Multiplicatively Weighted Voronoi diagram and Additively Weighted Voronoi diagram. Let *dis* represent the Euclidean distance and $w_i$ be the weight of each point $p_i$. A Multiplicatively Weighted Voronoi diagram is generated by using a distance function in Equation (1). An edge is generally a circular arc. An Additively Weighted Voronoi diagram is generated by using a distance function in Equation (2). An edge is generally a hyperbolic arc.

$$d(p, p_i) = dis(p, p_i)/w_i, \quad (1)$$
$$d(p, p_i) = dis(p, p_i) - w_i. \quad (2)$$

In the case where a sensor network has mobile nodes, a Voronoi diagram of moving points can be applied [18].

# 5 Incremental Deployment

Many nodes with communication, sensing, and processing abilities compose a WSN. These nodes have restricted energy and the lifetime of the system is affected by the energy on the nodes. To increase the lifetime of the network, new nodes can be deployed. Because the local observations made by the sensors depend on their position, the performance of the network is a function of the deployment.

A sensor node may not be covering the desired monitor area. Some reasons are the network density is low, nodes are not correctly deployed at the monitoring area, or sensor nodes lose their energy. Here we propose an algorithm to help incremental deployment of sensors. We define a scheme that examines the distribution of node density, its energy level, and sensing cover area; the algorithm indicates the quantity and which positions should have more nodes. The design utilizes ideas from the largest empty circle problem, which is explained next.

## 5.1 Largest Empty Circle

Following the definition in [15], the largest empty circle problem is: find the largest empty circle whose center is in the (closed) convex hull of a set of $n$ sites $S$, empty in that it contains no sites in its interior, and largest in that there is no other such circle with strictly large radius.

An advantage of our algorithmic approach is the fact that it can be computed outside the WSN or at the data sink, saving the network energy and not requiring the algorithm to be distributed.

## 5.2 The Problem

Given the localization and the approximate energy level of each node in a WSN, what is the minimum number of new nodes that should be added to

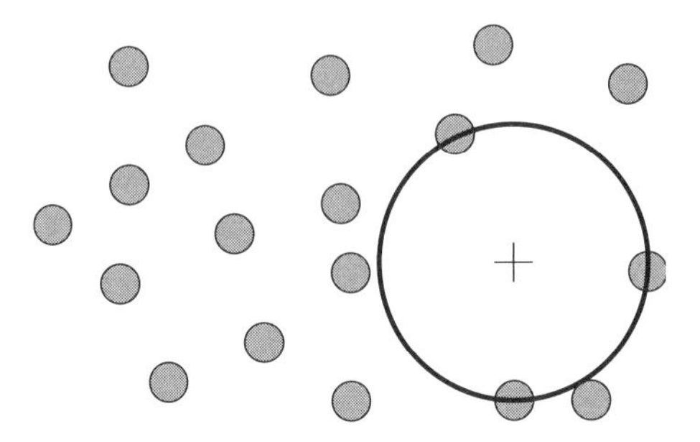

Figure 10: Largest empty circle example.

the network so that it does not lose any covering area? Also, where should the new nodes be positioned? The solution should be computed in reasonable time.

### 5.3 The Algorithm

The input of the algorithm consists of the position and the energy map of the network.

First, we classify the operational state of each node using the parameters defined by the MANNA architecture [2], as shown in the Table 1 classification.

Table 1: Operational state of sensor node.

| Operational State | Energy Level(%) |
|---|---|
| Active | >=30 |
| Critical | <30 and >15 |
| Inactive | <= 15 |

Only nodes with active states will become sites. Nodes on other operational states will be ignored. Other policies and other table values could be used instead. Using the set of sites, the algorithm calculates the largest empty circle. Regions with larger area are the ones that need new nodes because the nodes are not healthy (active) or there are few nodes at the region. The center of the circle indicates where the best position is to add a node. Repeating the process while the largest empty circle is larger than a threshold will indicate how many new nodes are necessary to guarantee that every circle area larger than the threshold is being covered and also where these new nodes should be placed.

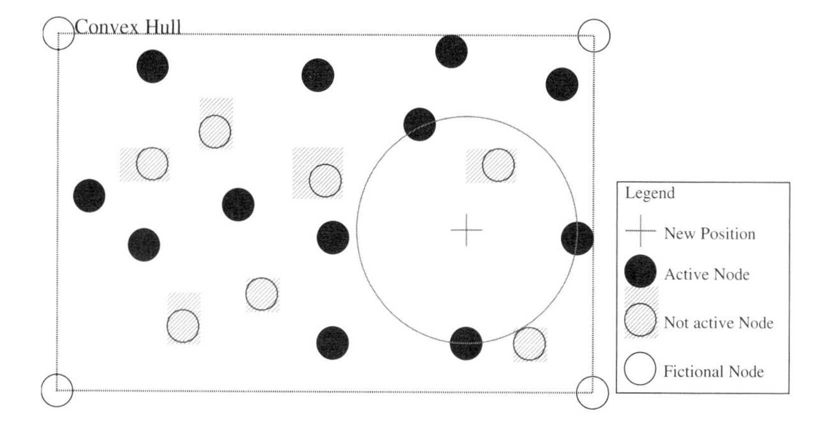


Figure 11: Algorithm example.

To calculate the largest empty circle it is necessary to have a convex hull, otherwise any point outside it could be the locale of the center of the circle. We choose four fictional nodes to define the convex hull of the rectangle surveillance area. The number of fictional nodes depends on the shape of the desired monitoring area. It does not affect the algorithm because it keeps looking for the largest empty circle until the area of the circle is larger than a threshold. If the region near the border were uncovered, the algorithm would, anyway, indicate that region.

Figure 11 shows the structure of the network from the design vision. At the border, fictional nodes were created to delimit the convex hull. Nodes not active were discarded. The algorithm calculates the largest empty circle considering only active nodes. The center of the circle indicates the best position to add a node. The worst-case complexity of the largest empty circle is $\Theta(n \log n)$. The algorithm can be extended to work with restricted areas (areas where we don't want to place sensors). Let's consider a real application of WSN, an application of monitoring and sensing a volcano. When the sensors are deployed, it is desirable that the sensors are not placed inside it or where the lava can reach them, otherwise, the sensors may be destroyed. Fictitious sites can be considered on the restricted area, so that the algorithm does not consider adding a sensor on the restricted area. We refer, in this chapter, to this algorithm, shown in pseudo-code in Algorithm 2, as the incremental algorithm (Figure 12).

```
Input: position and the energy level at each sensor node in the network.
Output: the quantity and the position of the sensor nodes.
for each active node begin
        set it as a site;
end for
count_new_nodes = 0;
Calculate Largest Empty Circle ();
do begin
        calculate largest_empty_circle_area();
        if(largest_empty_circle_area >threshold) then begin
                output the center of the Largest Empty Circle and set it as a
site;
                count_new_nodes = count_new_nodes + 1;
                update Largest Empty Circle ();
                keep_searching = true;
        end if
        else begin
                keep_searching = false;
        end else
while (keep_searching)
output count_new_nodes;
```

Figure 12: Incremental algorithm.

## 5.4 Guarantee Connectivity in the Graph

The network can be seen as a graph, where the sensor nodes are nodes in this graph. An edge connecting two nodes means that the two sensors are able to communicate. It is desired that the graph be connected, meaning that all sensor nodes are able to communicate with the data sink. The connectivity in the graph can be guaranteed by setting the value of threshold to radius/(2 * cos 30) , where radius is the radio range. This can be calculated from the case where there are three nodes that cannot communicate among themselves, which is shown in Figure 13.

Considering the sensing radius instead of the radio radius, it is possible to guarantee that the whole area among the sensor nodes is being sensed.

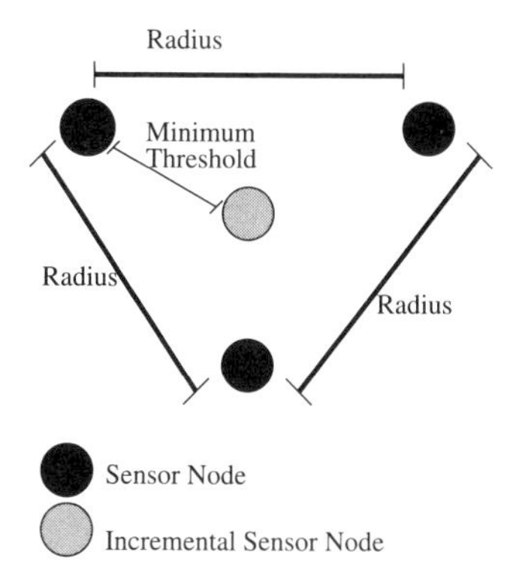


Figure 13: Minimum threshold to guarantee connectivity in the network.

# 6 Experimental Results

In order to evaluate our design, we created a Java application to simulate our experiments for large-scale sensor networks. The experiments were divided into two parts: one to demonstrate the scheduling schema and one for the incremental deployment. In this section, we present our results and discuss their implications.

## 6.1 Scheduling

### 6.1.1 Metrics

The key performance criterion in most wireless sensor networks is energy. We consider the energy saving as the number of off-nodes that our design outputs. Because there are many types of nodes with different types of energy levels, our energy-saving metric is the number of off-nodes, which is independent of sensor node types.

Another metric used is the percentage of the desired sensing area that is not being covered by the network. The idea is to verify if the sensing area is lost when a node is turned off.

A node that is turned off can also be described as a backup node. It is turned on when an operating node is not working properly, such as when it does not have enough energy or does not respond anymore.

### 6.1.2 Settings

Our experiments were conducted on a square sensing area. For all experiments, the network size is 100 nodes. To vary the density, we change the area. The position of each node is generated at random. We set up the threshold area as being a percentage of the sensing area of each sensor:

$$\text{Threshold Area} = \text{Percentage Coefficient} \times \text{Sensing Area} .$$

### 6.1.3 Results

Figure 14 shows the number of backup nodes as the density changes, for different threshold areas. Figures 14 and 15 differ from the sensor range, which is, respectively, 89 and 178 dm. As expected, when the density grows, the number of backup nodes also increases. This also happens when we compare the threshold of the Voronoi area and the number of backup nodes. Figure 16 illustrates when the network density is low and there is no backup node.

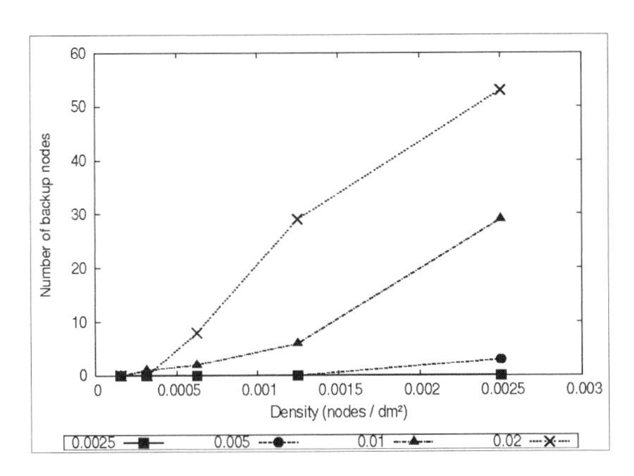


Figure 14: Number of backup nodes $\times$ density, varying the threshold area; sensor range is 89 dm.

## 6.2 Incremental Algorithm

Ideally, we would like to compare the coverage produced by this algorithm with that produced by an optimal solution. Unfortunately, finding the optimal solution for any nontrivial example is extremely difficult [19], even

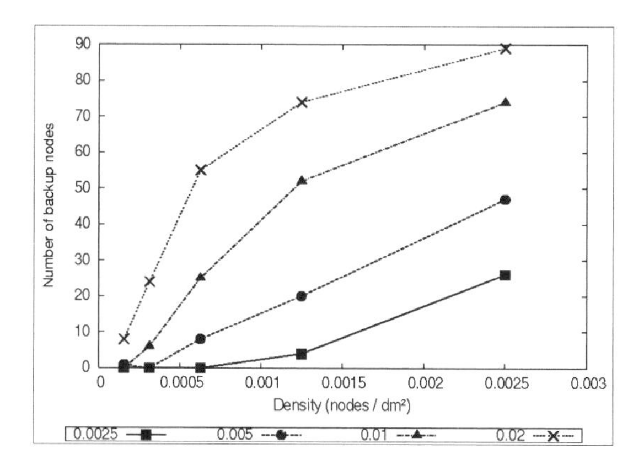


Figure 15: Number of backup nodes × density, varying the threshold area; sensor range (178 dm).

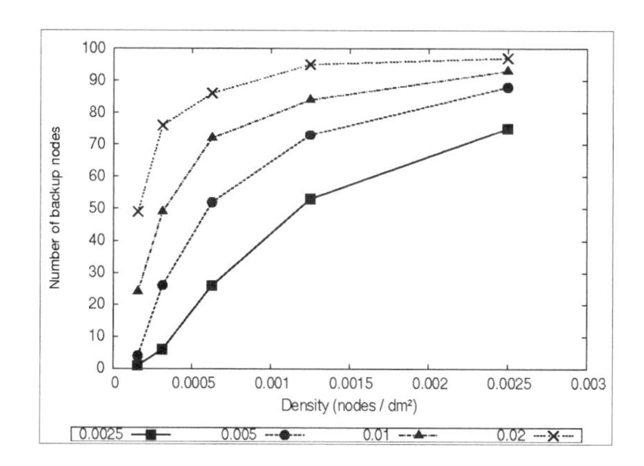


Figure 16: Number of backup nodes × density, varying the threshold area; sensor range (356 dm).

when we have good a priori maps of the environment. Consequently, in this chapter, we make no attempt to find such solutions.

In order to evaluate our design for adding new nodes to the network, we compare its performance to adding nodes randomly, considering different types of energy consumption. Although placing sensor nodes randomly is a naive approach, it serves as a base for comparison. In this section, we present our results and discuss their implications.

### 6.2.1 Metrics

The key performance criterion in adding new nodes to WSN is to keep the network active and sensing the desired area. Every point of the coverage area should be inside an active sensor. Our metric takes as relevant the area being covered and the number of nodes added.

### 6.2.2 Energy Dissipation Model

We used the models proposed by [8] to stress the network. The UNIFORM dissipation model captures uniformly distributed sensing activity. More precisely, during a sensing event, each node $n$ in the network has a probability $p$ of initiating a local sensing activity, and every node within a circle of $r$ centered at $n$ consumes a fixed amount of energy $e$. This latter feature of the model is inspired by a collaborative sensing algorithm.

In the uniform dissipation model, the energy at all nodes decreases at approximately the same rate. However, in a realistic environment, different regions in a sensor field may have different energy dissipation rates. To model this, there is the Hotspot model. In this model, there are $h$ hotspots uniformly distributed in random on the sensor field. Each node $n$ has a probability of $p = f(x)$ to initiate a local sensing activity, and every node within a circle of radius $r$ centered at $n$ consumes a fixed amount of energy $e$, where $f$ is a density function and $x = \forall \text{i} \{\min |h - h_i|\}$ is the distance from $n$ to the nearest hotspot. We used three density functions: $f(x) = ae^{-ax}$ is the density function for Exponetial distribution, where the hotspot effect drops quickly with increasing distance; the Pareto density $f(x) = a/(x+1)^{a+1}$, where the impact of the hotspot falls off more gradually than an exponential distribution with the same value $a$, and the Normal distribution.

### 6.2.3 Settings

Our experiments were conducted on a square desired sensing area, with area equal to 80,000 dm$^2$. The radius of the sensing area of each sensor is 20 dm. To simplify calculation, we define the area being covered as the inscribed square instead of a circle, so each node would have an area of 800 dm$^2$. With this number, a perfect placement of 100 nodes would cover 100% of the area. We compare the incremental algorithm against a random adding node algorithm using a random network using all the energy dissipation models described in Section 6.22.

In the experiment, we place 100 nodes on the desired area covering 100% of the area, and simulate using each of the energy models to find the nodes that would be considered inactive. Then we add nodes until we cover 100% of the area, using the incremental algorithm and the random insertion algorithm.

### 6.2.4 Results

Figures 17, 18, 19 and 20 show that the incremental algorithm performs close to the upper limit and independent of the energy model. It tries to cover the area that is not being covered and it also takes into account the energy information, covering the areas with low energy and consequently, being independent of the energy model.

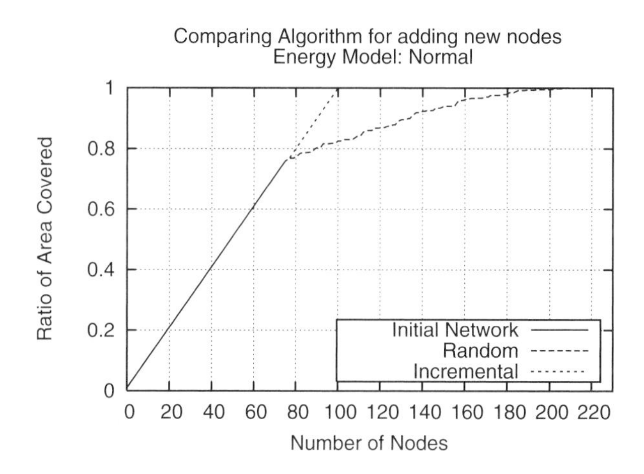


Figure 17: Adding new nodes to a network that lost energy as described by the normal model.

# 7 Concluding Remarks

Verifying if sensor nodes are actually monitoring a desired area is an important metric for wireless sensor networks. Energy savings in a wireless sensor network are critical, thus a management application is necessary to make the most of the network resource. Our design defines a schema for saving network resources. We presented the idea of a backup node, and

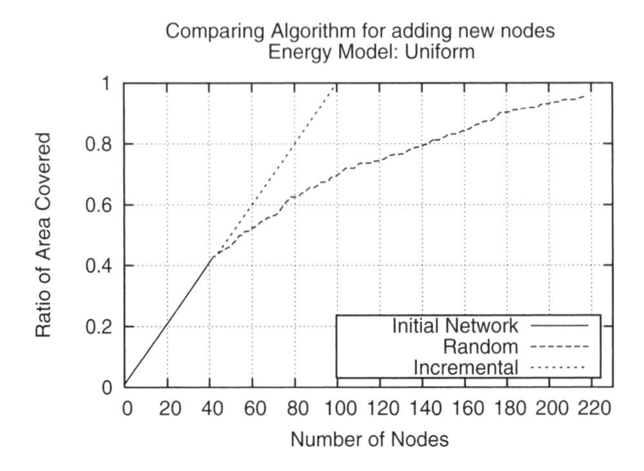


Figure 18: Adding new nodes to a network that lost energy as described by the uniform model.

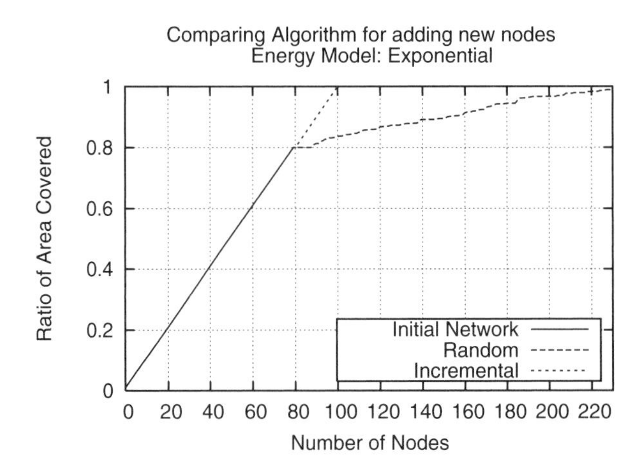


Figure 19: Adding new nodes to a network that lost energy as described by the exponential model.

defined and evaluated a criterion for determining which sensor nodes should be turned off.

Simulation results show that our approach is scalable and presents energy-efficient characteristics. The amount of backup nodes depends on the network density. It can save energy without losing sensing area.

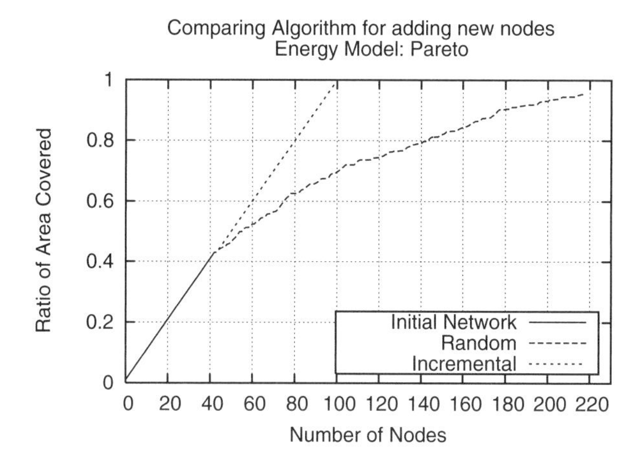


Figure 20: Adding new nodes to a network that lost energy as described by the Pareto model.

A good sensor deployment technique, in applications where it is possible to deterministically place sensors, is necessary to make the most of the network resource. Our design also defines a scheme that examines the distribution of node density, its energy level, and sensing cover area, and indicates the quantity and positions where new nodes should be deployed. This scheme is an efficient algorithm for incremental deployment of nodes in a wireless sensor network that combines results in computational geometry. It increases the lifetime of the network (as new nodes are added as needed) and minimizes the cost involved (as the minimum number of sensors that maintain the covering area is found). Furthermore, the running time is $\Theta(n \log n)$.

Experiments show that our algorithm is very close to the upper limit and performs 2.5 times better than the random insertion algorithm. It also works independently of the energy model.

Other computational geometry ideas can be applied in the wireless sensor network. Recently, Cheng et al. [20] proposed maintaining connectivity by introducing relay sensors and optimization techniques based on Steiner trees.

# References

[1] Badrinath, B.R., Srivastava, M.: Special issue on smart spaces and environments. *IEEE Pers. Commun.* **7** (2000).

[2] Ruiz, L.B., Nogueira, J.M., Loureiro, A.A.F.: Manna: A management architecture for wireless sensor networks. *IEEE Communication Magazine.* **41** (2003).

[3] K. Chakrabarty, S. S. Iyengar, H.Q., Cho, E.: Grid coverage for surveillance and target location in distributed sensor networks. *IEEE Transactions on Computers* **51** (2002) 1448–1453.

[4] S. S. Dhillon, K.C., Iyengar, S.S.: Sensor placement for grid coverage under imprecise detections. In: *Proceedings of the International Conference on Information Fusion* (2002), 1581–1587.

[5] Vieira, M.A.M., da Silva D.C., Jr., Coelho, C.N.: Survey on wireless sensor network devices. In: *IEEE International Conference on Emerging Technologies and Factory Automation (ETFA)* 2003 (2003).

[6] Vieira, M.A.M.: Embedded system for wireless sensor network. Master's thesis, Departamento de Ciência da Computação, Universidade Federal de Minas Gerais, Belo Horizonte-MG, Brasil (2004).

[7] Abrach, H., Bhatti, S., Carlson, J., Dai, H., Rose, J., Sheth, A., Shucker, B., Deng, J., Han, R.: MANTIS: System support for multimodal networks of in-situ sensors. In: *Second ACM International Workshop on Wireless Sensor Networks and Applications (WSNA)* (2003), 50–59.

[8] Zhao, J., Govindan, R., Estrin, D.: Residual energy scans for monitoring wireless sensor networks. In: *IEEE Wireless Communications and Networking Conference (WCNC 02)*, Orlando, FL (2002).

[9] Semiconductor, D.: Ds2438 datasheet. http://pdfserv.maxim-ic.com/arpdf/DS2438.pdf (2004).

[10] Tian, D., Georganas, N.D.: A coverage-preserving node scheduling scheme for large wireless sensor networks. In: *Proceedings of the First ACM International Workshop on Wireless Sensor Networks and Applications*, ACM Press (2002), 32–41.

[11] Clouquer, T., Veradej Phipatanasuphorn, P.R., Saluja, K.: Sensor deployment strategy for target detection. In: *Wireless Sensor Networks and Applications (WSNA)*, Atlanta,GA (2002).

[12] Aggarwal, A., Suri, S.: Fast algorithm for computing the largest empty rectangle. In: *Proceedings of the Third Annual Symposium on Computational Geometry.* (1987), 278–290.

[13] Meguerdichian, S., Koushanfar, F., Potkonjak, M., Srivastava, M.B.: Coverage problems in wireless ad-hoc sensor networks. In: *IEEE Infocom* (2001), 1380–1387.

[14] Bhardwaj, M., Chandrakasan, A., Garnett, T.: Upper Bounds on the Lifetime of Sensor Networks IEEE International Conference on Communications (2001) 785 – 790.

[15] O'Rourke, J.: Computational Geometry in C. Cambridge University Press (1993)

[16] Skiena, S.: *The Java Virtual Machine Specification.* Telos Pr. (1997).

[17] Heinzelman, W., Chandrakasan, A., Balakrishnan, H.: Energy-efficient communication protocol for wireless microsensor networks. In: *Proceedings of the Hawaii Conference on System Sciences* (2000).

[18] Albers, G., Guibas, L.J., Mitchell, J.S.B., Roos., T.: Voronoi diagrams of moving points. International Journal of Computational Geometry and Applications **8** (1998) 365–380.

[19] Howard, A., Mataric, M.J., Sukhatme, G.S.: An incremental deployment algorithm for mobile robot teams. In: *Proceedings of the IEEE/RSJ International Conference on Intelligent Robots and Systems* (2002).

[20] Cheng, X., Du, D., Wang, L., Xu, B.: Relay sensor placement in wireless sensor networks (2001).

# Chapter 9

## Message-Optimal Self-Configuration for Mobile Ad Hoc Networks

Han Namgoong
*Intelligent Robot Research Division*
*Electronics and Telecommunications Research Institute, Daejeon, 305-350, Korea*
E-mail: nghan@etri.re.kr

Dongman Lee and Dukyun Nam
*School of Engineering*
*Information and Communications University, Daejeon, 305-714, Korea*
E-mail: {dlee, paichu}@icu.ac.kr

## 1 Introduction

A Mobile Ad hoc NETwork (MANET) is a network in which no infrastructure for connectivity exists and participant hosts have a limited transmission range. Host mobility often leads to the instability of underlying network topology, which results in a new search for connectivity to other hosts and changes the roles of participating hosts in a logical configuration. Due to these intrinsic characteristics, MANETs require two separate phases for self-configuration: the initial search and logical connection (*initial configuration*), and the management of topology changes due to host mobility and failures (*mobility management*).

Hosts in a MANET usually exchange information via wireless communication. The wireless communication medium (i.e, the radio link) is a shared resource, thus the effective use of a link is a very important aspect in MANETs. A well-known scheme involves minimizing the number

M.X. Cheng, D. Li (eds.) *Advances in Wireless Ad Hoc and Sensor Networks.* Signals and Communication Technology, doi: 10.1007/978-0-387-68567-0_9.

of generated messages for self-configuration [7, 8]. To reduce the number of messages, a subset of hosts in a given network is selected in the initial configuration to form a set of gateways. The set is called a connected dominating set (CDS) [21], cluster heads [5, 12], or a virtual backbone [8] of the corresponding network. When a host wishes to send a message to another host, it simply transmits the message to one of the hosts in its corresponding CDS. This reduces the search space for the message destination into the hosts belonging to the CDS, not to all the hosts in a network. Clearly, this method saves a great number of generated messages in MANETs. There have been several research efforts on efficient construction (or finding) of CDS. Finding a minimum connected dominating set (MCDS) is an NP-hard problem [6, 13], thus approximation algorithms [8, 9, 19, 21] have been extensively studied.

These approximation algorithms focus mainly on how to find a near MCDS at the initial configuration. However, they do not effectively manage the impact of host mobility. In highly mobile networks, mobility causes frequent configuration changes. In this case, efficient mobility management affects the performance of self-configuration more than optimal initial configuration. That is, it is more crucial to reduce the number of messages to manage configuration in mobility management than in initial configuration.

To the authors' knowledge, there is no known self-configuration algorithm that provides both message-efficient initial configuration and mobility management.

The proposed scheme may spend more messages than Alzoubi's algorithm [3], which is known to be the best message-optimal CDS construction algorithm for initial configuration, but it consumes fewer messages for mobility management (to the ratio of 2.5). Our analysis shows that the proposed scheme outperforms Alzoubi's in terms of the total number of messages required for self-configuration.

This chapter is organized as follows. Sections 2 and 3 describe the preliminaries and the related work, respectively. Section 4 presents the proposed algorithm, and Section 5 offers our analysis. Finally, in Section 6, we present our conclusions.

# 2 Preliminaries

## 2.1 Definitions

D1. An undirected graph is referred to as connected if there is a communication path, that is, a sequence of a link, between every pair of distinct vertices of the graph. A connected graph is assumed to be a simple graph, where no two edges connect the same pair of vertices.

D2. A dominating set (DS) of a graph G = (V, E) is a subset v ⊂ V such that each node in (V – v) is adjacent to some node in v. DS may be connected or not connected. A connected dominating set (CDS) is a dominating set in which all members of a DS are connected in the corresponding graph G. A minimum connected dominating set (MCDS) is a CDS with a minimum number of members in G.

D3. A virtual backbone is a set of nodes in the CDS of the corresponding graph [8, 14, 21].

D4. A dominator is a node belonging to a CDS and a dominatee is a node not belonging to a CDS.

D5. A set of nodes is an independent set if it includes no pair of neighboring nodes. An independent set is maximal (MIS) if it can be a larger independent set.

## 2.2 Model

A mobile ad hoc network consists of a finite set of mobile hosts and a finite set of channels, that is, communication paths between mobile hosts. It is represented as a labeled and undirected graph, G = (V, E), where vertices (V) are for mobile hosts, and edges (E) represent bidirectional channels.[1] In Figure 1, the dotted circle represents transmission range and the bullet refers to a mobile host. Some possible CDSs and virtual backbones are as follows.

Possible DS = {A, E}, {C, F}, {C, D}, etc.
Possible CDS = {C, D}, {C, D, F}, etc.
Possible Virtual Backbone = {C, D}, etc.
MCDS = {C, D}.

## 2.3 Assumptions

A1. Message transmission delay is arbitrary but finite. A random mobility model is adopted, in which a node can move in any direction at any speed.

[1]The channel may be unidirectional due to different transmission ranges between hosts, which can be treated as link failure, but this does not affect the proposed algorithm.

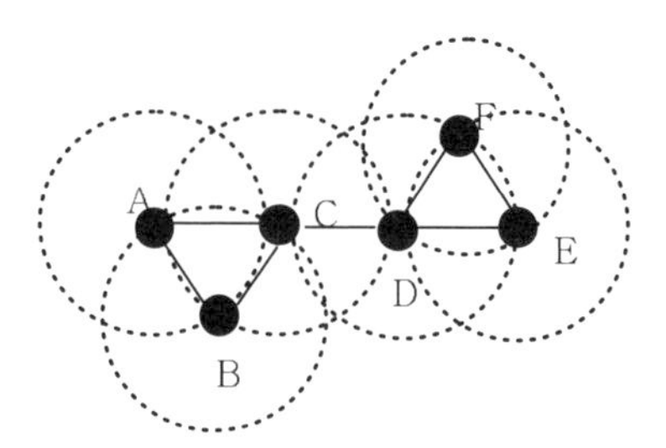


Figure 1: One instance of a mobile ad hoc network.

A2. All mobile hosts have the same transmission range,[2] r.

A3. The time gap between messages delivered from host to host is exponentially distributed, with the mean $1/\lambda$.

A4. The average number of packets per message is $\alpha$, and $\alpha \geq 1$.

A5. Inside a message, packets are separated by a constant time interval, $\tau_o$.

A6. Mobile hosts are uniformly distributed within a given network area.

A7. The time between location changes for each host is exponentially distributed, with the mean $1/\mu$.

Note that $\mu$ can be 0, in which case the hosts do not move.

# 3 Related Work

Traditional routing protocols of either link state [17, 15] or distance vector [10, 16] are impractical in a mobile ad hoc network because these protocols do not consider the effect of slow convergence of network topology, low bandwidth, low power, and host movements.

Virtual backbone-based routing [5] and spine-based routing [8, 18] use a similar approach to the cluster-based scheme [11, 12], and a virtual backbone (also called *spine*) consists of hosts similar to gateway hosts. Gateway hosts also form a dominating set [7, 21] of the corresponding mobile ad hoc network.

However, the virtual backbone proposed in [7, 8, 18] is expensive in terms of mobility maintenance and the algorithms in [21] and [19] also suffer from the inefficient approximation ratio [3]. The schemes in [2, 20] suffer from the same problem in relation to the maintenance of mobility.

[2]This assumption is actually the same case as Footnote 1.

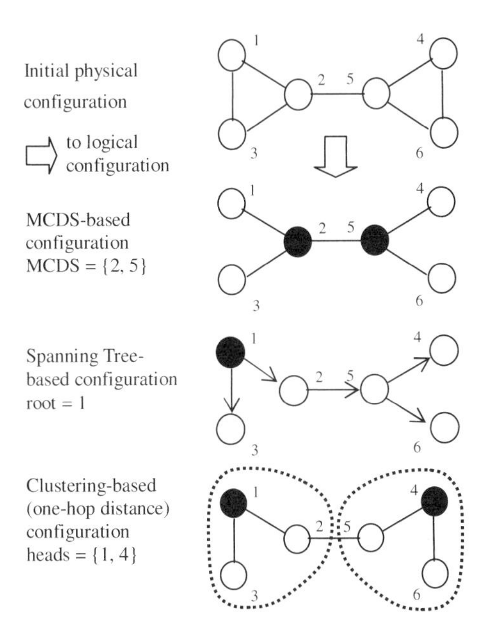


Figure 2: Some possible logical configurations.

Figure 2 represents many possible logical configurations that can be constructed at the initial configuration from the same physical configuration.

Table 1: Virtual backbone algorithm results

| | [7, 8, 18] | [21] | [5, 19] | [2, 20] |
|---|---|---|---|---|
| *Quality* | $\Theta(\log n)$ | $n$ | $n/2, n$ | $\leq 8$ |
| *Message* | $O(n^2)$ | $\Theta(m)$ | $O(n^2)$ | $O(n \log n)$ |
| *Time* | $O(n^2)$ | $O(\triangle^3)$ | $\Omega(n)$ | $O(n)$ |

Table 1 [3] shows some virtual backbone-based algorithms, where $n$ is the number of hosts, $m$ is the number of connections, and $\triangle$ is the maximum nodal degree. The *quality* of the virtual backbone is the ratio of its size to that of the MCDS and message and time represent the complexity.

Alzoubi's algorithm [1, 3] which improves [2, 20] with linear time and linear message complexity, is considered as the best in terms of message-optimal CDS construction, that is, at initial configuration. When a given

Table 2: The required number of messages for movement.

| Host Type | Min and Max = Number of Messages<br>($k$ : number of neighbors) |
|---|---|
| Dominator | max=($k$ - 1) * 2<br>=Number of (dominatee + connector) * 2<br>case : Only one dominator exists and it leaves |
| | min =0<br>case : One of dominators, i.e., from more than two, leaves |
| Connector | max=(Number of connector) + ($k$ - 1) * 2<br>=(Number of connector) + (number of dominator) * 2<br>+ { number of (dominatee + connector)} * 2<br>case : A connector leaves |
| | min =0<br>case : a connector joins to dominator/dominatee/connector |
| Dominatee or Candidate | max=1 + 3$k$<br>=one + {number of (dominatee + connector)} * 3<br>case : A dominatee comes in and no dominator exists |
| | min =0<br>case : A dominatee or candidate leaves |

network does not include mobile hosts, the algorithm is optimal. Alzoubi's algorithm manages the configuration using MIS-based CDS, which finds a maximal independent set (MIS), and builds the dominating tree from the MIS.

The algorithm has $O(n \log n)$ messages and $O(n)$ time complexity [1, 3]. The algorithm also maintains the status of each host, which is one of the following four: candidate, dominator, dominatee, or connector. The status of a host shows the role of the host in a given logical configuration and changes depending on the movement of the host itself and (or) neighboring hosts. Host movement means that a host first leaves (disconnects) from one location and joins (connects to) another location. Failure and recovery can be represented as leave and join, respectively.

The number of messages for mobility management is shown in Table 2.

We argue that the number of messages can be reduced if a CDS is constructed in the form of distributed spanning trees. Although the proposed scheme may spend more messages than Alzoubi's for message delivery, it consumes fewer messages for mobility management (to the ratio of 2.5).

The proposed scheme supports more efficient mobility management and outperforms the existing scheme in terms of total number of messages.

# 4 Proposed Algorithm

## 4.1 Overview

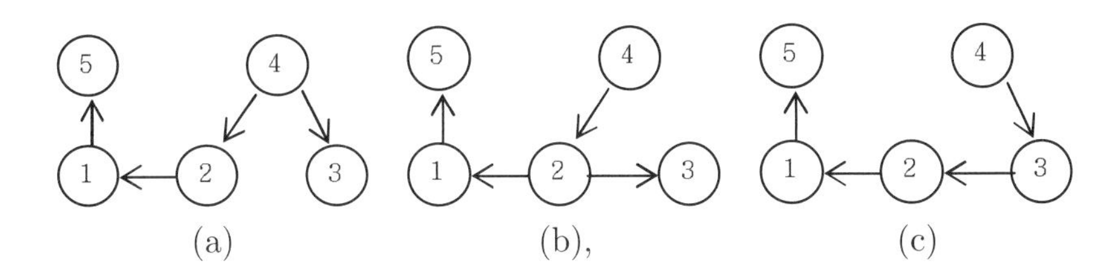


Figure 3: Possible STs (a), (b), and (c).

The proposed algorithm is similar to the minimum-weight spanning tree construction except that next edges are chosen randomly without considering the node's weight in breadth-first search style (i.e., select the neighbors as spanning edges as much as possible in the current round of propagation). Figure 3 shows three possible spanning trees (STs) when host 4 is a root.

The proposed scheme performs two steps in the initial configuration phase: construction request propagation and regional dominating set exchange.

In the construction request propagation step (lines 7~29, except 26), any node, also called an *initiator*, that is not currently involved in local ST construction, may initiate the initial configuration by issuing a construction request, *msg*(*request*, *n_set*, *n_id*, *n_id*, *n_state*), as shown in Figure 4. A construction request is propagated through the nodes establishing a region of the initiator (i.e., a root node in a local ST), where all nodes are not involved in other requests and receive a request from the same initiator (lines 7~20, except 16). The request is propagated to the end of a network (line 12) or to the border of other regions (line 16). When the request arrives at the end, the node returns a result to its parent (line 13). The result is a list of children (lines 20, 23, 27) of the same parent (lines 10 and 29) and the candidate parent nodes are stored in p_set (line 17 and note 2 in Figure 4). A regional dominating set consists of all nodes in a local ST and is compiled (line 23) at each node. This step ends (line 25) when all data are collected at the initiator.

In the regional dominating set exchange step (line 26), the initiator obtains another region's dominating set via a series of regional dominating set

```
Variables in each node
r_id : the nodes which initiate the initial configuration
n_id : the node which received and proceeds the message
p_id : the node which relays the message
n_state : the variable for node status, leaf or non_leaf, initially leaf
p_set : a set of neighbors which relay the message of the same r_id
i_set : a set of neighbors which relay the message
c_set : a set of children which have the same parent
n_set : a set of neighbors, all neighbors are constantly members.
r_set : a set of nodes with different r_id, i.e. different initiators
c_result : the combined result from children
All variables are initially empty except n_set and n_state.
  1. PROCEDURE msg(action, receiver, sender, text, role);
  2. ENUM action = ( request, result, notify, join, leave, reply ),
            role = ( leaf, non_leaf );
  3. TYPE n_state = role;
  4. SET r_id, n_id, p_id, receiver, sender, text, p_set, i_set, c_set, r_set, c_result;
  5. CONST n_set = all neighbors of each node as members;
  6. BEGIN
  7. case of (action = request)
  8.    if (r_id = ∅) then
  9.    { r_id := text; p_id := sender;
 10.       msg(notify, p_id, n_id, r_id, n_state);
 11.       i_set := i_set ∪ sender;
 12.       if (i_set = n_set) ∩ (c_set = ∅) then
 13.          msg(result, p_id, n_id, c_result ∪ n_id, n_state)
 14.       else msg(request, n_set-i_set, n_id, r_id, n_state); }
 15.    else
 16.    { if (r_id ≠ text) then r_set := r_set ∪ sender
 17.       else p_set := p_set ∪ sender;
 18.       i_set := i_set ∪ sender;
 19.       if (i_set = n_set) ∩ (c_set = ∅) then
 20.          msg(result, p_id, n_id, c_result ∪ n_id, n_state); }
 21. case of (action = result)
 22.    c_set := c_set-sender;
 23.    c_result := c_result ∪ text; i_set := i_set ∪ sender;
 24.    if (i_set = n_set) ∩ (c_set = ∅) then
 25.       if (r_id = n_id) then
 26.          for i ∈ r_set do get i-region's c_result except Note 4
 27.       else msg(result, p_id, n_id, c_result ∪ n_id, n_state);
 28. case of (action = notify)
 29.    c_set := c_set ∪ sender; n_state := non_leaf;
 30. case of (action = join)
 31.    msg(reply, sender, n_id, r_id, leaf);
 32. case of (action = reply)
 33.    if (r_id = ∅ ) then
 34.    { p_id := sender; n_state := leaf; r_id := text;
 35.       msg(notify, sender, n_id, r_id, n_state); }
 36. case of (action = leave)
 37.    if (sender ∈ c_set) then
 38.    { c_set := c_set-sender;
 39.       if (c_set = ∅) then n_state := leaf; }
 40.    else if (sender = p_id) then
 41.    { if (p_set = ∅) then handle the network-partition
 42.       else { select one node, i, from p_set; p_id := i;
 43.          msg(notify, i, n_id, r_id, leaf); }}
 44. END;
Note 1: After the execution of line 26, the node can make a single spanning tree or
        manage multiple regional spanning trees.
        The choice comes from configuration management policy.
Note 2: The first node which relays the request message is set as the parent (line 9)
        and other nodes of the same initiator are stored in p_set (line 17) for later
        use of the parent node's leave (line 42∼43).
Note 3: When a network partition occurs (line 41), the neighbor node changes its
        status depending on its position, i.e., child or parent, in the spanning tree
        and needs no extra messages.
Note 4: One initiator (r_set = ∅) or neighbors are initiators (r_set = n_set)
```

Figure 4: Proposed scheme.

exchanges. Every initiator sends to neighboring initiators in r_set the new information that has been obtained during the previous round. This step is done when all initiators receive no new dominating set information.

In the mobility management phase (lines 30∼43), when a node joins, the number of messages depends on the number of leaf and nonleaf nodes in the vicinity. A joining node broadcasts a join request, *msg(join, n_set, n_id, null, leaf)*, and waits for replies. When a neighbor receives the request, it sends its status (line 31). Upon receiving replies, the joining node decides to be a leaf of one of the responding nodes (line 34) and sends a message of a new parent (line 35). When receiving a leave request, *msg(leave, n_set, n_id, r_id, n_state)*, a node checks whether the leaving node is its child or parent (lines 37, 40). When a child node leaves, the parent node deletes the node from its children list (lines 38∼39). If a parent node leaves, the child node selects a candidate from p_set and notifies the parent (lines 42∼43).

**Example 1.**
In Figure 5, there exist three initiators (i.e., three distributed STs by three separate roots), thus three regions exist (regions 1, 3, and 5). Construction requests from different initiators, 1, 3, and 5, meet at the borders (two dotted lines) between the regions (regions 1 and 3, regions 3 and 5) where the adjacent nodes (nodes 1, 2, 4, and 5) will have different initiator values (i.e., different roots). Each node will have the initiator's identifier as the participated construction request identifier.

For example, node 2 has value 3 as the initiator's identifier, 3 as the parent of node 2, and 2 as the node identifier. A leaf node plays the role of both parent and child. A parent is responsible for sending the construction request to its children which, in turn, send the result to parents. Nodes on borders also pass on information of different initiators to their initiators.

## 4.2 Mobility Management

The proposed algorithm has two kinds of nodes: leaf nodes and nonleaf nodes. A leaf node is a node that is the end of a logical path in an ST. When it moves to another location and obtains new connectivity to some hosts, a host does two things: leave (disconnection) and join (connection).

When it joins (i.e., finds new neighbors after moving to some location), a host broadcasts its new join to all its neighbors and waits for replies from all its neighbors. If there is only one reply from a nonleaf neighbor, the joining host notifies that host of a new neighbor. If there is more than one nonleaf host, the joining host selects the one that has the smallest number

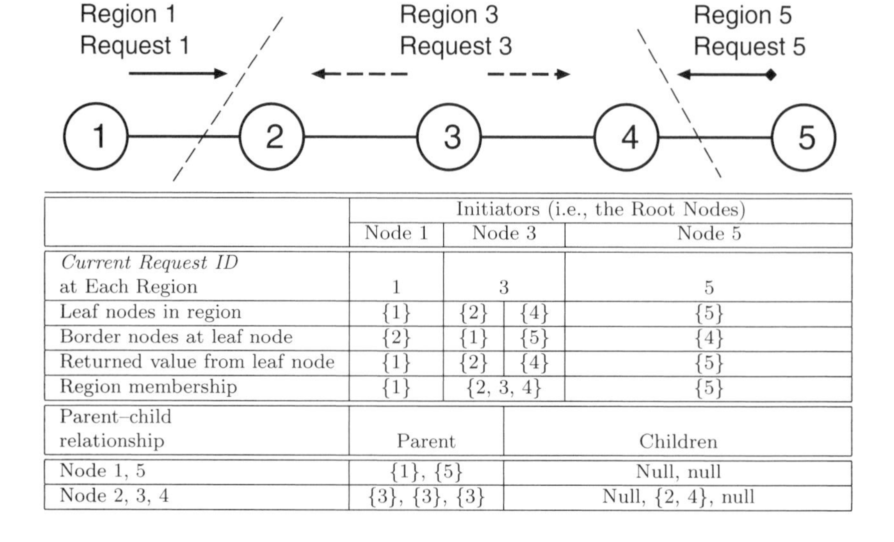


| | Initiators (i.e., the Root Nodes) | | | |
|---|---|---|---|---|
| | Node 1 | Node 3 | | Node 5 |
| *Current Request ID* at Each Region | 1 | 3 | | 5 |
| Leaf nodes in region | {1} | {2} | {4} | {5} |
| Border nodes at leaf node | {2} | {1} | {5} | {4} |
| Returned value from leaf node | {1} | {2} | {4} | {5} |
| Region membership | {1} | {2, 3, 4} | | {5} |
| Parent–child relationship | Parent | | Children | |
| Node 1, 5 | {1}, {5} | | Null, null | |
| Node 2, 3, 4 | {3}, {3}, {3} | | Null, {2, 4}, null | |

Figure 5: Concurrent initiators and their regions.

of neighbors.[3] In Figure 6 the dotted arrow line refers to the broadcast of "join" by host 7, and the arrow line (both (a) and (b)) is the path in the ST with the root of host 1. When only leaf hosts exist, a host is selected randomly (refer to Footnote 3) and notifies as a new neighbor.

If a host finds that one of its neighbors has departed from its vicinity (i.e., out of communication range, shown in Figure 6), the host checks whether any neighbors still remain connected. If there is no neighbor (i.e., the host is a leaf node), the remaining host changes its status as a single host tree (i.e., the network is separated, shown in Figure 6). Otherwise, the host deletes the disconnected host from its neighbors.

## 4.3 Complexity

### 4.3.1 Construction Request Cost

Transmission of a single message over a single communication link is a basic unit of complexity analysis. The proposed algorithm requires exactly one construction request message for each initiator which transmits the message

[3]The selection actually depends on the policy of topology management.

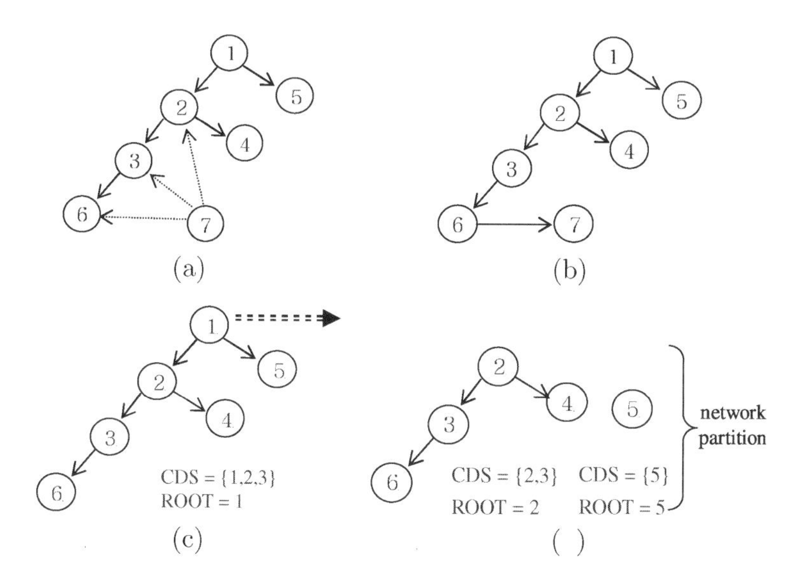


Figure 6: Host join and leave: (a) Before join; (b) after join; (c) before leave; (d) after leave.

to its neighbors. Construction request messages from the same initiators via neighbors are ignored; that is, only the construction request message that a node received first is relayed to its neighbors. When the result of a construction request is returned to the parent, only one message containing the combined result of all children is transmitted. The total number of messages for the construction request step depends on the network configuration and communication model such as broadcast, multicast, and point-to-point. In the regional dominating set exchange step, only the initiators take part directly in dissemination of their regional dominating set. Each initiator engages in a series of regional dominating set exchanges that enable all the initiators to obtain a complete dominating set. A noninitiating node may be indirectly involved in the regional dominating set exchange step as an interior node of the path for the transmission of a regional dominating set from an initiator to another initiator. During the regional dominating set exchange step, each initiator possesses its own regional dominating set that is sent to its neighboring initiators.

Let q be the multiple initiators, $n$ be the number of hosts in the network, and $((n/q) - 1)$ be the average number of noninitiator hosts in each region.

A regional dominating set may take one of these distances to a neighboring initiator: (1) one: initiators are all neighbors; (2) average: $((n/q) - 1)$; and (3) worst: $2^*((n/q) - 1)$. The average distance is computed assuming that the location of the initiator is in the middle of the region whereas the worst case assumes that initiators reside at the opposite end of each region. $q$ multiple initiators have maximum $q$ regions and each round requires maximum $(q-1)$ message exchanges (linearly connected) with neighboring initiators.

The work needed for single and multiple initiators is as follows.

**Case 1:** A Single Initiator

1. Construction Request Propagation Step
   The maximum distance for an initiator to go is $(n-1)$ in a row of $n$ nodes, and $(n-1)$ reply messages for the result from each node in ST, which results in $2 * (n-1)$. The minimum distance is 1 (all members within direct communication) and $(n-1)$ replies from $(n-1)$ children, thus $n$.

2. Regional Dominating Set Exchange Step
   The required cost is $(n-1)$ for each node should send exactly one result message, a combination of all results from its children.

3. All nodes want the Dominating Set
   Additionally, at most $(n-1)$ messages of the new dominating set will be sent from the initiator. When all nodes are in direct communication range, only one message is needed.

**Case 2:** Multiple Initiators

1. Construction Request Propagation Step
   Each region requires $((n/q)-1)$ messages for request propagation and $((n/q)-1)$ for request result replies at most, thus $2^*((n/q)-1)$ in each region and $2^*q^*((n/q)-1)$ in $q$ regions. When all nodes in a region are in direct communication range, only one message for request propagation is needed but still $((n/q)-1)$ reply messages are needed, which results in $((n/q)-1) + 1 = (n/q)$ in a region and $n$ in $q$ region.

2. Regional Dominating Set Exchange Step

   (a) Maximum $(q-1)$ rounds are required to collect all regional dominating sets for $q$ regions.

(b) Each round exchanges ($q$–1) messages.

(c) Each message in each round may traverse.

- maximum: 2*(($n/q$)–1)
- average: (($n/q$)–1)
- minimum: 1 (neighboring initiators case)

From (a), (b), and (c), this step requires ($q$–1)*($q$–1)*2*(($n/q$)–1) messages at most. The minimum case occurs when all the initiators are at the minimum distance, 1; only two rounds are required for regional dominating set exchange, and $q$ messages per round. The minimum number of messages is $2 * 1 * q = 2q$.

3. All nodes want the Dominating Set
Simply, ($n$–$q$) and 1 messages, the worst and minimum case (all nodes within transmission range), respectively, of the new dominating set are required for noninitiators; that is, each node relays the new dominating set to its neighbors.

Table 3: Complexity of initiators and steps.

| | Steps | | | |
|---|---|---|---|---|
| No. of Initiator | Construction Request | | Regional Dominating Set Exchange | |
| | Min. | Worst | Min. | Worst |
| Single | $n$ | 2($n$-1) | ($n$-1) | ($n$-1) |
| $q$ | $n$ | 2q(($n/q$)-1) | $2q$ | $2(q\text{-}1)^2((n/q)\text{-}1)$ |

Our analysis, shown in Table 3, shows that only the $q$ initiators will receive the dominating set.

### 4.3.2 Mobility Management Cost

From Figure 6, we can easily calculate the required number of messages for movement (shown in Table 4).

Table 4: Complexity of movement.

| The Required Number of Messages for Movement | |
|---|---|
| Node type | Min and Max Cost |
| Leaf | max cost = $k + 2$<br>case: A leaf joins to the vicinity of leaf and nonleaf nodes |
| | min cost = 1<br>case: A leaf leaves from the vicinity of nonleaf node |
| Nonleaf | max cost = $k + 2$<br>case: A nonleaf joins to the vicinity of nodes |
| | min cost = 1<br>case: A nonleaf leaves from the middle of linear connection |
| Note 1: $k$ is the number of nodes in the region where movements occur, thus the number of neighbors is ($k$ - 1).<br>Note 2: The proposed scheme generates explicit join/leave messages even though they are not mandatory. With heartbeats we can easily find the configuration changes, but explicit notices allow for a greater chance of more accurate configuration information. | |

# 5 Analysis

## 5.1 Initial Configuration Cost Comparison

Comparison factors in the initial configuration phase are (1) the number of messages; (2) the amount of space to keep topology information; and (3) the number of hosts that should keep topology information. The initial configuration phase complexity of Alzoubi's algorithm [1, 3] is shown in Table 1, $O(n \log n)$, and the proposed algorithm's complexity is presented in Table 3, $O(2(n-1))$. When $q$ nodes start initial configuration concurrently, the complexity will be $q$ times a single initiator complexity. The cost difference between the proposed algorithm and Alzoubi's algorithm becomes largest when all hosts are connected linearly. However, this maximum difference of initial configuration cost rapidly disappears through the difference of mobility management cost (see Section 5.3).

Clearly, the proposed algorithm takes a smaller amount of space in each host for a given topology because each host keeps only parent and child information in one-hop distance and maximally the number of neighbors. In Alzoubi's algorithm, each host should save the information of neighbors with a one-hop and two-hop distance, and other related topology information. The time complexity of the proposed algorithm is the same as the message complexity.

Algorithm [3] may have a smaller number of CDS than that of the proposed algorithm.

## 5.2 Mobility Management Cost Comparison

The total cost for mobility management ($C_{total}$) is the sum of (a) the cost for reconfiguration itself ($C_{reconf}$); and (b) the recovery cost for lost messages during reconfiguration ($C_{lost}$). $C_{reconf}$ is represented by $C_d * M_{reconf}$, where $C_d$ is the average cost for one message to traverse and $M_{reconf}$ is the expected number of messages to reconfigure. $C_{lost}$ can be calculated by $C_r * M_{loss}$, where $C_r$ is the recovery cost per lost message and $M_{loss}$ is the expected lost messages.
The total cost is

$$\begin{aligned} C_{total} &= C_{lost} + C_{reconf} \\ &= C_r * M_{loss} + C_d * M_{reconf}. \end{aligned} \tag{1}$$

$M_{loss}$ can be calculated as follows:

$$\begin{aligned} M_{loss} = \; & \text{the arrival rate}(\lambda) * \text{the probability of} \\ & ((\text{i}) \text{ receiver node failure,} \\ & (\text{ii}) \text{ message corruption, or} \\ & (\text{iii}) \text{ broken link due to mobility).} \end{aligned} \tag{2}$$

Because the three cases are independent, $M_{loss}$ is the product of the arrival rate by the sum of the probability of the three cases. Cases (i) and (ii) can be chosen from practical measurements. $M_{loss}$ depends on the probability of link breakage (i.e., Case (iii)), and is represented as follows [4]:

$$P_b = \left[\frac{1}{\alpha} - \left(\frac{\lambda}{\alpha(\lambda+\mu)}\right) - \left(\frac{\mu}{\lambda\alpha+\mu}\right)\right] e^{-(\lambda\alpha+\mu)\tau_o} + \left(\frac{\mu}{\lambda\alpha+\mu}\right). \tag{3}$$

$M_{loss}$ is a function of $\lambda$ (arrival rate), $\mu$ (mobility rate), $\alpha$ (packets per message) and $\tau_o$ (interpacket period). $M_{reconf}$ depends on the current configuration and the number of neighbors [1, 3, 7–9, 18]. Because node mobility is memoryless, the probabilities of different configurations are the same. The expected number of messages ($M_{reconf}$) for Alzoubi's scheme [1, 3] becomes

$$\begin{aligned}
\mathrm{E_{CDS}}[k] &= (1/7) * [(\text{dominator case of max and min}) \\
&\quad + (\text{connector case of max and min}) \\
&\quad + (\text{dominatee case of max and min})] \\
&= (1/7) * [(2k-2) + (2k - 2 + \text{the number of connector}) \\
&\quad + (3k+1)] \\
&= (1/7) * [(7k-3) + \text{the number of connectors}], \qquad (4)
\end{aligned}$$

where $k$ is the number of nodes in a region where movement occurs and should be equal to or greater than three because the minimum neighbors for message exchange except the end of configuration is two; that is, $k = 3$ for a linearly connected configuration. One-seventh is the probability of one of seven cases: the status of three nodes with maximum or minimum cost (six cases) and the not-moved case (Table 2). $\mathrm{E_{CDS}}[k]$ increases with the number of connectors. $\{(1/7) * (7k - 3)\}$ is the lower bound when the number of connectors is 0. The proposed scheme consumes $(k + 2)$ messages at maximum when leaf and nonleaf nodes move, and one at minimum. The expected value ($M_{reconf}$) is

$$\begin{aligned}
\mathrm{E_{ST}}[k] &= (1/5) * [(k+3) + (k+3)] \\
&= (1/5) * [2k+6]. \qquad (5)
\end{aligned}$$

$\mathrm{E_{ST}}[k]$ is less than $\mathrm{E_{CDS}}[k]$ when $k > 2.71$, that is, $k \geq 3$. Because $C_d$, $C_r$, and $M_{loss}$ are the same in both schemes, $C_{total}$ of the proposed scheme is always less than that of Alzoubi's. For a comparison of $C_{total}$ (Table 5) we fix related variables as follows.

$C_r$ is three from the (request/reply/acknowledgment) protocol and $C_d$ is one for any message traversing its neighbors. When $\alpha = 1$, $P_b$ becomes $(\tau/(\lambda + \mu))$ [4]. $\tau_o$ (0.01), $\lambda$ (5), a node failure probability (0.001), and the message corruption probability, (0), are fixed. The other variables are typical: i.e., a human being walks about 0.8 m/s and exchanges a few messages per second, and the message length is not that long (max $\alpha = 5$). The ratio between the two schemes converges on 2.5 when $k$ becomes large, 1.148 ($k = 4$), 1.313 ($k = 8$), and 1.496 ($k = 20$) under the same condition ($\mu = 0.5$, $\alpha = 5$).

## 5.3 Self-Configuration Cost Comparison

Because the cost of the initial configuration depends on a given configuration, we consider the worst case in terms of consumed messages. The

Table 5: Mobility management cost ($C_{total}$).

| $C_{total}$ by Mobility Rate ($\mu$), Neighbors ($k$ - 1) and Message Length ($\alpha$) | | | | | | | |
|---|---|---|---|---|---|---|---|
| $\alpha = 1$ | | | | $\alpha = 5$ | | | |
| $k$ | $\mu$ | $C_{total}$ [3] | Proposed | $k$ | $\mu$ | $C_{total}$ [3] | Proposed |
| 4 | 0.5 | 3.8472 | 3.0757 | 4 | 0.5 | 5.9771 | 5.2057 |
| 4 | 0.7 | 3.9428 | 3.1714 | 6 | 0.5 | 7.9771 | 6.0057 |
| 4 | 1 | 4.0744 | 3.3030 | 8 | 0.5 | 13.2932 | 10.1217 |
| 4 | 1.2 | 4.1551 | 3.3836 | 12 | 0.5 | 19.2932 | 13.7217 |
| 4 | 2 | 4.4316 | 3.6601 | 20 | 0.5 | 31.2932 | 20.9217 |
| 6 | 0.5 | 5.8472 | 3.8757 | 4 | 0.7 | 6.0009 | 5.2295 |
| 6 | 0.7 | 9.3932 | 7.4217 | 6 | 0.7 | 9.3932 | 7.4217 |
| 6 | 1 | 9.4159 | 7.4445 | 8 | 0.7 | 13.3932 | 10.2217 |
| 6 | 1.2 | 9.4373 | 7.4658 | 12 | 0.7 | 19.3932 | 13.8217 |
| 6 | 2 | 9.4402 | 7.4688 | 20 | 0.7 | 31.3932 | 21.0217 |
| 8 | 0.5 | 13.2932 | 10.1217 | 4 | 0.7 | 6.0343 | 5.2629 |
| 8 | 0.7 | 13.3932 | 10.2217 | 6 | 1 | 9.4159 | 7.4445 |
| 8 | 1 | 13.4159 | 10.2445 | 8 | 1 | 13.4159 | 10.2445 |
| 8 | 1.2 | 13.4373 | 10.2658 | 12 | 1 | 19.4159 | 13.8445 |
| 8 | 2 | 13.4402 | 10.2688 | 20 | 1 | 31.4159 | 21.0445 |
| 12 | 0.5 | 19.2932 | 13.7217 | 4 | 1 | 6.0552 | 5.2837 |
| 12 | 0.7 | 19.3932 | 13.8217 | 6 | 1.2 | 9.4373 | 7.4658 |
| 12 | 1 | 19.4159 | 13.8445 | 8 | 1.2 | 13.4373 | 10.2658 |
| 12 | 1.2 | 19.4373 | 13.8658 | 12 | 1.2 | 19.4373 | 13.8658 |
| 12 | 2 | 19.4402 | 13.8688 | 20 | 1.2 | 31.4373 | 21.0658 |
| 20 | 0.5 | 31.2932 | 20.9217 | 4 | 2 | 6.1295 | 5.3580 |
| 20 | 0.7 | 31.3932 | 21.0217 | 6 | 2 | 9.4402 | 7.4688 |
| 20 | 1 | 31.4159 | 21.0445 | 8 | 2 | 13.4402 | 10.2688 |
| 20 | 1.2 | 31.4373 | 21.0658 | 12 | 2 | 19.4402 | 13.8688 |
| 20 | 2 | 31.4402 | 21.0688 | 20 | 2 | 31.4402 | 21.0688 |

proposed scheme spends ($2n - 2$, maximum) in a linearly connected configuration and ($n$, minimum) when all nodes are within one transmission range. Alzoubi's scheme needs $cn$ ($c \geq 1$) from two steps [1]: distributed MIS construction; and MIS-based CDS construction. The minimum costs of the two steps are $n$ and a ceiling of $(n/2)$, respectively, in a row of nodes. Hence, the maximum difference is $[(2n - 2) - \{n + \lceil n/2 \rceil\}] \approx (n/2)$. Because the proposed scheme spends less $C_{total}$ than Alzoubi's to the ratio of 2.5, the maximum difference will rapidly disappear. Assume that $n$ is 200 nodes; only one node moves per second with ($\mu = 0.5$ and $\alpha = 5$), and each node has three neighbors (i.e., $k = 4$). Because the ratio of $C_{total}$ is 1.148, $\{(200/2)/1.148\} \approx 88$ seconds is enough to cover the maximum difference. If two nodes move per second, only 44 seconds will be necessary. Hence, we can conclude that the proposed scheme eventually consumes fewer messages than Alzoubi's.

### 5.4 Message Optimality

When a host moves into another region, it will generate one message for its intention to join that region. The neighbors should reply with each host's status information including the list of neighbors, and this step requires $k$–1 messages. Finally, the joining host decides its status and notifies all neighbors that sent their information regarding the configuration of the region. A minimum $(2 + (k - 1))$ messages are required in this two-round protocol. Instead of the two-round protocol, a one- (or three-) round protocol can be adopted depending on the policy. Note that Alzoubi's MIS-based CDS algorithm [1, 3] is not optimal in managing the configuration in any round protocol.

## 6 Conclusion

The proposed algorithm presents a message-optimal self-configuration algorithm for mobile ad hoc networks. Although the proposed algorithm does not result in a minimum number of members of a connected dominating set, it outperforms the well-known Alzoubi's algorithm in the total number of messages for self-configuration. The proposed algorithm also shows that mobility management affects the maintenance performance of a virtual backbone more than the virtual backbone construction method.

In future work we will further investigate fault-tolerance in a small region with a fixed number of hosts and consistency among hosts with some probability range. In a mobile ad hoc network with a large number of mobile hosts, reliability and consistency among hosts cannot be measured using a deterministic approach due to mobility. Instead, a small region, where hosts stay during some period with approximate probability, can be a good sample of the whole network.

## References

[1] K.M. Alzoubi, *Virtual backbone in wireless ad hoc networks*, Ph.D. dissertation, Dept. of Computer Science, Illinois Institute of Technology, Chicago, May 2002.

[2] K.M. Alzoubi, P.-J. Wan, and O. Frieder, Distributed heuristics for connected dominating sets in wireless ad hoc networks, *Journal of Communications and Networks* 4,1 (2002), pp. 22–29.

[3] K.M. Alzoubi, P.-J. Wan, and O. Frieder, Message-optimal connected dominating sets in mobile ad hoc networks. In *Proceedings of the Third ACM International Symposium on Mobile Ad Hoc Networking and Computing*, pages 157–164, Lausanne, Switzerland, June 2002.

[4] I.D. Aron and S.K.S. Gupta, On the scalability of on-demand routing protocols for mobile ad hoc networks: an analytical study, *Journal of Interconnection Networks* 2, 1 (2001), pp. 5–29.

[5] G. Chen and I. Stojmenovic, Clustering and routing in wireless ad hoc networks, Technical Report TR-99-05, Computer Science, SITE, University of Ottawa, June 1999.

[6] B.N. Clark, C.J. Colbourn, and D.S. Johnson, Unit disk graphs, *Discrete Mathematics,* 86, 1–3 (1990), pp. 165–177.

[7] B. Das and V. Bharghavan, Routing in ad-hoc networks using minimum connected dominating sets. In *Proceedings of International Conference on Communications*, pages 376–380, Montreal, Canada, June 1997.

[8] B. Das, R. Sivakumar, and V. Bhargavan, Routing in ad-hoc networks using a spine. In *Proceedings of the Sixth International Conference on Computer Communications and Networks*, pages 34–39, Las Vegas, September 1997.

[9] S. Guha and S. Khuller, Approximation algorithms for connected dominating sets, *Algorithmica*, 20, 4 (1998), pp. 374–387.

[10] C. Hedrick, Routing information protocol, RFC 1058, June 1988.

[11] P. Krishna, M. Chatterjee, N.H. Vaidya, and D.K. Pradhan, A cluster-based approach for routing in ad-hoc networks. In *Proceedings of the Second USENIX Symposium on Mobile and Location-Independent Computing*, pages 1–10, Ann Arbor, MA, April 1995.

[12] C.R. Lin and M. Gerla, Adaptive clustering for mobile wireless networks, *IEEE Journal on Selected Areas in Communications*, 15, 7 (1997), pp. 1265–1275.

[13] C. Lund and M. Yannakakis, On the hardness of approximating minimization problems, *Journal of the ACM*, 41, 5 (1994), pp. 960–981.

[14] M.V. Marathe et al., Simple heuristics for unit disk graphs, *Networks*, 25 (1995), pp. 59–68.

[15] J.M. McQuillan, I. Richer, and E.C. Rosen, The new routing algorithm for the ARPANET, *IEEE Transactions on Communications*, 28, 5 (1980), pp. 711–719.

[16] J.M. McQuillan and D.C. Walden, The ARPA network design decisions, *Computer Networks*, 1, 5 (1977), pp. 243–289.

[17] J. Moy, OSPF version 2, RFC 2328, April 1998.

[18] R. Sivakumar, B. Das, and V. Bharghavan, The clade vertebrata: Spines and routing in ad hoc networks. In *Proceedings of the 3rd IEEE Symposium on Computers and Communications*, pages 599–605, Athens, Greece, June 1998.

[19] I. Stojmenovic, M. Seddigh, and J. Zunic, Dominating sets and neighbor elimination-based broadcasting algorithms in wireless networks, *IEEE Transactions on Parallel and Distributed Systems*, 13, 1 (2002), pp. 14–25.

[20] P.-J. Wan, K.M. Alzoubi, and O. Frieder, Distributed construction of connected dominating set in wireless ad hoc networks. In *Proceedings of the 21st Annual Joint Conference of the IEEE Computer and Communications Societies (INFOCOM 2002)*, pages 1597–1604, New York, June 2002.

[21] J. Wu and H. Li, On calculating connected dominating set for efficient routing in ad hoc wireless networks. In *Proceedings of the Third ACM International Workshop on Discrete Algorithms and Methods for Mobile Computing and Communications*, pages 7–14, Seattle, WA, August 1999.

Chapter 10

# Energy-Efficient Two-Tree Multicast for Mobile Ad Hoc Networks

Sangman Moh
*Department of Internet Engineering*
*Chosun University, Gwangju, 501-759 Korea*
E-mail: smmoh@chosun.ac.kr

Chansu Yu
*Department of Electrical and Computer Engineering*
*Cleveland State University, Cleveland, OH 44115, USA*
E-mail: c.yu91@csuohio.edu

Ben Lee
*School of Electrical Engineering and Computer Science*
*Oregon State University, Corvallis, OR 97331, USA*
E-mail: benl@ece.orst.edu

Hee Yong Youn
*School of Electrical and Computer Engineering*
*Sungkyunkwan University, Suwon, 440-746 Korea*
E-mail: youn@ece.skku.ac.kr

## 1 Introduction

Wireless connectivity with mobility support will become an important enabling technology in future computing infrastructures. In particular, Mobile Ad hoc NETworks (MANETs) [1,2] have attracted a lot of attention

M.X. Cheng, D. Li (eds.) *Advances in Wireless Ad Hoc and Sensor Networks.* Signals and Communication Technology, doi: 10.1007/978-0-387-68567-0_10.

with the advent of inexpensive wireless LAN solutions such as IEEE 802.11 [3], HIPERLAN [4], and Bluetooth [5] technologies. MANETs have a great long-term economic potential but pose many challenging problems. Among them, energy efficiency may be the most important design criterion because one of the critical limiting factors for a mobile node is its operation time, restricted by battery capacity [6]. Energy consumption of wireless communication can represent more than half of the total system power consumption [7], therefore the key to energy efficiency is in energy-aware network protocols. Many research efforts have been devoted to energy-efficient MAC layer protocols [4,8,12], energy-conserving routing algorithms for MANETs [35–37], and measurement-based energy analysis of wireless network interfaces [7,38,39].

In this chapter, we consider an energy-efficient multicast for MANETs. Multicasting has been extensively studied for MANETs [9–11,19–31] because its operation is fundamental to many ad hoc network applications requiring close collaboration of multiple nodes in a multicast group. A multicast packet is delivered to multiple receivers along a network structure, such as a multicast tree or mesh, constructed at group creation time. However, the network structure is fragile due to node mobility, and thus some members may not be able to receive the multicast packet. In order to improve the packet delivery capability, multicast protocols for MANETs usually employ control packets to periodically refresh the network structure.

It has been shown that mesh-based protocols are more robust to mobility than tree-based protocols [9] due to many redundant paths between mobile nodes in the mesh. In contrast, tree-based protocols are much more energy efficient for the following reasons. First, the power-saving mechanism, such as the one defined in IEEE 802.11 wireless LAN standard [17], puts a mobile node into sleep-mode when it is not sending or receiving packets [4]. Second, a wireless Network Interface Card (NIC) typically accepts only two kinds of packets; unicast and broadcast packets (all ones). Because mesh-based protocols depend on broadcast flooding, every mobile node in the mesh must be ready to receive packets at all times during the multicast. In contrast, for unicast transmission along the multicast tree, only the designated receivers need to receive the transmitted data. Thus, a mobile node in tree-based protocols can safely put itself into a low-power sleep-mode to conserve energy if it is not a designated receiver.

Based on the aforementioned discussions, this paper proposes an energy-efficient, robust multicast scheme called Two-Tree Multicast (TTM). The proposed TTM is based on a multicast tree to save energy but maintains two maximally disjoint trees, called *primary* and *alternative trees*, to offer an improved packet delivery capability. A simulation study based on the QualNet simulator [15], which is the commercial version of GloMoSim [33], shows that the proposed TTM saves energy by a factor of $1.9 \sim 4.0$ compared to the mesh-based multicast without significant degradation in the packet delivery ratio. A combined performance metric, called energy per delivered packet, has also been measured to assess the general performance, such as delay or packet delivery ratio, together with energy, and results show that TTM outperforms the mesh-based and tree-based multicast by up to 80% and 40%, respectively.

The contribution of this chapter is two fold: it provides an in-depth comparison of tree-based and mesh-based multicasts with respect to energy consumption. Our analysis shows that mesh-based multicast consumes around $(f + 1)/2$ times more energy than tree-based multicast, where $f$ is average node connectivity. To the best of our knowledge, there is no such study in the literature evaluating the energy efficiency of tree-based and mesh-based multicasts. Second, the proposed TTM makes a prudent tradeoff between energy efficiency and packet delivery capability. Although the key idea of TTM is not new and it has been employed with routing algorithms to develop multipath routing schemes [13,14], the main thrust of the proposed TTM is to view the mesh structure as the superimposition of multiple trees and to utilize them at different time instances to not only improve energy efficiency but also offer better packet delivery capability.

The rest of the chapter is organized as follows: related work on network models and earlier multicast protocols for MANETs are described in the following section. In Section 3, the energy consumption of multicast protocols is analyzed in the context of a static ad hoc network. We assume the static network for simplicity in Section 3 but the following sections consider a dynamic network with node mobility. Section 4 presents the proposed TTM protocol. Section 5 discusses our simulation study, which shows the advantages of TTM compared to mesh-based and tree-based multicast. Finally, concluding remarks are given in Section 6.

# 2 Related Work

## 2.1 Power-Saving Mechanisms

Recent wireless LAN specifications usually provide power-saving mechanisms for energy-constrained applications. For example, the Bluetooth network interface operates in time-division multiplexing (TDM), where a master node controls up to seven neighboring slaves. Each slave node has a designated time slot for communication and can sleep in other time slots to conserve energy [34]. In the IEEE 802.11 standard, a master node, also called an Access Point (AP), periodically sends a beacon packet followed by a TIM (Traffic Indication Map) that indicates the desired receivers. Each slave wakes up when beacons are sent and checks whether it is the intended receiver. If it is not, it sleeps again; otherwise, it stays awake to receive data [4].

The IEEE 802.11 ad hoc power-saving mechanism operates in a similar fashion but without APs. Any node requiring communication sends beacons to synchronize with nodes in its vicinity. A beacon period starts with ATIM (Ad hoc TIM), during which all nodes listen, and the pending traffic is advertised. Each node turns itself on or off depending on the advertised traffic [4]. Unlike the AP-based mechanism, packets are buffered at the sender node and are directly transmitted to the receiver node. This power-saving mechanism reduces the available channel capacity because useful traffic cannot be transmitted during the ATIM window. In addition, it also suffers from longer packet delay because each intermediate node needs to buffer the packet until the next beacon period.

The abovementioned power-saving mechanisms favor unicast over broadcast communication. For unicast, all other neighbors do not need to wake up and thus can save energy. However, if a sender has more than one receiver, it must resort to broadcast resulting in many unnecessary receptions as well as wasted energy. As discussed in Section 1, mesh-based multicast protocols typically require each intermediate node to broadcast to multiple receivers and thus all its neighbors (not only the intended receivers) have to remain awake to receive the packet. For this reason, a multicast tree is a much more energy-efficient solution.

## 2.2 Multicast Protocols for MANETs

This section briefly overviews the previous multicast protocols developed for mobile networks. These protocols basically construct a network structure to deliver multicast messages. However, because it is difficult to maintain the network structure in the presence of node mobility, a common approach is for multicast group members to periodically exchange control packets (e.g., join messages [9,23]) to refresh the structure.

### Tree-based Multicast

As in wired and infrastructured mobile networks, tree-based multicast can be further classified as either *per-source tree multicast* or *shared tree multicast* [18]. In the per-source tree approach, each sender has to construct a separate multicast tree rooted at itself. Therefore, there will be as many trees as the number of senders and a significant amount of control overhead is required to maintain them. On the other hand, shared tree multicast has lower control overhead because it maintains only a single tree shared by all senders [16,19]. However, the path is not necessarily optimal, and the root node is easily overloaded due to the sharing of the single tree. *Associativity-Based Multicast Routing Protocol* (*ABAM*) [20] and *Multicast Routing Protocol based on Zone Routing* (*MZR*) [21] are per-source type multicast protocols. *Ad hoc Multicast Routing* (*AM-Route*) [22], *Ad-hoc On-demand Distance Vector Multicast Protocol* (*AODVM*) [23], *Ad hoc Multicast Routing protocol utilizing Increasing id-numberS* (*AMRIS*) [24], and *Lightweight Adaptive Multicast* (*LAM*) [25] are based on shared trees.

Figure 1 shows an example of a multicast tree. The tree consists of a root node ($r$), three nonmember intermediate nodes ($p$, $s$, and $t$), seven member nodes of a multicast group (shaded nodes in the figure), and ten

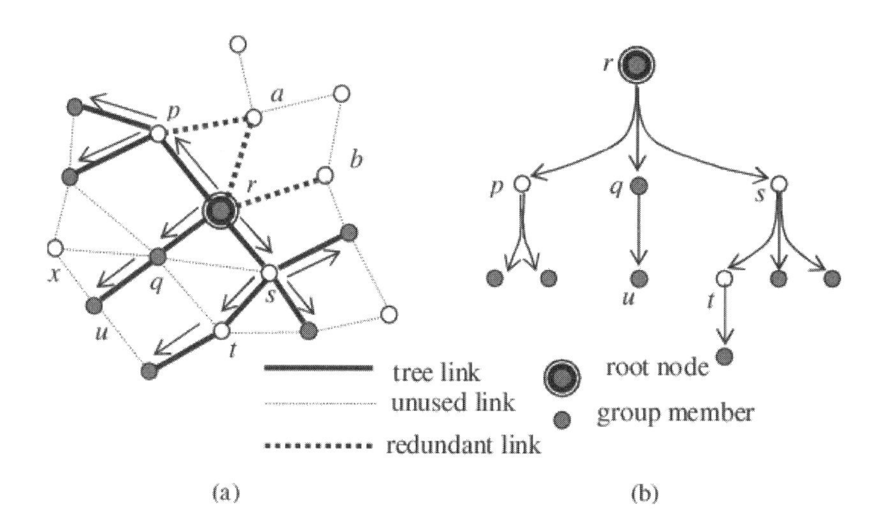


Figure 1: An example of tree-based multicast (5 transmissions and 17 receives). (a) Eight-node multicast group (including node $r$); (b) corresponding multicast tree.

tree links. A multicast packet is delivered from the root node $r$ to seven group members. For a member node $u$, for instance, the packet transmission is relayed through two tree links, that is, from $r$ to $q$ and then $q$ to $u$. Now consider the last transmission from $q$ to $u$. Even though all the nodes within node $q$'s radio transmission range (e.g., $s$, $t$, and $x$) can receive the multicast packet, only node $u$ will receive the packet and the rest of the nodes go into sleep-mode. Now, a single multicast packet requires five transmissions and 10 receives. However, because NIC typically accepts only unicast and broadcast addresses, nodes $r$, $p$, and $s$ must use broadcast address because they have more than one receiver. This increases the number of receives to 17 including 7 new receives ($r \rightarrow a, r \rightarrow b, p \rightarrow a, p \rightarrow r, s \rightarrow r$ and $s \rightarrow q$).

### Mesh-Based Multicast

The aforementioned tree-based protocols may not perform well under high node mobility because a multicast tree is fragile and needs to be frequently readjusted as its connectivity changes. A new approach unique to MANETs is the *mesh-based multicast*. A mesh is different from a tree because each node in a mesh can have multiple parents. When a single mesh structure spanning all multicast group members is used, multiple links exist and other links are immediately available when the primary link is broken due to node mobility. This avoids frequent network reconfigurations, which minimizes disruptions of on-going multicast sessions and reduces the control overhead to reconstruct and maintain the network structure. Note that these redundant links are usable because multicast packets are broadcast forwarded within the mesh.

Figure 2 shows an example of the mesh-based multicast for the MANET of Figure 1. Note that intermediate nodes $P$, $Q$, and $S$ are included in the mesh and participate in forwarding multicast packets. Thus, as in Figure 2a, a single multicast transmission requires data communication over 15 redundant links (dotted lines in the figure) in addition to 10 tree links. For example, sending a packet from $R$ to $U$ involves 3 transmissions (by $R$, $Q$, and $U$) and 14 receives (five neighbors of $R$, six neighbors of $Q$, and three neighbors of $U$). The transmission from node $Q$ is received not only by $U$ but also by neighbor nodes $R$, $S$, $T$, $W$, and $X$. The redundant link from $Q$ to $W$ may be useful when the path from $P$ to $W$ is broken, as shown in Figure 2b. However, most redundant links are not useful and waste energy. In summary, mesh-based multicast improves the packet delivery ratio by producing many redundant links, but as a result increases energy consumption compared to tree-based multicast.

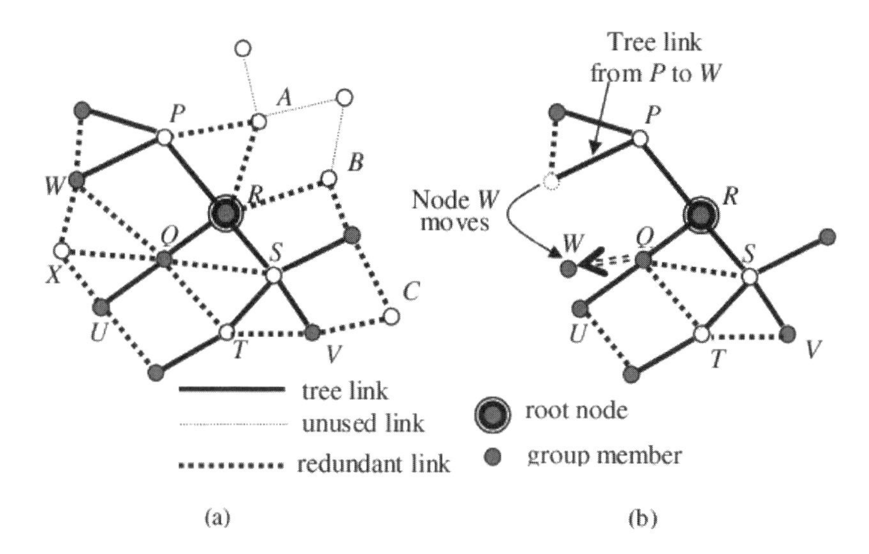


Figure 2: An example of mesh-based multicast (12 transmissions and 40 receives). (a) Eight-node multicast group ($P \rightarrow W$); (b) corresponding multicast mesh ($Q \rightarrow W$).

Protocols such as On-Demand Multicast Routing Protocol (ODMRP) [26], Neighbor-Supporting Multicast Protocol (NSMP) [27], Core-Assisted Multicast Protocol (CAMP) [28], Multicast Core-Extraction Distribution Ad hoc Routing (MCEDAR) [29], and Clustered Group Multicast (CGM) [30] fall into this category.

# 3 Energy Efficiency of Multicast Protocols

This section evaluates the tree-based and mesh-based multicast protocols in terms of energy efficiency. Section 3.1 presents an example study to show the benefit of multicast trees over multicast meshes using a static ad hoc network. Section 3.2 generalizes this to provide an analytical result of energy consumption of multicast protocols.

## 3.1 Energy Model and Example Study

### Energy Model (First-Order Radio Model)

Let the total energy consumption per unit multicast message be denoted $E$, which includes the transmission energy ($E_{TX}$) as well as the energy required to receive the transmission ($E_{RX}$). This chapter only considers data packets for simplicity. According to the first-order radio model [32], $E = E_{TX} + E_{RX} = N_{TX} \cdot e_{TX} + N_{RX} \cdot e_{RX}$, where $N_{TX}$ and $N_{RX}$ are the number of transmissions and the number of receives, respectively, and $e_{TX}$ and $e_{RX}$ are the energy consumed to transmit and receive a unit multicast message via a wireless link, respectively. If we assume that $e_{TX}$ and $e_{RX}$ are the same[1] and denoted by $e$, the total energy consumption is simply $E = (N_{TX} + N_{RX})e$.

Let $\Gamma_+$, $\Gamma_1$, and $\Gamma_0$ be the set of tree nodes with more than one receiver, with exactly one receiver, and with no receiver, respectively. Thus, the set of all tree nodes is $\Gamma = \Gamma_+ + \Gamma_1 + \Gamma_0$. It is straightforward to show that, in a multicast tree, $N_{TX}$ is the number of tree nodes except the leaf receiver nodes (i.e., root and intermediate nodes) and $N_{RX}$ is $\sum_{i \in \Gamma_+} f_i + |\Gamma_1|$, where $f_i$ is the number of neighbors of node $i$. In a multicast mesh, $N_{TX}$ is the number of tree nodes (i.e., root, intermediate, and receiver nodes) for the multicast group and $N_{RX}$ is the sum of the number of neighbors of all tree nodes ($\sum_{i \in \Gamma} f_i$).

### Example Network Model (Static Ad Hoc Network)

Consider a static ad hoc network consisting of 64 nodes placed in an 88 grid, and three multicast groups consisting of 4, 10, and 25 members in the groups, respectively, where the group members are selected randomly. Figures 3a–c show the multicast trees with the shortest paths from the root to receiver nodes for the three group sizes, and Figures 3d–f show the multicast meshes with one-hop redundant links. Note that the figure does not include discarded links through which a packet is transmitted and received but ultimately dropped by the receiver. Node connectivity[2] is assumed to be 4; that is, the radio transmission range is limited to a node's four direct neighbors in the east, west, south, and north. In Figure 4, node connectivity is assumed to be 8 so that a node can communicate with all its eight

[1] In reality, $e_{TX}$ and $e_{RX}$ are slightly different. For example, $e_{TX} = 300\ mA$ and $e_{RX} = 250\ mA$ for WaveLAN-II from Lucent [3].

[2] Node density, defined as the number of nodes per unit area, does not indicate the connectivity between mobile nodes. Node connectivity is a relative measure of the node density compared to the radio transmission range of the underlying wireless network interface. (For simplicity, we assume that the radio transmission range as well as transmission power is fixed and cannot be dynamically controlled.) With low node connectivity, a mobile ad hoc network can be partitioned. Most research on ad hoc networks implicitly assumes a relatively large node connectivity to avoid network partitioning.

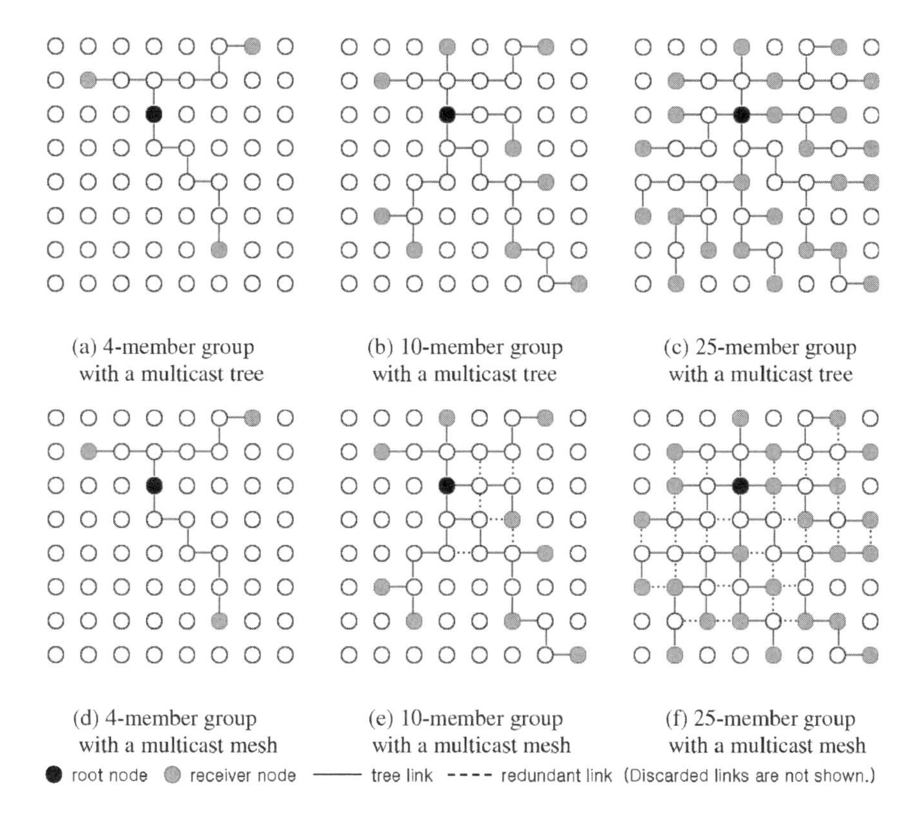


Figure 3: Examples of multicast groups in a lowly populated network (node connectivity = 4).

surrounding neighbors. In Figures 3 and 4, each solid line denotes a tree link, and each dotted line denotes a redundant link as explained in Section 2.

### Energy Comparison of the Example Networks

Based on the previously introduced energy model, we compare energy efficiency of tree-based and mesh-based multicast schemes. For example, consider Figures 3b and e for 10-member group multicast. $E_{tree}$ (Figure 3b) can be estimated as $E_{tree} = 53e$ because $N_{TX}$ is 19 (27 tree nodes, 8 leaf receiver nodes) and $N_{RX}$ is 34 ($4 \times 5 + 14$). $E_{mesh}$ for a 10-member group with a multicast mesh (Figure 3e) can be estimated as $E_{mesh} = 129e$ because $N_{TX}$ is 27 (27 tree nodes) and $N_{RX}$ is 102 (sum of the number of neighbors of the 27 nodes). Therefore, it can be concluded that mesh-based multicast

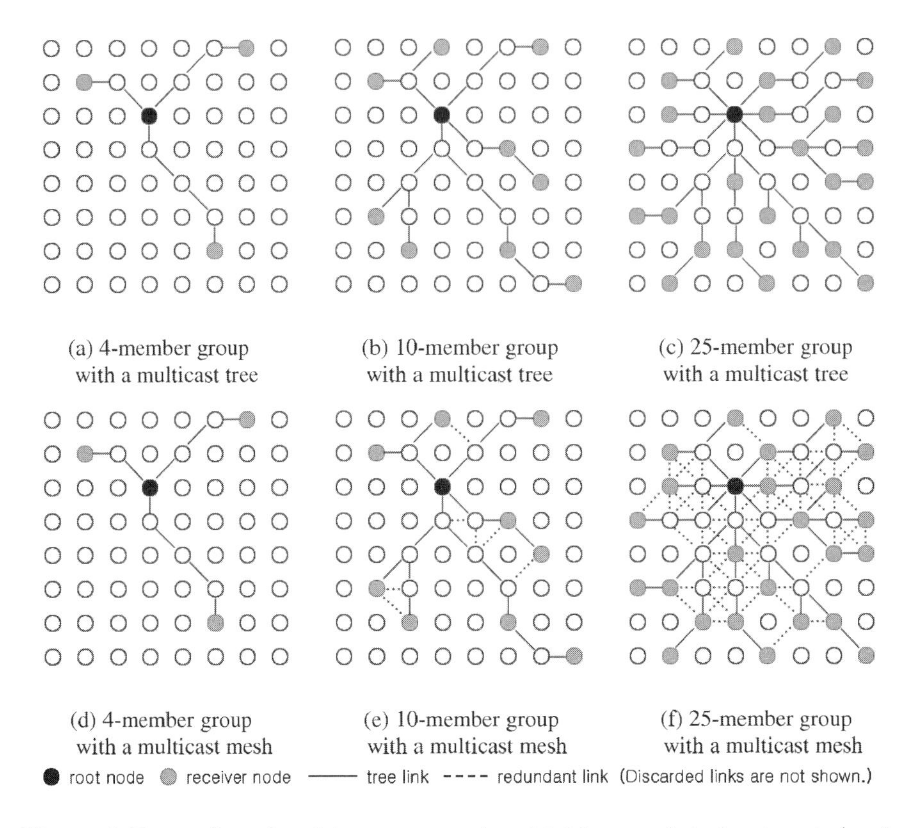


Figure 4: Examples of multicast groups in a highly populated network (node connectivity = 8).

consumes 2.4 times ($E_{mesh}/E_{tree} = 129e/53e$) more energy than tree-based multicast for a 10-member group with node connectivity of 4.

Similarly, the energy efficiency for 4-member and 25-member groups with node connectivity of 4 and 8 is shown in Figure 5. In summary, multicast meshes consume about 2$\sim$4 times more energy than multicast trees. Another important observation is that high node connectivity or a dense network makes the mesh-based multicast protocol suffer more because of more redundant links.

## 3.2 Quantitative Analysis of Energy Consumption

In this subsection, total energy consumption of a multicast mesh and a multicast tree is analyzed quantitatively. Consider a static ad hoc network

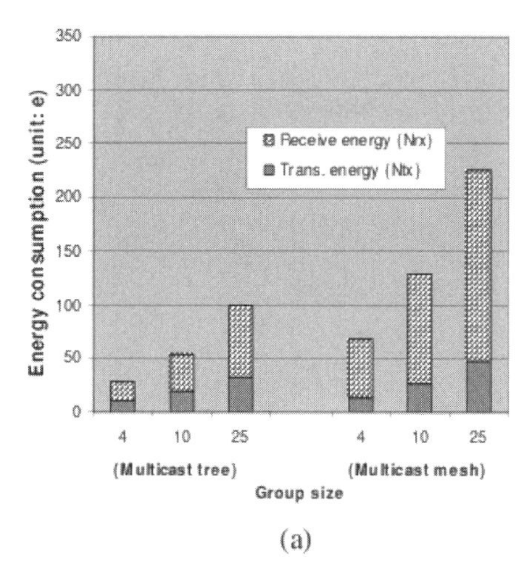


(a)

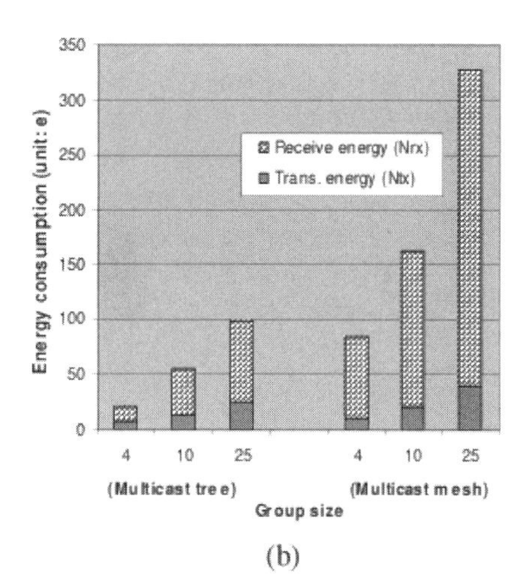


(b)

Figure 5: Comparison of multicast tree and multicast mesh. (a) Lowly populated network ($f = 4$); (b) highly populated network ($f = 8$).

consisting of $k^2$ nodes placed in a $k \times k$ grid. Figure 6 shows examples of tree-based multicast with the shortest paths from the root to receiver nodes on an $8 \times 8$ grid network with node connectivity of 4 and 8. Figure 7 shows examples of mesh-based multicast with one-hop redundant links on an $8 \times 8$ grid network. For upper bound analysis, we focus on complete multicast, where all the nodes in a network are member nodes as in Figures 6a, 6d, 7a, and 7d. Figures 6b, 6e, 7b, and 7e show the worst cases where the total energy consumption is about the same as the complete multicast but with fewer member nodes. This happens when the member nodes reside at the edges of the network. Figures 6c, 6f, 7c, and 7f show the best cases where a multicast tree or mesh consists of only the member nodes and thus the total energy consumption is the lowest with the given number of member nodes.

The following two theorems formally analyze the upper and lower bounds of total energy consumption in a static ad hoc network based on the tree-based multicast and the mesh-based multicast.

**Theorem 3.1** *For a static ad hoc network of $k \times k$ grid topology with node connectivity of $f$, the total energy consumed to transfer a multicast message in a tree-based multicast method, $E_{tree}$, is bounded by $(2n - O(n^{1/2}))e \leq E_{tree} \leq (2k^2 - O(k))e$, where $n$ is the number of member nodes and $e$ is the energy consumed to transmit or receive a multicast message via a link.*

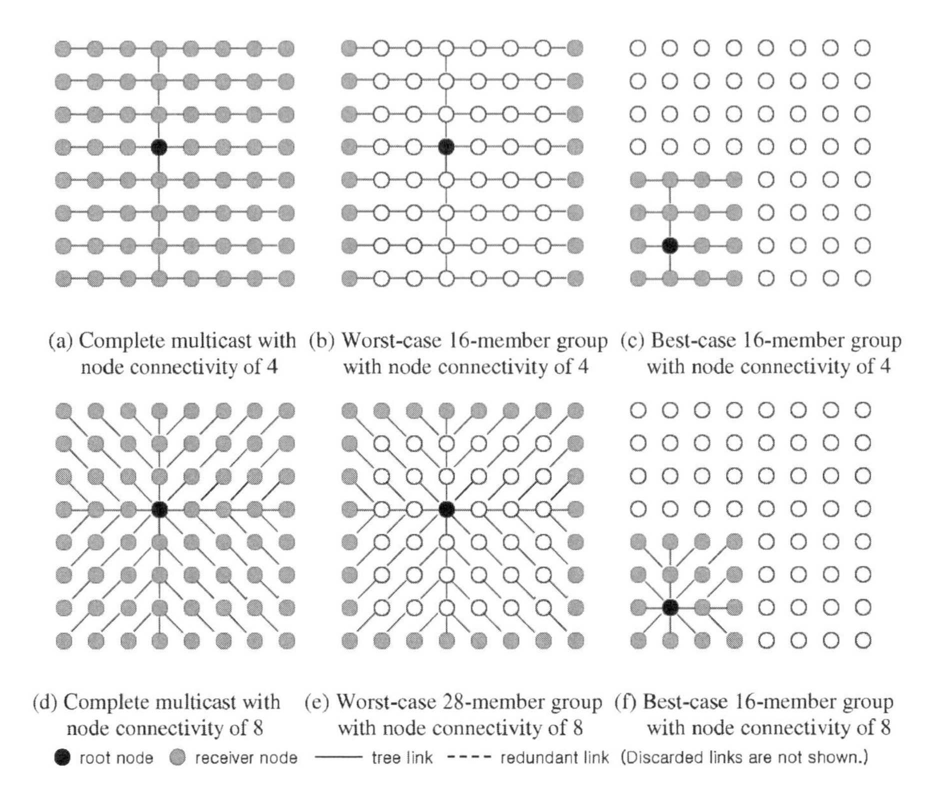


Figure 6: Examples of tree-based multicast on an $8 \times 8$ grid network.

*Proof.* Given a static ad hoc network of $k \times k$ grid topology with node connectivity of $f$, the total energy consumption of a tree-based multicast method for complete multicast can be regarded as the upper bound. In a complete multicast, $N_{TX} = k^2 - O(k)$, where $O(k)$ is mainly due to the boundary nodes having less node connectivity than $f$, and $N_{RX} = k^2 - 1 + O(k)$, where $O(k)$ is due to the tree nodes that send to more than one receiver. Hence, $E_{tree} = (N_{TX} + N_{RX})e \leq (2k^2 - O(k))e$. In the best case, where a multicast tree consists of only member nodes, $N_{TX} = n - O(n^{1/2})$ and $N_{RX} = n - 1 + O(k)$. Hence, $E_{tree} \geq (2n - O(n^{1/2}))e$. □

**Theorem 3.2** *For a static ad hoc network of $k \times k$ grid topology with node connectivity of $f$, the total energy consumed to transfer a multicast message in a mesh-based multicast method, $E_{mesh}$, is bounded by $((f + 1)n -$*

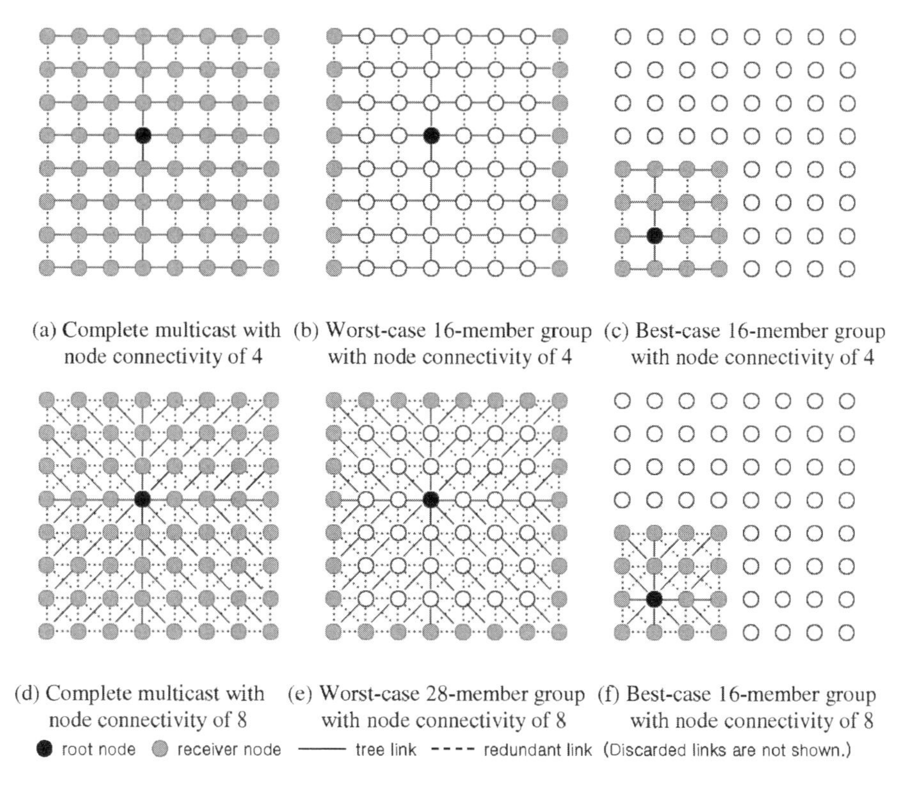


Figure 7: Examples of mesh-based multicast on an $8 \times 8$ grid network.

*$O(n^{1/2}))e \leq E_{mesh} \leq ((f+1)k^2 - O(k))e$, where $n$ is the number of member nodes and $e$ is the energy consumed to transmit or receive a multicast message via a link.*

*Proof.* Given a static ad hoc network of $k \times k$ grid topology with node connectivity of $f$, the total energy consumption of a mesh-based multicast method for complete multicast can be regarded as the upper bound. In the complete multicast, $N_{TX} = k^2$ and $N_{RX} = fk^2 - O(k)$ because the mesh-based multicast protocol uses broadcast-style communication ($O(k)$ is due to the boundary nodes having smaller node connectivity than $f$). Hence, $E_{mesh} = (N_{TX} + N_{RX})e \leq ((f+1)k^2 - O(k))e$. In the best case, where a multicast mesh consists of only member nodes, $N_{TX} = n$ and $N_{RX} = fn - O(n^{1/2})$. Hence, $E_{mesh} \geq ((f+1)n - O(n^{1/2}))e$. □

According to Theorems 3.1 and 3.2, we can conclude that $E_{mesh}/E_{tree} \approx (f+1)/2$ because the most energy-consuming scenario for multicast tree renders the situation that multicast mesh also consumes the most energy (and the same argument applies to the least energy-consuming scenario). Because node connectivity, $f$, is usually much larger than 2 (otherwise, network partitioning occurs), mesh-based multicast protocols consume more energy compared to tree-based multicast protocols by a factor of $(f+1)/2$. The above analysis is based on the assumption that all nodes are located in a grid fashion. Even when the nodes are located in an arbitrary manner, the above analysis applies because the node connectivity is directly related to the tree structure and the number of transmissions.

# 4 Energy-Efficient Two-Tree Multicast

This section introduces a new multicast protocol, *Two-Tree Multicast* (*TTM*), which not only reduces the total energy consumption but also alleviates the energy balance problem without having an adverse effect on the general performance. TTM is a tree-based multicast protocol and thus consumes less energy than mesh-based protocols. It uses a shared tree rather than per-source trees in order to minimize the tree construction and maintenance overhead.

Unique to TTM is the use of two trees called primary and alternative trees for a multicast group. When the primary tree becomes unusable or overloaded, the alternative tree takes on the responsibility of the primary tree and a new alternative tree is immediately constructed. A group member with the largest remaining battery energy is selected as the root node of the new alternative tree. This is similar to the *root relocation scheme*, where the root node is periodically replaced with the one near the center of the network to achieve the shortest average hop distance from the root to all the receiver nodes [16]. Two trees can reduce the latency problem when a link error occurs on the primary tree by immediately switching to the alternative tree. Tree replacement is also useful for alleviating the energy balance problem inherent in shared tree multicast.

### TTM Procedures

Using the same examples shown in Figures 1 and 2, Figure 8 shows the two trees constructed for the same multicast group of eight members. The primary tree consists of a primary root ($r_p$), three intermediate nodes, and

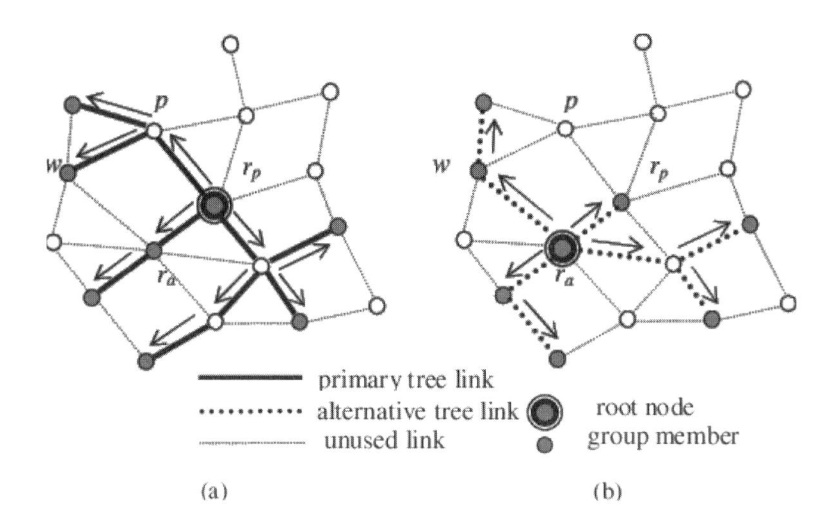


Figure 8: An example of two trees in TTM for an eight-node multicast group. (a) Primary tree; (b) alternative backup tree.

seven receiver nodes. On the other hand, the alternative tree consists of an alternative root ($r_a$), one intermediate node, and seven receiver nodes including $r_p$. As in tree-based and mesh-based multicast protocols, TTM reconstructs two trees periodically (e.g., every three seconds [9]) using periodic join messages sent by all the receiver nodes to $r_p$ and $r_a$. Note that the join message includes information about the remaining battery energy of the corresponding member node, which will be used to select a root node for a new alternative tree. The two root nodes independently construct multicast trees based on the forwarding paths that the join messages traverse (called the tree construction and maintenance procedure). When a sender node intends to send a multicast packet, it forwards the packet to $r_p$; then $r_p$ delivers the message to the member receivers along tree links of the primary tree [25,26] (called multicast message delivery procedure).

When a tree connection is broken due to node mobility during the join interval or $r_p$'s residual energy reduces to a predetermined threshold, the primary tree yields its responsibility to the alternative tree; that is, $r_p$ sends a control message to $r_a$ notifying that the alternative tree will take the role of the primary tree. Upon receiving the control message, the alternative root ($r_a$) selects a new alternative root ($r'_a$) that has the largest remaining battery energy among the member nodes. Then, $r_a$ informs the sender(s) and all the members including $r'_a$ of the replacement tree. When each member receives a control message from $r_a$, it sends a join request message to $r_a$ (i.e., the new $r_p$) and $r'_a$ (tree replacement procedure). The shared tree multicast protocol described in [19] is used as the basic multicast protocol in our implementation. Table 1 summarizes the operations for implementing the proposed TTM protocol.

Table 1: Host operation for the TTM protocol. (Messages in each procedure are described in time sequence.)

| Sender | Primary Root ($r_p$) | Alternative Root ($r_a$) | Member Nodes |
|---|---|---|---|
| **Tree construction and maintenance procedure** | | | |
| | Receive join messages from the member nodes. Construct a multicast tree based on the forwarding paths traversed by the join messages. | | Periodically send a join message to $r_p$ and $r_a$. |
| **Multicast message delivery procedure** | | | |
| Send a multicast message to $r_p$. | Send a multicast message to the member nodes. | | Receive a multicast message from $r_p$. |
| **Tree replacement procedure** | | | |
| | Send a control message to $r_a$ about the tree replacement. | Receive a control message from $r_p$ and select a new alternative root ($r_a'$). Send a control message to both the sender(s) and all the members including $r_a'$ about the tree replacement. | Receive a control message from $r_a'$. Send a join request message to $r_a$ (*i.e.*, a new $r_p$) and $r_a'$. |

## Tree Construction: Maximally Disjoint Trees

Upon receiving join messages from member nodes, both $r_p$ and $r_a$ refresh the tree structure for use in subsequent multicast packets. In order to provide high tolerance to failure, TTM constructs maximally disjoint trees with minimum overlapping of links or nonmember nodes. When a primary tree breaks, the chance that the alternative tree will still work is higher if the two trees are constructed as maximally disjoint as possible.

The problem of finding disjoint trees is not trivial. Our approach is to use multipath routing algorithms proposed for unicast routing [13,14]. When a member node, say $w$, sends join messages to $r_p$ and $r_a$, it includes path information from itself to either $r_p$ or $r_a$. The idea is to make the two paths

($w \rightarrow r_p$ and $w \rightarrow r_a$) disjoint as much as possible. When routing paths from member nodes to $r_p$ are maximally disjoint from those from member nodes to $r_a$, we can obtain two trees that are maximally disjoint. For example, consider a member node $w$ in Figure 8. When it sends a join message to $r_a$ with path information, the path is most probably $w \rightarrow r_a$. However, when it sends one to $r_p$, it has two choices: $w \rightarrow p \rightarrow r_p$ and $w \rightarrow r_a \rightarrow r_p$. TTM selects the first path because it does not include the $w \rightarrow r_a$ link.

## 5 Performance Evaluation

In this section, performance of the proposed TTM is evaluated via simulation. The simulation environment is described in Section 5.1 including a physical communication model, node mobility, multicast traffic, and parameters used in multicast protocols. Section 5.2 discusses the simulation results.

### 5.1 Simulation Environment

Our simulation study is based on the *QualNet* simulator [15]. QualNet is a scalable network simulation tool for wireless and wired networks and supports a wide range of ad hoc routing protocols. QualNet simulates a realistic physical layer that includes a radio capture model, radio network interfaces, and the IEEE 802.11 Medium Access Control (MAC) protocol using the distributed coordination function (DCF). The radio network interface card model also simulates collisions, propagation delay, and signal attenuation.

Our evaluation is based on the simulation of 40 mobile nodes moving over a square area of $1000 \times 1000\ m^2$ for 15 minutes of simulation time. The radio transmission range is assumed to be 250 m and a free space propagation channel is assumed[3] with a data rate of 2 Mbps. Mobile nodes are assumed to move randomly according to the random waypoint model [9,19,35]. Two parameters, maximum node speed and pause time, determine the mobility pattern of the mobile nodes. Each node starts moving from a randomly selected initial position to a target point, which is also selected randomly in the simulated area. Node speed is chosen between 0 and the specified maximum speed (2 or 20 m/s). When a node reaches the target point, it

[3]Each mobile node can communicate with each another if it is within the transmission circle, defined by the radio transmission range. The area of the transmission circle is $250^2 \times \pi$ m$^2$. Because 40 nodes are located in the network of area size $1000^2$ m$^2$, it results in a node connectivity of 7.9 ($= 40 \times (250^2 \times \pi)/1000^2$).

stays there for the duration of the pause time (30 s) and then repeats the movement.

A recently released QualNet version 2.9 includes a mesh-based multicast protocol ODMRP [9] but does not support any tree-based multicast protocols. The proposed TTM protocol is implemented within the QualNet simulation framework. We use the shared tree multicast protocol described in [19] as the basic operation principle for each tree. The periodic join message is transmitted every 3 seconds for all the simulated protocols. (It is JOIN DATA in case of ODMRP.) For ODMRP, the acknowledgment timeout for JOIN TABLE and the maximum number of JOIN TABLE retransmissions are given by 25 ms and 3, respectively. The overhead due to the control messages, which include the control messages transferred during the tree switching, is considered in the simulation.

A separate application file specifies the traffic as well as the application type: FTP, HTTP, Telnet, or Constant Bit Rate (CBR). In our simulation, a multicast CBR (MCBR) source and its corresponding destinations are randomly selected among 40 mobile nodes, where the number of destinations is varied from 4 to 40 to see the effect of the group size on the performance. The MCBR source sends a 512-byte multicast packet every 100 ms. For simplicity, we assume a multicast message consists of one data packet.

## 5.2 Simulation Results and Discussion

In this chapter, we are specifically interested in total energy consumption and energy balance across all mobile nodes. For each node, energy consumption is measured at the radio layer during simulation. According to the specification of IEEE 802.11-compliant WaveLAN-II [3] from Lucent, the power consumption varies from 0.045 W (9 mA $\times$ 5 V) in sleep-mode to 1.25 and 1.50 W (230 and 250 mA $\times$ 5 V) for receiving and transmitting modes, respectively. The instantaneous power is multiplied by time duration to obtain the energy consumed. For example, data transmission of a 512-byte packet consumes 3.1 milli-Joules (1.50 W $\times$ 512 $\times$ 8 bits/2 Mbps). In order to measure the energy balance, we observed the peak-to-mean ratio; that is, the energy consumption of the most utilized node divided by the average energy consumption over all nodes. In the ideal case, the ratio becomes one when the total energy consumption is evenly distributed. In practice, the ratio is larger than one but a smaller ratio indicates better energy balance.

General performance measures such as packet delivery ratio are also an important metric. Because the data transmission in our simulation is based on UDP rather than TCP, some data packets may be lost. For this reason, we introduce a new performance metric, energy per delivered packet, to combine the energy performance with general performance. It is measured as the ratio of the total energy consumption over the total number of successfully delivered packets.

Figures 9a and b show the total energy consumption for the mesh-based multicast, STM (Single Tree Multicast) and TTM with low node speed (0~2 m/s) and with high node speed (0~20 m/s), respectively. As shown in the two graphs, the total energy consumption increases linearly with the group size. Energy consumption of STM and TTM is almost the same, but they consume less energy than the mesh-based multicast by factors of 2.4~4.0 and 1.9~3.1 at low and high node speed, respectively. Even at high node speed, STM and TTM consume almost the same amount of energy as in the case of low node speed, but the mesh-based multicast consumes less energy at high node speed compared to that at low node speed. Therefore, STM and TTM are less sensitive to node mobility in terms of total energy consumption compared to the mesh-based multicast.

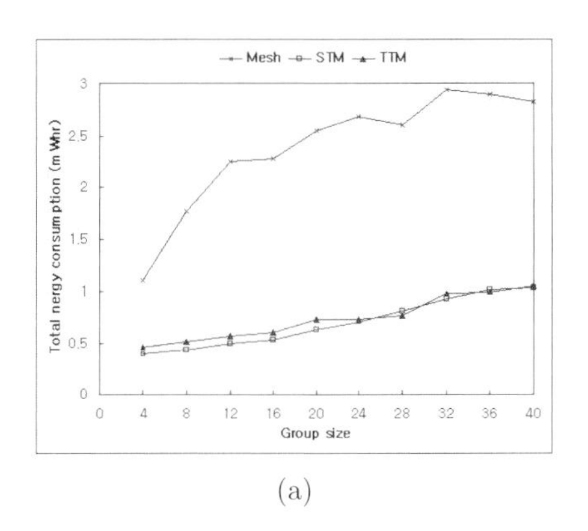


(a)

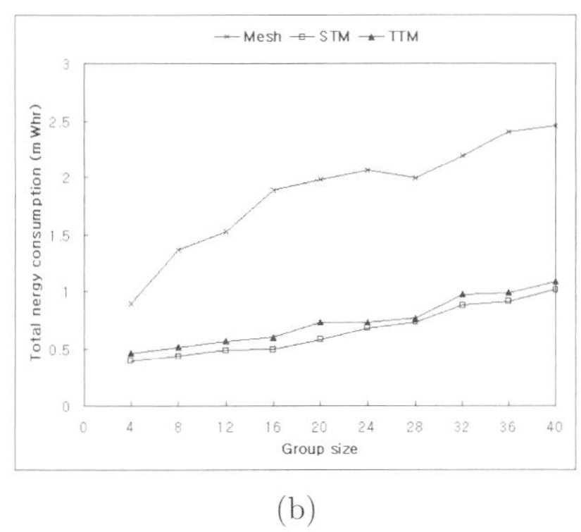


(b)

Figure 9: Total energy consumption (mesh: mesh-based multicast ODMRP, STM: shared, single-tree multicast, TTM: shared, two-tree multicast). (a) At low node speed ($0 \sim 2\ m/sec$); (b) at high node speed ($0 \sim 20\ m/sec$).

Figures 10a and b show the *energy balance* (i.e., peak-to-mean ratio of energy consumption) for the mesh-based multicast, STM, and TTM with low node speed (0~2 m/s) and with high node speed (0~20 m/s). As can be seen in the two graphs, the mesh-based multicast and TTM are almost the

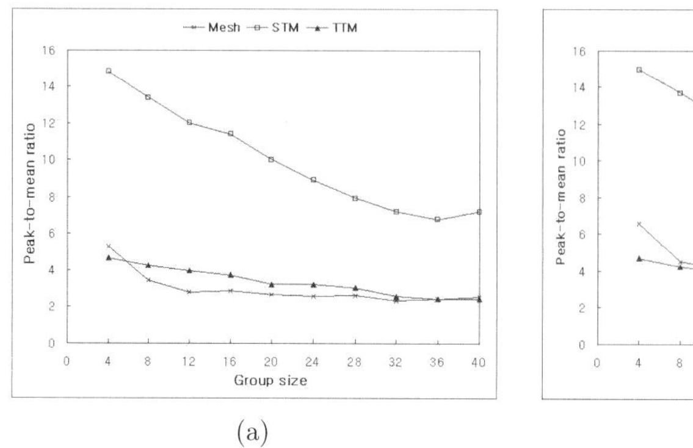


(a)

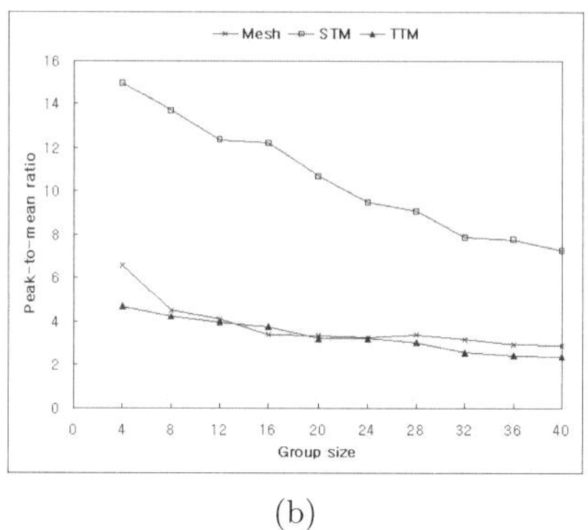


(b)

Figure 10: Peak-to-mean ratio (Mesh: mesh-based multicast ODMRP, STM: shared, single-tree multicast, TTM: shared, two-tree multicast). (a) At low node speed ($0 \sim 2\ m/sec$); (b) at high node speed ($0 \sim 20\ m/sec$).

same in terms of energy balance and show less peak-to-mean ratio than STM by factors of 2.6~3.2 and 3.0~3.4 at low and high node speed, respectively. At high node speed, all three methods become slightly worse (i.e., peak-to-mean ratios increase) compared to those at low node speed. Notice here that the energy balance improves with the group size.

Figures 11a and b show the energy per delivered packet for the mesh-

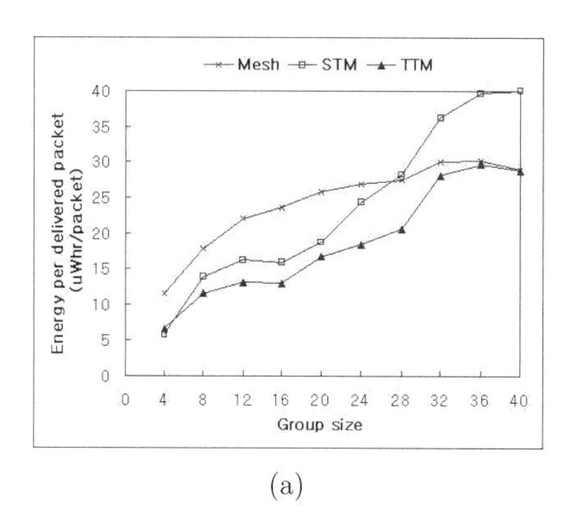


(a)

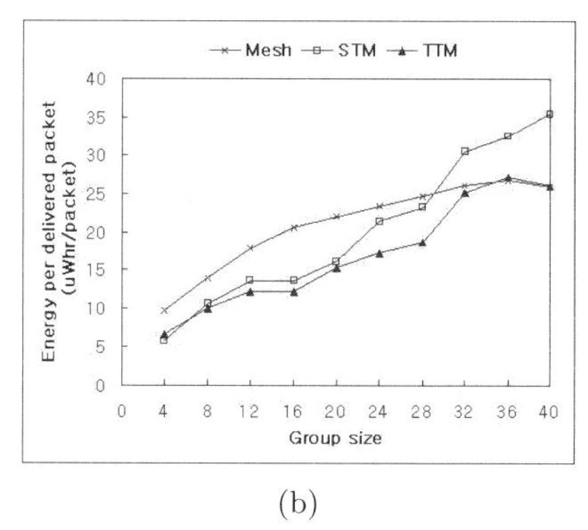


(b)

Figure 11: Energy per delivered packet (mesh: mesh-based multicast ODMRP, STM: shared, single-tree multicast, TTM: shared, two-tree multicast). (a) At low node speed ($0 \sim 2\ m/sec$); (b) at high node speed ($0 \sim 20\ m/sec$).

based multicast, STM, and TTM with low node speed (0∼2 m/s) and with high node speed (0∼20 m/s). As described in the previous section, the energy per delivered packet is a combined metric of total energy consumption and packet delivery ratio. At low node speed, TTM is better than the mesh-based multicast and STM by factors of 1.0∼1.8 and 1.0∼1.4, respectively. Even at high node speed, TTM outperforms the mesh-based multicast and STM by factors of 1.0∼1.7 and 1.0∼1.4, respectively. With a large group size, the mesh-based multicast and TTM show almost the same value. At high node speed, all three methods become slightly worse than those at low node speed. Note also that the energy per delivered packet increases with the group size.

## 6 Conclusion

The motivation of this chapter is to reevaluate the multicast protocols proposed for MANETs in terms of energy efficiency, which may be the most critical factor restricting the lifetime of each mobile node as well as the overall lifetime of MANETs. With the power-saving mechanisms in effect, multicast trees are more energy efficient than multicast meshes, although it is just the opposite with respect to general performance. This is because mesh-based protocols depend on broadcast flooding and, therefore, mobile nodes must be awake at all times to receive packets during multicast communication. An analytical result showed that mesh-based protocols consume around $(f + 1)/2$ times more energy than tree-based protocols, where $f$ is the node connectivity ($f \gg 2$).

The proposed two-tree multicast always constructs and maintains a pair of trees, called primary and alternative trees. When the primary tree becomes unusable, the alternative tree takes the responsibility of the primary tree and a new alternative tree is constructed immediately. According to our performance evaluation, TTM saves energy consumption by a factor of 1.9∼4.0 compared to the mesh-based multicast without significantly affecting the general performance. In terms of a combined performance metric, energy per delivered packet, TTM results in up to 80% and 40% better performance than the mesh-based multicast and STM, respectively.

## References

[1] Internet Engineering Task Force (IETF) Mobile Ad Hoc NETworks (MANET) Working Group Charter, http://www.ietf.org/html.

charters/manet-charter.html, (2000).

[2] J. Jubin and J. D. Tornow, The DARPA Packet Radio Network Protocols, *Proceedings. of the IEEE*, 75, 1 (1987), pp. 21–32.

[3] A. Kamerman and L. Monteban, WaveLAN-II: A high-performance wireless LAN for the unlicensed band, *Bell Labs Technical Journal*, (1997), pp. 118–133.

[4] H. Woesner, J. Ebert, M. Schlager, and A. Wolisz, Power-saving mechanisms in emerging standards for wireless LANs: The MAC level perspective, *IEEE Personal Communications*, 5, 3 (1998), pp. 40–48.

[5] Complete Bluetooth Tutorial, http://infotooth.tripod.com/tutorial/complete.htm (2000).

[6] C. E. Jones, K. M. Sivalingam, P. Agrawal, and J. C. Chen, A survey of energy efficient network protocols for wireless networks, *Wireless Networks*, 7, 4 (2001), pp. 343–358.

[7] R. Kravets and P. Krishnan, Power Management Techniques for Mobile Communication. In *Proc. of Int. Conf. on Mobile Computing and Networking (MobiCom '98)*, (1998), pp. 157–168.

[8] B. Prabhakar, E. U. Biyikoglu, and A. E. Gamal, Energy-efficient transmission over a wireless link via lazy packet scheduling. In *Proc. of the IEEE Infocom 2001*, Vol. 1 (2001), pp. 368–394.

[9] S.-J. Lee, W. Su, J. Hsu, M. Gerla, and R. Bagrodia, A performance comparison study of ad hoc wireless multicast protocols. In *Proc. of the IEEE Infocom 2000*, Vol. 2 (2000), pp. 565–574.

[10] J. E. Wieselthier, G. D. Nguyen, and A. Ephremides, On the construction of energy-efficient broadcast and multicast trees in wireless networks. In *Proc. of the IEEE Infocom 2000*, Vol. 2 (2000) pp. 585–594.

[11] J. Ryu, S. Song, and D.-H. Cho, A power-saving multicast scheme in 2-tier hierarchical mobile ad-hoc networks. In *Proc. of the 52nd IEEE Vehicular Technology Conference (VTC 2000)*, Vol. 4 (2000), pp. 1974–1978.

[12] S. Singh and C. S. Raghavendra, Pamas — Power aware multi-access protocol with signaling for ad hoc networks, *ACM Computer Communication Review*, Vol. 28, Issue 3 (1998) pp. 5–26.

[13] A. Nasipuri and S. R. Das, On-demand multipath routing for mobile ad hoc networks. In *Proc. of Int. Conf. on Computer Communication and Network (ICCCN '99)*, (1999), pp. 64–70.

[14] M. R. Pearlman, Z. J. Hass, P. Sholander and S. S. Tabrizi, On the impact of alternate path routing for load balancing in mobile ad hoc networks. In *Proc. of the First Annual Workshop on Mobile Ad Hoc Networking and Computing (MobiHoc 2000)*, (2000), pp. 3–10.

[15] Scalable Network Technologies, Inc., QualNet: Network simulation and parallel performance, http://www.scalable-networks.com/products/qualnet.stm (2001).

[16] C.-C. Chiang, M. Gerla, and L. Zhang, Adaptive shared tree multicast in mobile wireless networks. In *Proc. of the IEEE Global Telecomm. Conference (GlobeCom '98)*, Vol. 3 (1998), pp. 1817–1822.

[17] IEEE Std 802.11-1999, Local and Metropolitan Area Network, Specific Requirements, Part 11: Wireless LAN Medium Access Control (MAC) and Physical Layer (PHY) Specifications (1999).

[18] G. Xylomenos and G. C. Polyzos, IP multicast for mobile hosts, *IEEE Communications Magazine*, Vol. 35, Issue 1 (1997) pp. 54–58.

[19] M. Gerla, C.-C. Chiang, and L. Zhang, Tree multicast strategies in mobile, multihop wireless networks, *Baltzer/ACM Journal of Mobile Networks and Applications (MONET)*, 3, 3 (1999), pp. 193–207.

[20] C. Toh, G. Guichal, and S. Bunchua, ABAM: On-demand associativity-based multicast routing for ad hoc mobile networks. In *Proc. of the 52nd IEEE Vehicular Technology Conference (VTC Fall 2000)*, Vol. 3 (2000), pp. 987 – 993.

[21] V. Devarapalli, and D. Sidhu, MZR: A multicast protocol for mobile ad hoc networks. In *Proc. of IEEE International Conference on Communications*, Vol. 3 (2001), pp. 886–891.

[22] E. Bommaiah, M. Liu, A. McAuley, and R. Talpade, AMRoute: Adhoc Multicast Routing Protocol, *Internet-Draft*, draft-talpade-manet-amroute-00.txt (1998).

[23] E. Royer, and C. Perkins, Multicast operation of the ad-hoc on-demand distance vector routing protocol. In *Proc. of the Int. Conf. on Mobile Computing and Networking (MobiCom '99)* (1999), pp. 207–218.

[24] C. Wu, Y. Tay, and C. Toh, Ad hoc Multicast Routing protocol utilizing Increasing id-numberS (AMRIS) Functional Specification, *Internet-Draft*, draft-ietf-manet-amris-spec-00.txt (1998).

[25] L. Ji, and M. Corson, A lightweight adaptive multicast algorithm. In *Proc. of IEEE Global Telecomm. Conference (GlobeCom '98)*, Vol. 2 (1998), pp. 1036–1042.

[26] S. Lee, M. Gerla, and C. Chiang, On-demand multicast routing protocol. In *Proc. of IEEE Wireless Communications and Networking Conference (WCNC '99)* (1999), pp. 1298–1302.

[27] S. Lee, and C. Kim, Neighbor supporting ad hoc multicast routing protocol. In *Proc. of the First Annual Workshop on Mobile Ad Hoc Networking and Computing (MobiHoc 2000)* (2000), pp. 37–44.

[28] J. Garcia-Luna-Aceves, and E. Madruga, The core assisted mesh protocol, *IEEE Journal on Selected Areas in Communications*, 17, 8 (1999), pp. 1380–1394.

[29] P. Sinha, R. Sivakumar, and V. Bharghavan, MCEDAR: Multicast core extraction distributed ad-hoc routing. In *Proc. of IEEE Wireless Communications and Networking Conference (WCNC '99)*, Vol. 3 (1999), pp. 1313-1317, Sep. 1999.

[30] C. Lin, and S. Chao, A multicast routing protocol for multihop wireless networks. In *Proc. of IEEE Global Telecomm. Conference (GlobeCom '99)*, (1999) pp. 235-239.

[31] J. E. Wieselthier, G. D. Nguyen, and A. Ephremides, Algorithms for energy-efficient multicasting in ad hoc wireless networks. In *Proc. of Military Communication Conference (MILCOM '99)*, Vol. 2 (1999), pp. 1414–1418.

[32] W. R. Heinzelman, A. Chandrakasan, and H. Balakrishnan, Energy-efficient communication protocols for wireless microsensor networks. In *Proc. of the Hawaii Int. Conf. on System Science* (2000), pp. 3005–3014.

[33] L. Bajaj, M. Takai, R. Ahuja. K. Tang, R. Bagrodia, and M. Gerla, GloMoSim: A scalable network simulation environment, *Technical Report*, No. 990027 (Computer Science Dept., UCLA, 1999).

[34] M. Frodigh, P. Johansson, and P. Larsson, Wireless ad hoc networking — The art of networking without a network, *Ericsson Review*, 4 (2000), pp. 248–263.

[35] J.-C. Cano and P. Manzoni, A performance comparison of energy consumption for mobile ad hoc network routing protocols. In *Proc. of Int. Symp. on Modeling, Analysis and Simulation of Computer and Telecomm. Systems (MASCOTS 2000)* (2000), pp. 57–64.

[36] J.-H. Chang and L. Tassiulas, Energy conserving routing in wireless ad-hoc networks. In *Proc. of the IEEE Infocom 2000* (2000), pp. 22–31.

[37] S. Singh, M. Woo, and C. S. Raghavendra, Power-aware routing in mobile ad hoc networks. In *Proc. of the Int. Conf. On Mobile Computing and Networking (MobiCom '98)* (1998), pp. 181–190.

[38] M. Stemm and R. H. Katz, Measuring and reducing energy consumption of network interfaces in hand-held devices. In *Proc. of Int. Workshop on Mobile Multimedia Communications (MOMUC-3)*, (1996).

[39] L. M. Feeney, Investigating the energy consumption of a wireless network interface in an ad hoc networking environment. In *Proc. of the IEEE Infocom 2001*, Vol. 3 (2001), 1548–1557.

Chapter 11

# SPLAST: A Novel Approach for Multicasting in Mobile Wireless Ad Hoc Networks

Yehuda Ben-Shimol
*Communication Systems Engineering Department*
*Ben-Gurion University of the Negev, Beer-Sheva 84105, P.O.B. 653, Israel*
E-mail: benshimo@bgu.ac.il

Amit Dvir
*Communication Systems Engineering Department*
*Ben-Gurion University of the Negev, Beer-Sheva 84105, P.O.B. 653, Israel*
E-mail: azdvir@bgu.ac.il

Michael Segal
*Communication Systems Engineering Department*
*Ben-Gurion University of the Negev, Beer-Sheva 84105, P.O.B. 653, Israel*
E-mail: segal@bgu.ac.il

## 1 Introduction

A Mobile Ad hoc NETwork (MANET) is a network architecture that can be rapidly deployed without relying on preexisting fixed network infrastructure [1]. Wireless communication is used to deliver information between nodes, which may be mobile and rapidly change the network topology. The wireless connections between the nodes (which later are referred to as links or edges) may suffer from frequent failures and recoveries due to the motion of the

M.X. Cheng, D. Li (eds.) *Advances in Wireless Ad Hoc and Sensor Networks.* Signals and Communication Technology, doi: 10.1007/978-0-387-68567-0_11.

nodes and due to additional problems related to the propagation channels (e.g., obstructions, noise) or power limitations.

Group communication is the basis for numerous applications in which a single source delivers concurrently identical information to multiple destinations. This is usually obtained with efficient management of network topology in the form of a tree having specific properties. For example, multicast routing refers to the construction of a spanning tree rooted at the source and spanning all destinations [2–5]. Delivering the information only through edges that belong to the tree generates an efficient form of group communication which uses the smallest possible amount of network resources. In contrast, with unicast routing from the source to each destination, one needs to find a path from the source to each destination and generate an inefficient form of group communication where the same information is carried multiple times on the same network edges and the communication load on the intermediate nodes may significantly increase.

Generally, there are two well-known basic approaches to construct multicast trees: the Minimal Steiner Tree (SMT) and the Shortest Path Tree (SPT). The Steiner tree (or group-shared tree) tends to minimize the total cost of a tree spanning all group nodes with possibly additional nongroup member nodes. The construction of the SMT is known to be a NP-hard problem [6,7]. Some heuristics that offer efficient solutions to this problem are given in [8–10]. To the best of our knowledge the best solution was derived by [11].

SPT tends to minimize the cost of each path from the source to each destination. This can be achieved in polynomial time by using one of the two well-known algorithms by Bellman [12] or Dossey et al. [13]. The goal of a SPT is to preserve the minimal distances from the root to the nodes without any attempt to minimize the total cost of the tree.

The problem considered in this chapter is a distributed construction and maintenance of a good multicast tree with properties that are suited for mobile ad hoc networks. This tree is required to balance between the minimization properties of the SMT and the SPT by defining two constraints. The first one states that the cost of each path from the source to any terminal in the multicast tree does not exceed a given constant factor $\alpha$ from the corresponding shortest-path cost in the original graph. The second constraint states that the total cost of the multicast tree does not exceed a given constant factor $\beta$ from the total cost of the Minimum Spanning Tree (MST) with truncating all nongroup members of degree one in a recursive fashion. The truncated tree is known to be the Minimum Spanning Tree Heuristic

(MSTH) of the SMT problem.

The proposed algorithm is based on a combination of the Light Approximate Shortest-Path Tree (LAST) algorithm given by Khuller et al. [14] and on the concept of spider spanning graphs explained in [15]. Our novel approach utilizes these two concepts with additional ideas in order to construct and maintain a new multicast tree (which we call SPLAST) under node mobility.

This chapter is organized as follows. Section 2 presents essential definitions and key terms and cites related work. Section 3 presents the building blocks of the SPLAST algorithm. Section 4 presents a centralized static solution that is extended to a distributed solution in Section 5. Section 6 presents and analyzes a new comprehensive SPLAST solution for wireless mobile scenarios. Average-case properties of SPLAST are explored by thorough simulations in Section 7. Finally, conclusions and recommendations are given in Section 8.

## 2 Definitions and Key Terms

In this section we briefly introduce all the definitions and theorems that are used in the rest of the chapter and are important for the understanding of the present problem and its solution. We use notations and terms of graph theory and refer to other research and results that are relevant to our work.

A *communication network* is usually defined as an undirected, connected weighted graph $G(V, E)$ where $V$ is the set of $n$ nodes and $E$ is the set of undirected edges of cardinality $m$. Each edge $e(u, v) \in E$ connects two nodes $u, v \in V$. Every edge $e(u, v) \in E$ is assigned a nonnegative real value (cost) $c(e)$. A sequence of edges (path) that connects two nodes $u, v \in V$ is represented by $P(u, v)$ with total path cost $|P(u, v)| = \sum_{e_i \in P(u,v)} c(e_i)$. The cost of a spanning tree $T$ of $G$ is defined as $|T| = \sum_{e_i \in T} c(e_i)$. *A Minimum Spanning Tree (MST(G))* of graph $G$ is defined as a tree spanning all nodes in the graph with a minimum total cost [16]. Let $s \in V$ be the source node of the graph (named *root* ) and let $M \subseteq V$ be a subset of nodes that are called *terminals* (or multicast group). *Multicast tree* refers to any tree spanning the root, all multicast members, and possibly additional nonmulticast members of degree larger then one that serve as intermediate nodes.

The *Steiner Minimal Tree* (SMT(G)) is the multicast tree with the minimal total cost. One difference between the *SMT* and the multicast tree is the special communications role of the source node in the multicast tree.

Usually one would limit the distance between the source and multicast members in the multicast tree, a property that is not considered in the construction process of the Steiner tree. Therefore, in addition to the computational problems of constructing the SMT, the worst-case end-to-end path length of a SMT is not bounded ([6,7]) and it may be as long as the longest path within the graph.

Let $d_G(v, u)$ be the minimal path cost from node $v \in V$ to node $u \in V$ in a given graph $G$ and let $d_G(v)$ be the minimal path cost from node $v \in V$ to the root $s \in V$ in $G$. A *Shortest-Path Tree (SPT(G))* [12] $T_{sp}$ of graph $G$ with source $s \in V$ is defined as a tree spanning all nodes in the graph with a minimum path cost to the root such that for each node $v \in V$ in $T_{sp} : d_{T_{sp}}(v) = d_G(v)$ and the total cost of $T_{sp}$ is not constrained. A *Short-Path-Tree* $T_s$ of graph $G$ with source $S \in V$ is defined as a tree spanning all nodes in $G$ with $d_{T_{sp}}(v)/d_{T_s} \leq \delta$.

Several researchers have been working on multicast tree problem. We briefly discuss some previous work here.

Kortsarz and Peleg [17] considered a *d-MST* problem, which finds a minimum-weight spanning tree of a given subset of the vertex set, with diameter no more than $d$. Khuller et al.[14] construct a *Light Approximate Shortest Paths Tree* ($LAST(G)$) $T$ which is a spanning tree of $G$ rooted at $s$. For $\alpha > 1$ and $\beta \geq 1$ $T$ is called an $(\alpha, \beta)$-$LAST$ rooted at $s$ if: (a) for each vertex $v$, the distance between $s$ and $v$ in $T$ is at most $\alpha$ times the shortest distance from $s$ to $v$ in $G$; that is, $d_T(v)/d_G(v) \leq \alpha$; and (b) the total weight of $T$ is at most $\beta$ times the weight of a minimum spanning tree of $G$ (i.e., $|T|/|MST(G)| \leq \beta$).

Wu et al. [18] construct a *Light Approximate Routing cost spanning Tree* ($LART(G)$), which is at most a constant factor larger than the MST(G). In addition the path cost to any vertex $u$ from the source $s$ is not larger than a constant factor of the routing distance of $G$ which is defined as $c(G) = \sum \alpha_{ij} \cdot d_G(i, j)$, where $\alpha_{ij}$ is the requirement between nodes $i$ and $j$. Only the special case where all the requirements are set to one was discussed in [18].

We formulate the problem of multicast routing as follows. Given an undirected, simple, weighted graph $G(V, E)$, a group of terminals $M \subseteq V$ and a multicast root $s \in M$, find and maintain in a distributed fashion a multicast tree $T'(V', E')$, $T' \subseteq G$ and $V' \subseteq V$, $E' \subseteq E$ such that $T'$ spans all the nodes in $M$ and satisfies the following conditions: (a) the path length in the multicast tree between the source and each node from the multicast group is as small as possible; and (b) the total weight of the multicast tree is also as small as possible. This property should be preserved efficiently under dynamical changes in edge weights. The length of unicast routing paths between the root and any other node from the multicast group is minimized in $T'$ (under the other constraint given above). Therefore, our multicast tree efficiently preserves energy for both unicast and multicast transmissions.

In practice the weight of an edge $e(u, v)$ is defined as a function of the power transmission level of the nodes $u$ and $v$ that is required to establish a connection between them. Proximity changes between nodes may lead to an increase/decrease in transmission power which is then treated as a change in the edge's weight.

# 3 Fundamental SPLAST Techniques

We use several techniques and algorithms in order to accomplish our task of building the multicast tree. First we build a tree that is at most larger by a constant factor than the MST(G), with the length of the path to any vertex $u$ from the source $s$ not larger by a constant from $d(u)$. Next, we decompose the obtained tree into smaller parts (which we call *spiders*) and connect them in an efficient way. In what follows we describe each step of our algorithm in more detail.

## 3.1 Light Approximate Shortest Paths Tree (LAST)

One of the key components of our proposed algorithm is the light approximate shortest-paths tree algorithm [14,19]. The results given in [14,19] show that a single tree or graph can balance between the minimizations of both MST and SPT trees. The LAST algorithm consists of several steps: (a) build the MST(G) and the SPT(G). (b) Perform a Depth First Search (DFS(G,s)) walk on graph $G$ starting with the root $s$. During this walk construct another graph $H$ that holds the required solution and is set initially to MST(G). The DFS walk ensures that the distance from each node $v$ to the root $s$ is compared against $\alpha \cdot d(v, s)$. If this distance is higher than $\alpha \cdot d(v, s)$, edges from $SPT(G)$ are added to $H$. (c) At the last step the minimal spanning tree of $H$ is calculated and gives the $(\alpha, \beta)$-**LAST** approximation T.

**Theorem 1 [14].** *Let $G$ be a graph with $n$ nodes and $m$ edges of nonnegative edge weights. Let $s$ be a vertex of $G$ and $\alpha, \beta \geq 1$ and $\beta = 1 + 2/(\alpha - 1)$.*

*Then $G$ contains an $(\alpha, \beta)$ – LAST tree rooted at $s$. LAST can be computed in linear time given a MST(G) and a SPT(G), and in $O(m + n \log n)$ time otherwise.*

### 3.2 Spider Decomposition

***Definition* 1 [15].** A *spider* is a tree with at most one node of degree greater than two. A *center* of a spider is a node from which there are edge disjoint paths to the leaves of the spider. A *foot* of a spider is a leaf, or, if the spider has at least three leaves, the spider's center.

Note that if a spider has at least three leaves, its center is unique and every spider contains disjoint paths from its center to all its leaves. A *nontrivial spider* is a spider with at least two leaves.

Let $G$ be a graph, and $M$ be a subset of its nodes. A *spider decomposition* of $M$ in $G$ is a set of node disjoint nontrivial spiders in $G$ such that the union of the feet of the spiders in the decomposition contains $M$.

A spider decomposition of $M$ in $G$ may be found as follows [15]. Let $T$ be any rooted spanning tree of $G$. The depth of a node in $T$ is defined as the distance to the node from the root. Choose a node $v$ of maximum depth in the tree such that the subtree rooted at $v$ contains at least two nodes in $M$. Ties are broken arbitrarily. By choice of $v$, all the paths from the nodes in $M$ to the node $v$ in the subtree of $v$ are node-disjoint, and together with $v$ form a nontrivial spider centered at $v$. We now delete the subtree rooted at $v$ from the tree (or the graph). If no node in $M$ remains in the tree, we are done. If the tree contains two or more nodes in $M$, then we can find a spider decomposition of these nodes recursively. Otherwise, there is exactly one node in $M$ remaining in the tree. In this case, we add the path in the tree from this node in $M$ to the spider centered at $v$. This leaves a spider centered at $v$ and we are done.

## 4 The Static Algorithm

In this section our spider-based LAST (SPLAST) multicast tree is presented.

The construction of the static SPLAST algorithm is given in Algorithm 1. Notice that we use a slightly modified LAST algorithm in step 1; we don't iterate through all nodes of $G$, but rather from one described in [14,19] only through the nodes of $M$.

**Algorithm 1: SPLAST Construction**

| |
|---|
| ***Input:*** *Graph $G(V, E)$, set of terminals $M \subseteq V$, source node $s \in V$ and two real numbers $\alpha$ and $\beta$ such that $\alpha > 1$ and $\beta \geq 1 + 2/(\alpha - 1)$.* |
| ***Output:*** *SPLAST Graph $H'(\alpha, \beta)$, SPLAST Tree $T'(\alpha, \beta)$.* |
| **Step 1:** *Run an $(\alpha, \beta)$–LAST algorithm on $G$ (with our minor change). Produce graph $H(\alpha, \beta)$ and tree $T(\alpha, \beta)$.* |
| **Step 2:** *Find a spider decomposition $D$ of $M$ in $T(\alpha, \beta)$* |
| **Step 3:** Connect the spiders from D with the source $s$ based on the paths of $T$. *Produce tree $T'$.* |
| **Step 4:** *Delete from graph $H$ all the nodes that do not belong to $R$, where $R$ is the set of nodes of $T'$. Produce Graph $H'$* |

**Theorem 2.** *For a given graph $G$, a multicast group $M$, root $s$, $\alpha \geq 1$, and $\beta \geq 1 + 2/(\alpha - 1)$ the algorithm correctly produces subgraph $H'(\alpha, \beta)$ and tree $T'(\alpha, \beta)$ such that: for node $v$, the distance between $s$ and $v$ in $T'(\alpha, \beta)$ and in $H'(\alpha, \beta)$ is at most $\alpha$ times the shortest distance from $s$ to $v$ in $G$; the total weight of $T'(\alpha, \beta)$ and the total weight of $H'(\alpha, \beta)$ is at most $\beta$ times the weight of a minimum spanning tree of $G'$ where $G'$ is the subgraph of $G$ only with the nodes that belongs to $R$.*

*Proof.* The *LAST* algorithm [14,19] and step 3 in Algorithm 1 ensure that the desired $\alpha$ factor is achieved. Let $T$ be any spanning tree of $G$ with a root $s$ and let $z_1, z_2, z_3, \ldots, z_k$ be any $k$ nodes of $T$. From [19] it is known that

$$\sum_{i=1 \ldots k} d_T(z_{i-1}, z_i) \leq 2 \cdot c(T). \tag{1}$$

Let $T'$ be the MST of $H'$. Every node that belongs to $T'_m$ also belongs to minimum spanning tree $T_m$ of $G$, so the path $p(z_{i-1}, z_i)$ between $z_{i-1}$ to $z_i$ in $T'_m$ $(z_{i-1}, z_i \in T'_m)$ is the same in $T_m$ as well. Theorem 2 [19] proves that $\sum_{i=1 \ldots k} (\alpha - 1) \cdot d(z_i) \leq d_{T_m}(z_{i-1}, z_i) = d_{T'_m}(z_{i-1}, z_i)$. Therefore, from *Equation* (1) it follows that $\sum_{1 \ldots k} d(z_i) \leq 2 \cdot c(T'_m)/(\alpha - 1)$ and the total cost of $H'$ is $|H'| = |T'_m| + \sum_{1 \ldots k} d(z_i) \leq |T'_m| \cdot (1 + 2/(\alpha - 1)$ From [14] we can see that if $H'$ is $(\alpha, \beta)$ LAST so $T'$ is $(\alpha, \beta)$ LAST too.

# 5 The Distributed Algorithm

Notice that the SPLAST algorithm can be viewed as a combination of several algorithms, for example MST, SPT, and DFS. First, we briefly explain the distributed version of the above algorithms. Then we show how to combine them into a unified distributed SPLAST algorithm.

### 5.1 Distributed LAST Algorithm

We begin our discussion with the distributed LAST algorithm. At the first stage we use the algorithm proposed in [20,21] that builds an MST (named $T_m$) in a distributed fashion (we assume that all nodes have unique IDs). Next we proceed to compute SPT by running an improved distributed Bellman–Ford algorithm that is presented in [22]. All these operations can be done in $O(n^2)$ time and messages in the worst case. In what follows, we use the distributed DFS algorithm given in [23,24] in order to produce the DFS walk. We note that the output of each of the algorithms (MST, SPT and DFS) is produced in a distributed fashion; that is, each node knows its neighbors in the corresponding tree. In addition, after MST construction each node $v$ knows its distance from the root $s$, $d_{T_m}(v,\ s)$, and after the SPT computation each node $v$ knows its $d(v)$. This can be accomplished by a simple broadcast process from root $s$ towards the leaves in the corresponding tree. In Algorithm 2 we present a pseudo-code for distributed LAST followed by additional explanations.

**Algorithm 2: Distributed $(\alpha,\ \beta)$ Light Approximate Shortest-Paths Tree**

***Input:*** *Weighted graph $G(V, E)$, root $s$ and $\alpha > 1, \beta \geq 1 + 2/(\alpha - 1)$.*
***Output:*** *Graph $H(\alpha,\ \beta), T(\alpha,\ \beta)$-LAST.*
**Step 1**: *Find a MST$(G)$ $T_m$ by employing distributed algorithm [20,21];*
*Find a SPT$(G)$ $T_{sp}$ by employing distributed algorithm [22] with start node $s$;*
**Step 2:** *Find a preorder numbering of $T_m$ using $s$ as the start node by employing the distributed DFS algorithm [23,24];*
**Step 3:** $H = T_m$;
*For each node $v \in M$ in the preorder sequence of $T_m$ do*
*Find a shortest $s - v$ path $P$ in $H$;*
*If $c(P) > \alpha \cdot d(v)$*
*Then add all the edges in a shortest $s$–$v$ path in $G$ to $H$;*
*Update $d_H(v,\ s)$ and send to next node in the walk;*
*End; (if)*
*End; (for)*
**Step 4:** *Find a SPT of $H$ with start node $s$ by the employing the distributed algorithm in [22]. Produce $T$.*

At the beginning of Algorithm 2 each node is aware of its distance from the root $s$ in the MST $T_m$ and the distance from $s$ in the SPT $T_{sp}$. We construct the required graph $H$ when initially $H = T_m$. We perform a distributed DFS walk on $H$ [23,24]. When node $v$ has been activated during this walk, $v$ checks whether it belongs to the multicast group. If not, $v$ sends its distance $d_H(v, s)$ to the next node on the walk. If node $v$ is a member of the multicast group, $v$ checks if $d_H(v, s) > \alpha \cdot d(v)$. If so, $v$ sends a "*connect*" message to its parent $u_1$ in $T_{sp}$ (shortest-path tree) and adds the edge $(v, u_1)$ to $H$. Node $u_1$ sends the same message to its parent $u_2$ in $T_{sp}$ and adds the edge $(u_1, u_2)$ in $H$. This process propagates upwards towards the root $s$ and stops after the "*connect*" message reaches $s$. In parallel, $v$ sends its new distance from the root in $H$ to the next node in the DFS walk. Each node $w$ upon receiving a message with the updated distance $d_H(v, s)$ from its neighbor, updates $d_H(w, s)$ if necessary. Note that every node may participate in the DFS walk several times.

## 5.2 Distributed Spider Algorithm

Here we explain the distributed spider decomposition algorithm. The algorithm works as follows: every leaf in the LAST-T initiates the process by sending to its parent in $T$ a message *spider(k, prune)*, where $k$ plays a role of an accumulator: initially $k = 1$ if node $v \in M$ or $k = 0$ otherwise, and *prune* is a Boolean variable: initially *prune* $= 0$ if $v \in M$ or *prune* $= 1$ otherwise. When a node $w$ receives messages from all its children in $T$ it sums the total value of $k$ values derived from the received messages plus its own $k$ value, obtaining $k_w$. If the sum $k_w$ is equal to or greater than 2, $w$ declares itself as a center of a spider and sends a message *spider(0, 0)* to its parent $u$ in $T$. Otherwise, it sends a message *spider*$(k_w, k_w \oplus 1)$. Each node $w$ removes the children that sent the message *spider*(0,1) from its neighborhood list. If a node $z$ receives a message *spider*(1/0 , 0) from one of its children it must propagate this message to its parent with "0" in the *prune* field. The algorithm terminates when the root receives messages from all its children and removes nodes if needed. After the algorithm terminates each center is connected to the root and only the relevant paths connecting terminals to the root $s$ remain in the tree.

## 5.3 Distributed SPLAST Algorithm

The steps of the distributed SPLAST algorithm are based on the corresponding steps of the static SPLAST algorithm. Each step follows a distributed approach analogous to the corresponding centralized approach as presented in Algorithm 1.

# 6 Mobile SPLAST Algorithm

Our main challenge is to maintain the resulting SPLAST under dynamic changes of the network. We assume that each node has limited battery power, and therefore a required increase in transmission power that cannot be supported by the battery corresponds to edge failures. Similarly edge recoveries may be related to a decrease in the required transmission power which now can be supported by the remaining energy of the battery. Recoveries and failures of network edges may also happen due to obstructions and other phenomena related to physical properties of the radio channels. Weight modifications are functions of changes in the required transmission power that can be supported by the battery. In the following we describe how to deal with each one of these updates and how to adjust the underlying algorithms (i.e., MST, DFS, and SPT) to fit mobile scenarios.

## 6.1 Maintaining the Minimum Spanning Tree

A primary requirement from a distributed MST algorithm for mobile networks is a quick response to topological changes that are caused by failures, recoveries, and weight modification of the edges. Without loss of generality, we assume that at any given time only one topological change may occur. The solution presented in [25] is suitable for cases of edge failures and recoveries but not for modifications of edge weights. We modify that algorithm in order to provide a comprehensive solution for this case as well. The idea is that each node periodically checks the weight of its adjacent edges. If the weight of any edge in the MST $T_m$ is increasing or the weight of some nontree edge is decreasing, both endpoint nodes of this edge will update their neighborhood list by deleting this edge with its old weight and follow the failure delete algorithm presented in [25]. Afterwards these nodes insert the delete edge with modified weight by updating their neighborhood list and follow the recovery algorithm of [25]. This technique provides a way to maintain a distributed mobile MST.

The algorithm uses $O(n+m)$ messages. In contrast, if one uses an algorithm (such as GHS [20]) to reconstruct the tree after every failure or recovery, the message complexity changes to $O(m+n\log n)$.

## 6.2 Maintaining the Shortest-Path Tree

Changes in the network topology trigger the execution of the algorithm that produces an updated Short-Path-Tree where $\delta$ equals 3. In the proposed algorithm we utilize the observation, given in [26].

**Observation 1.** When an edge $e(v, u)$ fails, a new SPT may be found by replacing the failed edge with a new edge of graph $G$, thus producing a new tree with path lengths up to three times longer than the optimal.

### 6.2.1 Edge Failure

When an edge $e(v, u) \in T_{sp}$ fails, $T_{sp}$ is split into two disjoint subtrees $S_u$ and $S_v$ such that, without loss of generality $S_u$ contains root $s$, and node $u$, and $S_v$ contains node $v$ where $d(u) \leq d(v)$. First, each node in the graph should be informed about the failed node and mark itself to the appropriate subtree. In order to deal with this case, node $u$ sends to root $s$ a $failed(e(u, v), up)$ message regarding the failed edge. When root $s$ receives $failed(:, :)$ message it broadcasts to all nodes in $S_u$ a $failed(e(u, v), down)$ message.

Each node in $S_u$ that receives this message marks itself as part of $S_u$. Node $v$ broadcasts to all nodes in $S_v$ a $failed(e(v, u), up)$ message. Each node in $S_v$ that receives this message marks itself as part of the $S_v$ tree and sends a *search* message on its outgoing non-SPT edges. Each node in $S_v$ that receives a *search* message replies with an *ignore* message on the same edge.

Each node $w$ in $S_u$ that receives a *search* message via edge $e(x, w)$, $x \in S_v$ sends back on the same edge a message $distance(value)$, where $value$ is equal to $d(w) + c(x, w)$ such that e(x, w) connects between $S_u$ and $S_v$. If a leaf $z$ in an $S_v$ tree receives $distance(value)$ messages from all its non-SPT edges, it chooses the minimum $value(min\_val)$ and sends a *minimum_length(min_val, z)* to its parent in the $S_v$ tree. Every node $k$ in $S_v$ that receives a *minimum_length(min_val, z)* message from all its children in $S_v$ and also receives a message $distance(value)$ from all its non-SPT edges sends to its parent a message $minimum\_length(min\_val, d)$ where $min\_val$ is the minimum between all $min\_val$ and $value$ that node $k$ receives and $d$ is the node that sent this value. This process ends when node $v$ receives *minimum_length(min_val, d)* messages from all its children in $S_v$ and $distance(value)$ messages from its non-SPT edges between $S_u$ and $S_v$.

Afterwards, node $v$ chooses the minimum between all *min_val* and *value* that it receives and sends a $connect(d)$ message to node $d$ that corresponds to this minimum distance. When node $d$ receives the $connect(d)$ message from $v$ it sends a *connect_trees* message to the node in $S_u$ thus reconnecting the two trees $S_u$ and $S_v$. Node $d$ broadcasts the new distance from the root to all nodes in $S_v$ in order to provide a consistent update of distance from nodes of $S_v$ to root $s$.

### 6.2.2 Edge Recovery

When an edge $e(u, v)$ recovers, a cycle appears in the SPT $T_{sp}$. The proposed LCA mechanism from [25] enables finding the heaviest edge $e(x, y)$ in the cycle in a distributed fashion. After we isolate the heaviest edge $e(x, y)$, we delete it from the tree and start the process of a failed edge as explained above. Notice that we are using the recovered edge $e(u, v)$ in the failed process as a non-SPT edge.

### 6.2.3 Edge Modification

In the case where edge $e(u, v)$ changes its weight, notice that we determine the type of $e(u, v)$ before we perform any update. The changes we deal with are a non-SPT edge that decreased its weight or an edge in the SPT that changed its weight (notice that in all cases $d(u) \leq d(v)$). In all cases we follow the same procedure below. First we check whether the new edge improves the path length from $v$ to the root $s$. If not, this process is terminated, otherwise three cases may occur:

*Case 1:* A non-SPT $e(u, v)$ edge weight decreased. Node $v$ sends a *search* message to node $u$ to check whether the new weight improves its path length. Node $u$ receives via $e(u, v)$ the *search* message and replies to $v$ with a *distance*(*value*) where $value = d(u) + c(u, v)$. If *value* improved $d(v)$, node $v$ switches between the improved edge $e(u, v)$ and the edge that connected node $v$ to its parent in the SPT.

*Case 2:* An edge weight increase in $e(u, v) \in$ SPT. Node $v$ deletes the edge $e(u, v)$ with the old weight and starts the failed edge process for this edge.

*Case 3:* An edge weight decrease in $e(u, v) \in$ SPT. Node $v$ broadcasts a message $new\_dist(d(v))$ to all its children in the SPT regarding the new distance from the root.

## 6.3 Maintaining the Depth-First-Search Tree

The main idea of this algorithm relies on the distributed approach of constructing the DFS tree [23].

### 6.3.1 Edge Failure

When edge $e(u,v)$ fails, the original DFS tree is split into two disjoint subtrees $S_u$ and $S_v$ such that, without loss of generality $S_u$ contains root $s$, and node $u$, and $S_v$ contains node $v$ where in the DFS walk starting from $s$, $u$ is reached before v. Each node in the graph should be informed about the failed node and mark itself to the appropriate subtree. This task can be accomplished by simple broadcast from each one of the roots of $S_u$ and $S_v$ where $v$ is the root of $S_v$ and knows about the failure and $s$ is the root of $S_u$ and is not aware of the failure. Therefore node $u$ should inform $s$. Before $s$ issues an instruction to $u$ to start reconnecting $S_u$ and $S_v$ we perform two convergecast processes (one in $S_u$ and one in $S_v$) in order to guarantee that each node in the converged subtrees has marked itself as belonging to the correct subtree. Afterward, root $s$ starts the connection processes by choosing node $w$ in $S_u$ with the smallest ID and then sends it a *connect* message. When node $w$ receives this message it tries to connect $S_u$ with $S_v$ with its outgoing non-DFS tree edges. In the case of success this process terminates; otherwise, node $w$ sends to the root an *unsuccessful* message. When root $s$ receives this message it picks the next smallest node in $S_u$ and sends it a *connect* message. This node in turn tries to connect in the same way as $w$ did. This process continues until one of the nodes is successful in connecting the two subtrees.

### 6.3.2 Edge Recovery

When edge $e(u,v)$ recovers from a failure, node $u$ sends to its parent in the DFS tree a message *failed*$(u)$. Node $v$ sends to its parent in the DFS tree a *failed*$(v)$ message. Each node $w$ that receives one of these messages appends itself to the message and passes to its parent *failed*$(v,w)$ or *failed*$(u,w)$, accordingly. When the root $s$ receives both messages it compares them. If they are not equal it means that $u$ and $v$ are not from the same branch of the DFS tree, therefore the root deletes the edge that connects the root to the first node in the $v$ branch (the last node in $v$ message). If some nodes in the two sequences are equal, it means that $u$ and $v$ are from the same branch of the DFS tree, therefore the root sends a *delete*$(z)$ message where

$z$ is the first node that is not equal in the sequence (the first node in the sequence of $u$ that is not contained in the $v$ sequence when searching from right to left). When $z$ receives the *delete* message it deletes the edge that connects it to $v$. If node $u$ receives a message from $v$, it means that $u$ and $v$ are from the same branch of the DFS tree and the path from $v$ to the root pass in $u$ so the new edge does not change the DFS tree.

### 6.4 Maintaining SPLAST

The distributed SPLAST algorithm has four steps. The first two steps of the algorithm were explained in Section 6.1 and 6.2. The rest of the algorithm is essential for the maintenance of SPLAST, when the network experiences topology changes. Whenever the tree is updated, the $\alpha$ factor of the corresponding tree is deterioratedy. Therefore, after a number of changes in the SPT we may need to reconstruct a new SPT from scratch.

## 7 Simulation Results

The theoretical bounds given before provide information on the worst-case properties of the SPLAST tree and motivate its usefulness for multicast applications in wireless ad hoc networks. However, average properties are also important from a practical networking point of view. The average properties of SPLAST trees were investigated with thorough simulation experiments. For many given graphs with a various number of nodes and different topologies we compared the efficiency of the SPLAST tree against the minimum spanning tree, the shortest-path tree, the minimal Steiner tree, and one heuristic of the Steiner tree constructed with the shortest-path heuristic (SPH) [6] which is known to have good average results.

The $\alpha$ parameter measures the effectiveness of each tree with relation to the shortest path of each multicast destination to the source, and the $\beta$ parameter measures the effectiveness with relation to the total tree cost. As was mentioned previously, the SPLAST algorithm tries to balance between these two requirements.

Simulation details are as follows. Each point on each graph represents the average parameter of several experiments on random graphs with a fixed number of nodes. The total number of nodes varies from 4 to 20 and the size of multicast group is one half of the total number of nodes.

Figure 1 shows the properties of the $\alpha$ parameter. In order to compare the results for a different number of nodes, the distance from the source to

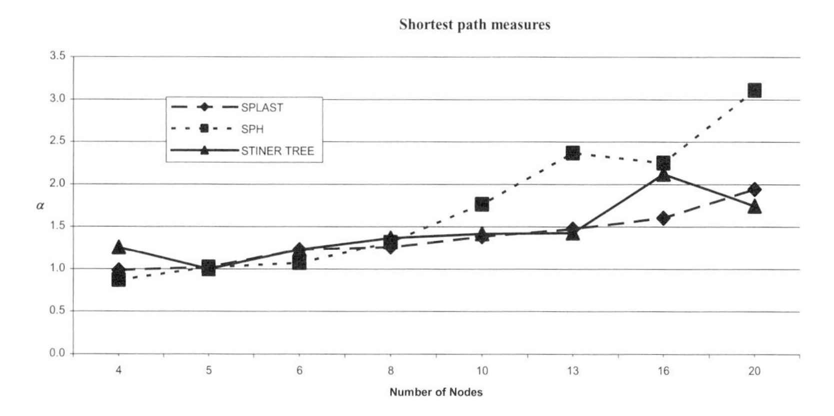


Figure 1: Comparison among the shortest distance properties of the SPLAST, SMT, and SPH trees.

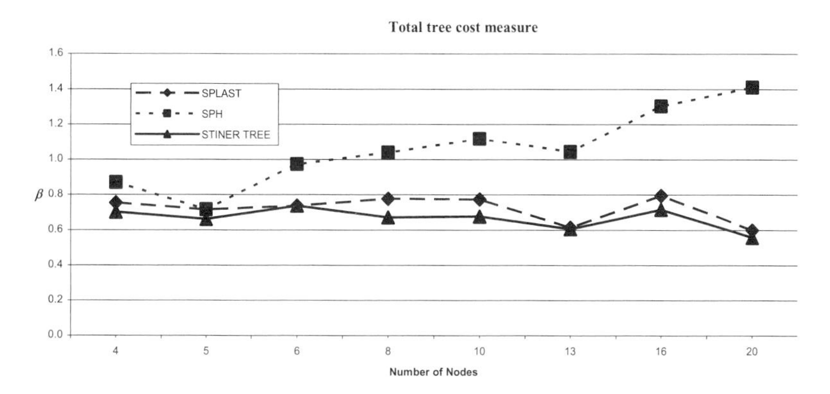


Figure 2: Comparison among the total tree cost property of the SPLAST, SMT, and SPH trees.

each multicast destination was normalized by the shortest distance obtained from the SPT. The SPLAST tree shows good performance compared with the SMT and the SPH trees.

Figure 2 shows the properties of the $\beta$ parameter. In order to compare the results for a different number of nodes, the distance from the source to each multicast destination was normalized by the total weight of the MST. Again the SPLAST tree shows good performance on the average compared against the SMT and the SPH trees.

## 8 Conclusion

In this chapter we presented a novel approach for the construction of multicast tree in a wireless ad hoc network: the SPLAST algorithm which combines the LAST algorithm and spider decomposition. SPLAST trees are proved to keep all the properties of the LAST trees expressed by the $\alpha$ and $\beta$ parameters while offering a complete solution for multicast applications in wireless ad hoc networks.

The analysis of the SPLAST algorithm started from a discussion of its underlying components: the LAST algorithms, shortest-path and minimal spanning trees, and spider decomposition.

A centralized static solution was presented and later extended to a distributed implementation, which is crucial for communication networks. The applicability of SPLAST to wireless ad hoc networks must be tested with various conditions that are unique for this type of network: rapid topology changes caused by edge failures, edge recovery, and changes to edge weights. We developed a detailed solution to wireless ad hoc scenarios, and analyzed the overhead of maintaining its various components.

Supported by good worst-case theoretical bounds and good average results explored by thorough simulations, the SPLAST algorithm suggests a good balance between the minimum spanning tree and shortest-path tree with low complexity in computation time and messages. This suggests that SPLAST trees may be used as an uniform algorithm for both unicast and multicast routing in wireless ad hoc networks.

## References

[1] Z.J. Haas, J. Deng, P. Papadimitratos, S. Sajama. Wireless ad hoc networks. In *Wiley Encyclopedia of Telecommunications*, John G. Proakis, Ed. Wiley, 2002.

[2] S. Deering, *Scalable multicast routing protocol*, Ph.D. dissertation, Stanford University, 1989.

[3] B. Awerbuch, Y. Azar. Competitive multicast routing, *Wireless Networks*, 1995; 1: 107–114.

[4] J. Nonnenmacher, M. Lacher, M. Jung, G. Carl, EW Biersack. How bad is reliable multicast without local recovery? In *Proceedings IEEE Infocomm 98*, San Francisco, CA, April 1998; 972–979.

[5] S. Paul, K. Sabnani, J. Lin, S. Bhattacharyya. Reliable multicast transport protocol (RMTP). *IEEE J. Selected Area Comm.*, 1997; **15**(3); 407–421.

[6] F.K. Hwang, D.S. Richards, P. Winter. *The Steiner Tree Problem*, North-Holland, 1992; 93–202.

[7] M.R. Garey, R.L Graham, D.S. Johnson. The complexity of computing Steiner minimal trees, *SIAM J. Appl. Math.*, 1977; 32: 835–859.

[8] P. Winter. Steiner problem in networks: A survey, *Networks*, 1987, **17**(2): 129–167.

[9] F. Hwang, D. Richards. Steiner tree problems, *Networks*, 1992; **22**(1), 55–89.

[10] R. Ravi. *Steiner trees and beyond: Approximation algorithms for network design*, Ph.D. Dissertation, Brown University, 1993.

[11] G. Robins, A. Zelikovsky. Improved Steiner tree approximation in graphs. In *Proceedings of the Eleventh Annual ACM-SIAM Symposium on Discrete Algorithms*, 2000; 770–779.

[12] R. Bellman. *Dynamic Programming*, Princeton University Press, 1957.

[13] J. Dossey, A. Otto, L. Spence, C. Eynden. *Discrete Mathematics*, Harper Collins College, 1993.

[14] S. Khuller, B. Raghavachari, N. Young. Balancing minimum spanning and shortest path trees, *Algorithmica*, 1994; **120**(4): 305–321.

[15] P.N. Klein, R. Ravi. A nearly best-possible approximation algorithm for node-weighted Steiner tree, *Algorithms*, 1995; **19**: 104–115.

[16] J.B. Kruskal. On the shortest spanning subtree of a graph and the traveling salesman problem, *Proc. Am. Math. Soc.*, 1956; **7**: 48–50.

[17] G. Kortsarz, D. Peleg. Approximating shallow-light trees. In *Proceedings of the Eighth Symposium on Discrete Algorithms*, 1997; 103–110.

[18] B.Y. Wu, K.M. Chao, C.Y. Tang. Light graphs with small routing cost, *Networks* 2002; **39**(3): 130–138.

[19] J. Cheriyan, R. Ravi. Lecture Notes on Approximation Algorithms for Network Problems, http://www.math.uwaterloo.ca/~jcheriya/PS_files/ln-master.ps, 1998; 78–83.

[20] R.G. Gallager, P.A. Humblet, P.M. Spira. A distributed algorithm for minimum weight spanning trees, *ACM Trans. on Program. Lang. and Systems*, 1983; **5**(1): 66–77.

[21] P.K. Mohapatra, Fully Sequential and Distributed Dynamic Algorithms for Minimum Spanning Trees, http://www.arxiv.org/PS_cache/cs/pdf/0002/0002005.pdf, 2000.

[22] L. Brim, I. Cerna, P. Krcal, R. Pelanek. Distributed shortest path for directed graphs with negative edge lengths, *Technical Report FIMU-RS-2001-01*, Faculty of Informatics, Masaryk University Brno, 2001.

[23] M.B. Sharma, N.K. Mandyam, S.S. Iyangar. An optimal distributed depth-first-search algorithm. In *Proceedings of the Seventeenth Annual ACM Conference on Computer Science: Computing Trends in the 1990s*,1989; 287–294.

[24] D. Kumar, S.S. Iyengar, M.B. Sharma. Corrections to a distributed depth-first search algorithm. *Inform. Process. Lett.*, 1990, **32**(4): 183–186.

[25] C. Cheng, I.A. Cimet, S.P.R. Kumar. Protocol to maintain a minimum spanning tree in a dynamic topology, *Comput. Commun. Rev.*, 1988, **18**(4): 330–338.

[26] E. Nardelli, E. Proietti, P. Widmayer. Swapping a failing edge of a single source shortest paths tree is good and fast, *Algorithmica*, 2003, **35**(1): 56–74.

Chapter 12

# A Simple and Efficient Broadcasting Scheme for Mobile Ad Hoc Networks

Shihong Zou
*State Key Lab of Networking and Switching*
*Beijing University of Posts and Telecommunications, Beijing 100876, China*
E-mail: zoush@bupt.edu.cn

Shiduan Cheng
*State Key Lab of Networking and Switching*
*Beijing University of Posts and Telecommunications, Beijing 100876, China*
E-mail: chsd@bupt.edu.cn

Wendong Wang
*State Key Lab of Networking and Switching*
*Beijing University of Posts and Telecommunications, Beijing 100876, China*
E-mail: wdwang@bupt.edu.cn

## 1 Introduction

Mobile Ad hoc NETworks (MANETs) [1] consist of a collection of mobile hosts without a fixed infrastructure. Due to limited wireless power a host may not communicate with its destination directly. It usually requires other

M.X. Cheng, D. Li (eds.) *Advances in Wireless Ad Hoc and Sensor Networks.*
Signals and Communication Technology, doi: 10.1007/978-0-387-68567-0_12.

hosts to forward its packets to the destination through several hops. So in MANET every host acts as a router when it is forwarding packets for other hosts. Because of mobility of hosts and time variability of the wireless medium, the topology of MANET varies frequently. Therefore the routing protocol plays an important role in MANET. There has been extensive research on routing protocols, such as DSR [2], AODV [3], ZRP [4], LAR [5], and so on. A common feature of these routing protocols is that their route discovery all relies on networkwide broadcasting to find the destination. In addition, networkwide broadcasting is essential in some applications, such as intrusion alert in military fields, abnormality alarm of environmental monitoring, and so on. In a word, networkwide broadcasting is a basic operation in MANET.

The simplest broadcasting scheme is flooding, which is used by most existing routing protocols. It is very costly and often results in a serious broadcast storm [6]. Tseng et al. [6] proposed five schemes to relieve broadcast storms, called probabilistic, counter-based, distanced-based, location-based, and cluster-based schemes. Williams and Camp [8] classify current broadcasting schemes into four families: simple flooding, probability-based methods (including the pure probabilistic scheme and counter-based scheme), area-based methods (including the distance-based scheme and location-based scheme) and neighbor knowledge methods. Both distance-based and location-based schemes need the support of GPS (Global Positioning System) and make the mobile hosts more complex. The neighbor knowledge scheme needs periodic hello messages to maintain neighbors information. It is only suited for static networks or networks with low mobility because it is difficult to get accurate neighbor knowledge in a highly mobile network. In contrast, the counter-based scheme is relatively simple. It does not rely on GPS or a hello message, although its efficiency is much higher than simple flooding. Therefore, the counter-based scheme is best suited for the highly mobile scenario in MANETs. However, the performance of the counter-based scheme proposed in [6] (denoted as CBB) is not satisfying. This chapter proposes the enhanced counter-based broadcasting scheme CBB+. CBB+ makes several improvements on CBB. First, it performs backoff only at the MAC layer, instead of at both the network layer and MAC layer. Second, it introduces priority in rebroadcasting and assigns higher priority to the host located at the radio border of a broadcast. Third, it tunes the threshold of the counter dynamically. Analyses and simulation results show that these approaches improve the performance of the counter-based schemes greatly.

In this chapter we assume that the MAC layer protocol is IEEE 802.11 DCF (Distributed Coordination Function) [10] because it is widely used in MANETs.

The rest of the chapter is organized as follows. Section 2 introduces IEEE 802.11 DCF, the counter-based broadcast scheme and related work. Section 3 proposes the enhanced counter-based broadcast scheme CBB+. Section 4 shows simulation and analysis results. Section 5 concludes the chapter.

## 2 Preliminaries and Related Work

The Distributed Coordination Function (DCF) supports asynchronous data transfer on a best-effort basis. It is the fundamental access method of IEEE 802.11 [10] and frequently used in MANET as the MAC protocol. As specified in the standards, DCF must be supported by all the hosts in a Basic Service Set (BSS). The DCF protocol is based on Carrier Sense Multiple Access with Collision Avoidance (CSMA/CA). In DCF carrier sense is performed both at the physical layer, which is also referred to as physical carrier sensing, and at the MAC layer, which is known as virtual carrier sensing. The goal of collision avoidance is to avoid the case where all the hosts transmit data immediately after the medium has been idle for DIFS and so collision happens. Collision avoidance is implemented by a backoff procedure. DCF demands a host start a backoff procedure right after the host transmits a message, or when a host wants to transmit but the medium is busy and the previous backoff procedure has been ended. To perform a backoff procedure, a counter is first set to an integer picked randomly from its current contention window. When the medium is detected to be idle for a slot (a fixed period), the counter is decreased by one. Only when the counter reaches zero can the host transmit data.

There are two techniques used for packet transmission in DCF. The default one is the two-way handshaking mechanism, also known as the basic access method. A positive MAC acknowledgment is transmitted by the destination host to confirm the successful packet transmission. The optional one is a four-way handshaking mechanism, which uses a Request-To-Send/Clear-To-Send (RTS/CTS) technique to reserve the medium before data transmission. This technique has been introduced to reduce the performance degradation caused by hidden terminals. However, RTS/CTS and ACK cannot be used in broadcasting because the receivers of a broadcast are all the hosts in the coverage. Without RTS/CTS the transmission of a broadcast is more vulnerable to hidden terminals. Without ACK the sender cannot make sure if a broadcast is successfully conducted. It cannot retransmit the lost broadcast packets. Therefore, the collision is especially harmful to the transmission of a broadcast.

The counter-based scheme (CBB) works as follows. When a host receives a new broadcast, its rebroadcast may be blocked due to a busy medium, running backoff procedure, or buffered packet ahead in the queue. Therefore, the host may receive the same message again and again from other rebroadcasting hosts before it starts rebroadcast. To avoid the broadcast storm, each host is required to maintain a counter for every broadcast it received. When it receives a redundant broadcast it adds 1 to the corresponding counter. When the counter reaches a threshold (C) the host cancels its own rebroadcast. Threshold C is suggested to be 3 or 4.

An improvement on the pure probabilistic scheme has been proposed in [9]. Its major idea is to privilege the rebroadcast by hosts that are located at the radio border of the sender. The distance between two hosts can be approximated by comparing their neighbor lists. It requires hosts to periodically send a hello message.

# 3 Proposed CBB+ Scheme

CBB+ improves the performance of CBB through the following measures.

## 3.1 Backoff Only at MAC Layer

DCF demands that each host backoff in order to avoid collision. Now consider the scenario where several neighbor hosts hear a broadcast from host X. If the surrounding medium of X has been quiet for enough of a long time, all of X's neighbors may have passed their backoff procedures. Thus, after hearing the broadcast message, they may all start rebroadcasting at around the same time [6]. To avoid this kind of collision, it is necessary to introduce a Random Assessment Delay (RAD) before rebroadcasting [8]. All the current broadcast schemes implement RAD at the network layer [6–8]. However, when the medium is busy or the broadcast origination rate is too high, the rebroadcast packets still need to backoff again at the MAC layer. These two backoff procedures will increase the probability of collision greatly. The statement is proved as follows.

Because in DCF there are no retransmissions for broadcast, there is no binary exponential backoff for broadcast. All broadcasts are sent with the initial contention window. Let $x1, x2$ be the backoff time (in slot) of two hosts at the network layer, $y1, y2$ be the backoff time (in slot) of two hosts at the MAC layer. And $x1, x2 \in [0, A-1]$, $y1, y2 \in [0, B-1]$, where $B$ is the contention window of the MAC layer; $A$ can be thought of as the contention window of the network layer. It is clear that $x1, x2, y1, y2$ are all integers and all independent of each other. Therefore, the probability of two rebroadcasts colliding in CBB is the probability of $x1 + y1 = x2 + y2$. First we prove the following theorems.

**Theorem 3.1** *The sum of two random integers with uniform distribution does not conform to uniform distribution.*

*Proof.* Let $x, y$ be the random integer with uniform distribution; their distribution functions are as follows:

$$\begin{aligned} p(x=i) &= \tfrac{1}{Q-P+1} \quad , \quad P \leq i \leq Q \\ p(y=j) &= \tfrac{1}{S-R+1} \quad , \quad R \leq j \leq S, \end{aligned}$$

where $P, Q, R, S$ are all integers.

Then we have

$$\begin{aligned} p(x+y=P+R) &= \tfrac{1}{(Q-P+1)(S-R+1)} \\ p(x+y=P+R+1) &= p(x=P)p(y=R+1)+p(x=P+1)p(y=R) \\ &= \tfrac{2}{(Q-P+1)(S-R+1)}. \end{aligned}$$

Hence, the random integer $x + y$ does not conform to uniform distribution. □

**Theorem 3.2** *Given $K$ independent random integers $x_1, \ldots, x_K$ with the same probability distribution, when the distribution is a uniform distribution the probability that at least two of them are equal minimizes.*

*Proof.* Let the distribution function of $x_j, 0 \leq j \leq K$ be

$$p(x_j = i) = p_i \qquad 0 \leq i \leq N \qquad N > K.$$

Let $\omega$ denote the probability that none of them are equal. Then it is left to prove that $\omega$ is maximized when $x_j$ conforms to uniform distribution.

$$\begin{aligned} \omega &= p\{x_1 = a_1, x_2 = a_2, \cdots, x_K = a_K\} \\ &= p(x_1 = a_1)p(x_2 = a_2)\ldots p(x_K = a_K) \\ &= p_{a_1} p_{a_2} \cdots p_{a_K}, \end{aligned}$$

where $a_1 \dots a_K$ are different from each other.

According to the theorem that the geometric mean is less than or equal to the average mean, we can obtain that when $p_{a_1} = p_{a_2} = \cdots = p_{a_K}$, $\omega$ is maximized.

Note that $a_j$ is picked randomly from $[0, N]$, so $\omega$ is maximized when the distribution is a uniform distribution.

Hence the probability that at least two of them are equal minimizes when the distribution is a uniform distribution. □

According to Theorem 3.1 we know that when backoff is conducted at the network layer and MAC layer, respectively, the sum of backoff time does not conform to the uniform distribution. From Theorem 3.2 we obtain that under the same condition, the colliding probability of two-layer backoff is greater than that of backoff only at the MAC layer. The colliding probability of these two methods when there are two hosts is calculated hereinafter.

Without loss of generality, assume $A \geq B$. Let $z1, z2$ be the backoff time when there is backoff only at the MAC layer. In order to compare the colliding probability of the two methods under the same max backoff time, we obtain $z1, z2 \in [0, A + B - 2]$. Moreover, according to the definition of backoff time, $x1, x2, y1, y2, z1, z2$ are all integers in their corresponding ranges and conform to uniform distribution.

So the colliding probability of two hosts when backoff occurs at two layers is $p(x1 + y1 = x2 + y2)$.

$$\begin{aligned} p(x1 + y1 = x2 + y2) &= \sum_{k=0}^{A+B-2} p(x1 + y1 = k) \times p(x2 + y2 = k) \\ &= \sum_{k=0}^{A+B-2} (p(x1 + y1 = k))^2; \end{aligned}$$

$$p(x + y = k) =$$

$$\begin{cases} \sum_{j=0}^{k} p(x = j)p(y = k - j) = \frac{k+1}{AB}, & 0 \leq k \leq A - 1 \\ \sum_{j=k-A+1}^{B-1} p(x = k - j)p(y = j) = \frac{A+B-1-k}{AB}, & A \leq k \leq A + B - 2; \end{cases}$$

$$\begin{aligned} p(x1 + y1 = x2 + y2) &= \sum_{k=0}^{A-1} \left(\frac{k+1}{AB}\right)^2 + \sum_{k=A}^{A+B-2} \left(\frac{A+B-1-k}{AB}\right)^2 \\ &= \frac{(A+B)[(A+B-1)(2A+2B-1)-A(B-1)]}{6(AB)^2}. \end{aligned} \quad (1)$$

The colliding probability of backoff only at the MAC layer is $p(z1 = z2)$.

$$p(z1 = z2) = \frac{1}{A + B - 1}. \qquad (2)$$

When $A, B$'s values are both 32, from (1) and (2) we obtain

$$\begin{cases} p(x1 + y1 = x2 + y2) & = & 0.0713 \\ p(z1 = z2) & = & 0.0159. \end{cases}$$

The former is 4.5 times the latter.

Besides removing the backoff operation at the network layer, CBB+ needs a little modification on DCF to provide a function of canceling some queueing packet at the MAC layer at the request of the network layer. The function improves performance of the counter-based broadcast scheme. Because wireless bandwidth is very limited and all the hosts contend for the wireless medium, there may be a long delay before the transmission of a packet. While a packet is waiting for transmission, the host may receive the same broadcast many times, which causes the counter to reach the threshold. Without a cancellation function, the packet still has to be transmitted, which only wastes the scarce wireless bandwidth.

The colliding probability is decreased greatly by backoff only at the MAC layer. To get the same colliding probability, the max backoff time of CBB+ is much less and the broadcast delay can be reduced enormously. The simulation results of Section 4 show that backoff for 64 * 20 us in CBB+ can obtain a better effect than backoff at the network layer for 10 ms and at the MAC layer for 32 * 20 us in CBB.

### 3.2 Privilege the Border Hosts

The expected additional coverage [6] is illustrated in Figure 1. After B receives a broadcast from A, the shadowed area will be additionally covered if B rebroadcasts. The smaller the distance between the receiver and the sender, the smaller the expected additional coverage by the rebroadcast. According to this observation, we assign priority to the receivers according to their distance from the sender. The receiver near the sender is assigned lower priority and starts a longer backoff procedure; the receiver far from the sender has higher priority and starts a shorter backoff procedure. Hence the hosts far away from the sender can rebroadcast first. The chance of rebroadcasting by the hosts near the sender is reduced. Because near hosts

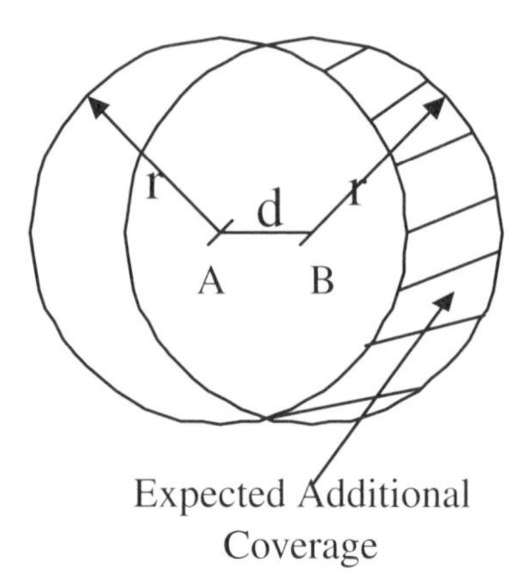


Figure 1: Illustration of expected additional coverage.

have a longer backoff procedure, it is highly possible for them to receive $C$ redundant broadcasts during the backoff procedure.

Let $r$ be the wireless transmission radius. We divide the coverage area into $m$ annuluses according to the distance to the sender. In the $i$th ($1 \leq i \leq m$) annulus, the distance to the sender is in the range $[r(i-1)/m, ri/m]$. The range of backoff_time is also divided into $m$ intervals, whose length is proportional to the number of hosts in the corresponding annulus. Assume the density is even; the number of hosts in an annulus is proportional to its area. Which backoff_time interval a host uses is determined by the annulus at which it is located. For example, $m = 3$. The ratio of the area of the three annuluses is 1:3:5, and the corresponding ratio of the number of the hosts is also 1:3:5. So, we divide the range of backoff_time into 3 intervals in proportion to the corresponding area. Let $MB$ be the max backoff time; then the interval of the backoff_time of the hosts whose distance to the sender is located in $[2r/3, r]$ is set to $[0, 5MB/9 - 1]$, that of $[r/3, 2r/3]$ is set to $[5MB/9, 8MB/9 - 1]$, and that of $[0, r/3]$ is set to $[8MB/9, MB - 1]$. The distance to the sender can be estimated by the signal power of the received packet, which can be provided by current wireless interfaces. So GPS is not required in CBB+.

Hereinafter we prove that, given the same number of hosts, the mechanism of dividing the range of backoff time into several intervals will not increase the colliding probability.

**Theorem 3.3** *Given the same number of hosts, the mechanism of dividing* *the range of backoff time into several intervals does not increase the colliding probability.*

*Proof.* First we consider the case of dividing the range into two intervals.

Let the number of backoff slots be $N = (a + b)\times n$, and the number of the hosts be $K = (a + b) \times k$, where $a, b, n, k$ are all integers, and $a, b > 0, n > k > 1$.

Let $p1$ be the probability without collision when the range is not divided; then

$$\begin{aligned} p1 &= \frac{(a+b)n((a+b)n-1)((a+b)n-2)\cdots((a+b)n-(a+b)k+1)}{((a+b)n)^{(a+b)k}} \\ &= \frac{n\left(n-\frac{1}{a+b}\right)\left(n-\frac{2}{a+b}\right)\cdots\left(n-\frac{a+b-1}{a+b}\right)\times(n-1)\left(n-1-\frac{1}{a+b}\right)\cdots\left(n-(k-1)-\frac{a+b-1}{a+b}\right)}{n^{(a+b)k}}. \end{aligned}$$

Let $p2$ be the probability without collision after dividing $N$ and $K$ into two parts: $a \times n$ slots and $a \times k$ hosts, $b \times n$ slots and $b \times k$ hosts; then:

$$\begin{aligned} p2 &= \left(\frac{an(an-1)\cdots(an-ak+1)}{(an)^{ak}}\right)\left(\frac{bn(bn-1)\cdots(bn-bk+1)}{(bn)^{bk}}\right) \\ &= \frac{n\left(n-\frac{1}{a}\right)\left(n-\frac{2}{a}\right)\cdots\left(n-\frac{a-1}{a}\right)n\left(n-\frac{1}{b}\right)\left(n-\frac{2}{b}\right)\cdots\left(n-\frac{b-1}{b}\right)\times(n-1)\cdots}{n^{(a+b)k}}. \end{aligned}$$

We can prove the following inequality (the proof is given in the appendix).

$$\begin{gathered} (n-i)\left(n-i-\tfrac{1}{a+b}\right)\left(n-i-\tfrac{2}{a+b}\right)\cdots\left(n-i-\tfrac{a+b-1}{a+b}\right) \\ < \\ (n-i)\left(n-i-\tfrac{1}{a}\right)\left(n-i-\tfrac{2}{a}\right)\cdots\left(n-i-\tfrac{a-1}{a}\right) \\ \times \\ (n-i)\left(n-i-\tfrac{1}{b}\right)\left(n-i-\tfrac{2}{b}\right)\cdots\left(n-i-\tfrac{b-1}{b}\right), \end{gathered} \tag{3}$$

where $i$ is an integer with the range of 0 to $k - 1$.

Varying $i$ from 0 to $k-1$ we can get $k$ inequalities. Then we multiply the left expressions and the right expressions, respectively, and so obtain a new inequality, whose left expression is the numerator of $p1$ and right expression is the numerator of $p2$. So $p1 < p2$ holds. Hence $(1 - p2) < (1 - p1)$.

So, after dividing the colliding probability becomes smaller.

Given the case where the range is divided into more than two intervals, we can obtain that the colliding probability is smaller by applying the above conclusion repeatedly. □

### 3.3 Dynamically Tune the Threshold

In CBB the counter threshold is constant for every host, regardless of the distance from the receiver to the sender. However, based on the same observation that the rebroadcast by the host near the sender has low probability to cover new hosts, CBB+ dynamically tunes the counter threshold. When a host receives a message from a neighbor within distance of D, the counter threshold of this broadcast is decreased by 1. It can be implemented by adding 2 (instead of 1 in the normal CBB procedure) to the counter when receiving a message from a neighbor within distance of D.

If the threshold D is set too large, for example, D equals $r$, then the counter threshold C of all the hosts will be one less than the normal value, which will reduce the reachability of the broadcasting scheme. If the threshold D is set too small, then this mechanism will not be effective because the number of hosts that are located within the distance D of the sender is too small. In addition, the setting of D should take the counter threshold C into account. When C is larger, D can also be larger to some extent; when C is smaller the value of D must also be smaller because the dynamic tuning will have a significant impact on the reachability of the broadcast.

According to [6], the average expected additional coverage of a neighbor rebroadcasting is $0.41\pi r^2$ when neighbors are uniformly distributed in the transmission range. In simulations the threshold D is set to be $r/4$. The expected additional coverage by a neighbor within distance $r/4$ from the sender is $0.0005\pi r^2$, which equals 0.1% of the average value.

After introducing the above-mentioned three improvements to CBB, we describe the detailed procedure of CBB+ as follows.

S1. When a broadcast is received in a host for the first time, the corresponding counter is initialized to one (if the distance to the sender is smaller than D, the counter is initialized to two). CBB+ estimates the distance to the sender according to the signal power of the received packet and produces a random integer as backoff_time in the corresponding interval, then passes the rebroadcast packet together with the backoff_time to the MAC layer.

S2. When the MAC layer is to transmit the rebroadcast, if the current backoff procedure has ended, the MAC layer starts a backoff procedure according to the backoff_time parameter.

S3. During the backoff procedure, if a redundant message is received, the counter is increased by one (if the distance to the sender is smaller than

D, then the counter is increased by two). If the counter reaches C, CBB+ notifies the MAC layer to cancel the rebroadcast. Otherwise it does nothing.

## 4 Simulation Results and Analysis

To validate the performance improvement of CBB+, We have implemented CBB and CBB+ on NS-2 [11]. We do simulations with various numbers of hosts, various broadcast origination rates, and various mobile speeds.

In all simulations the MAC layer is DCF, and the physical layer is IEEE 802.11 DSSS. Wireless bandwidth is set to 2 Mbps. The network area is 350 m * 350 m. The transmission radius of hosts is 100 m; the number of simulated hosts is 20, 30, 40, 50, 60, 70, 90, 110, respectively. The broadcast origination rate is increased from 10 packets/second to 80 packets/second. The broadcast payload is 64 bytes. Simulation time is 100 s. In CBB+, the max backoff time (MB) is 64 slots and the range of backoff time [0, 63] is divided into three intervals [0, 35], [36, 56], [57, 63]. The RAD range of CBB is 0–10 ms. The threshold of counter (C) is 3. D (the distance threshold) is 25 m. Every simulation result is averaged from 10 simulation trials. In all the simulations we use the random waypoint mobility model [12].

In the description of simulation results, CBB-BK denotes the CBB with the first improvement. CBB-BK-PRI denotes the CBB with the first two improvements. CBB-BK-PRI-DTH denotes the CBB with all three improvements, namely, CBB+. To show the effect of each of the individual improvements on the performance of the broadcasting scheme, first we do a simulation with various numbers of hosts and various broadcast origination rates to examine the performance of CBB, CBB-BK, CBB-BK-PRI, and CBB+. The maximum moving speed of hosts is 2 m/s with zero pause time. Then we especially examine the effect of mobility on the performance of CBB and CBB+. In these simulations the maximum moving speed is set to 2,5,10,15, . . . , 50 m/s, respectively. The pause time is 0 s.

We evaluate the performance of the broadcast schemes according to the following metrics:

*REachability (RE):* the number of mobile hosts receiving the broadcast message divided by the total number of mobile hosts that are reachable, directly or indirectly, from the source host [6]. The closer to 100% RE is, the better the performance is.

*Delay*: the interval from the time when the broadcast was initiated to the time when the last host finishes its rebroadcasting [6].

*Packet sent per broadcast per host (PPB):* The number of packet transmissions caused by a broadcast divided by the total number of the hosts. In simple flooding, PPB is one. The smaller the PPB is, the better the performance is.

The comparison of the above three metrics among the above four schemes is shown in Figures 2–4. The $X$ coordinate has two dimensions: in fact, one is the number of hosts, which is marked on the top of the figure $(20, 30, 40, \ldots, 110)$, and the other is the broadcast origination rate, which is marked on the bottom of the figure $(10, 20, 30, \ldots, 80)$. For every number of hosts, the broadcast origination rate is varied from 10 packets/second to 80 packets/second.

Figure 2 shows that when the number of hosts increases, the reachability increases because the probability for all the neighbors not to rebroadcast decreases. Figure 3 presents that delay increases when the number of hosts increases due to the increasing number of collisions. Figure 4 shows that when the number of hosts increases, the PPB decreases. The reason is that in a constant network area, when the number of hosts increases, the average number of neighbors also increases. The greater the number of neighbors is, the higher the probability for a host to receive C redundant broadcasts is. Hence the probability for a host not to rebroadcast is lower when the number

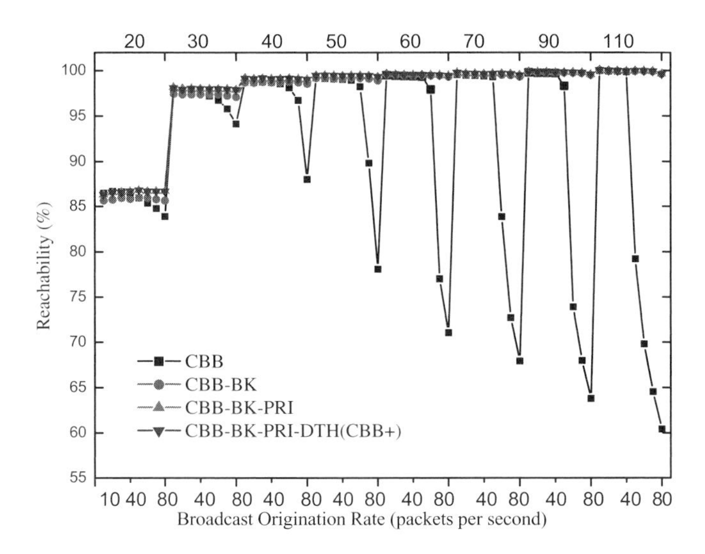


Figure 2: Reachability.

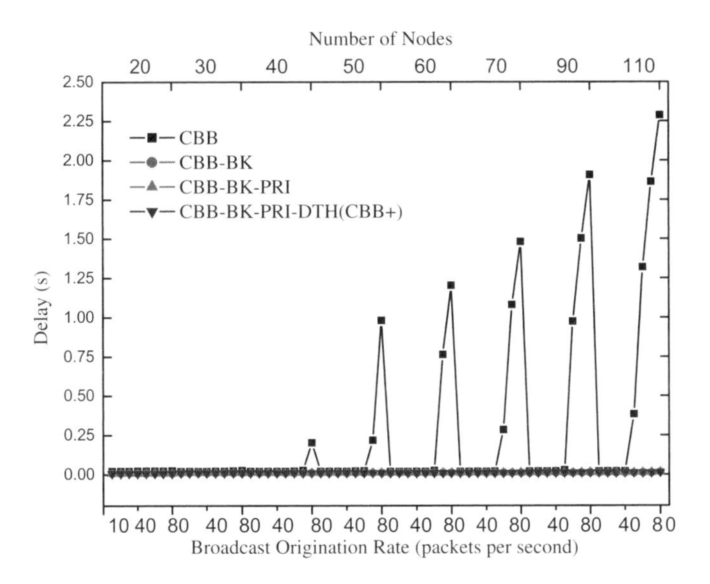


Figure 3: Delay.

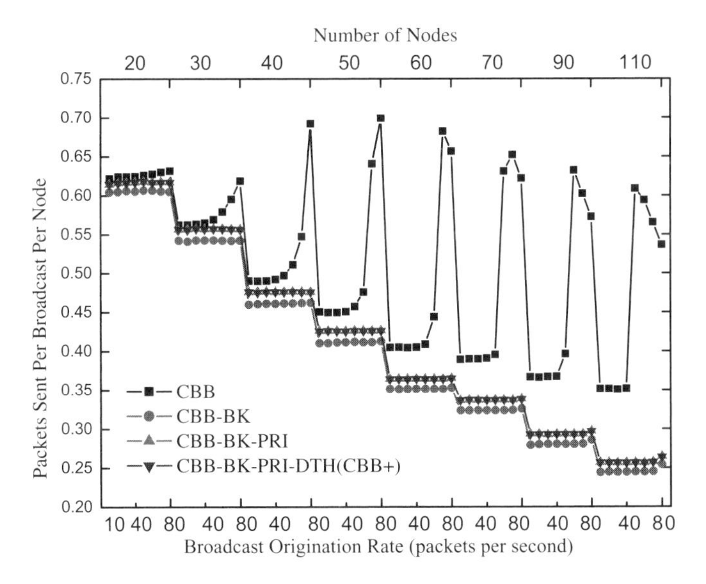


Figure 4: Packets sent per broadcast per host.

of hosts is greater. And so PPB decreases with the increasing number of hosts.

From Figures 2–4 we can also see that the performance of CBB degrades seriously when the number of hosts increases or the broadcast origination rate increases. This is due to the increasing number of collisions. On the contrary, given the same scenario, reachability of CBB-BK, CBB-BK-PRI, and CBB+ are much higher than that of CBB; their delays and PPBs are much lower than those of CBB. It is due to the effects of the carefully designed collision avoidance with the first improvement, the heuristic to improve the spatial diffusion with the second improvement, and the heuristic algorithm to decrease the number of rebroadcasts with the third improvement.

Due to the poor performance of CBB compared with the three improved schemes, we can't see clearly the effect of the individual improvement on the performance of the broadcast scheme in Figures 2–4. To examine the effect of the three individual improvements, the performance comparisons of CBB-BK, CBB-BK-PRI, and CBB+ are shown in Figures 5–7. From Figures 5 and 6, we can see that the performance of CBB-BK-PRI and CBB+ is superior to that of CBB-BK in terms of reachability and delay. In Figure 7, the PPB of CBB-BK is less than that of CBB-BK-PRI and CBB+. This is because the reachability of CBB-BK is lower than that of

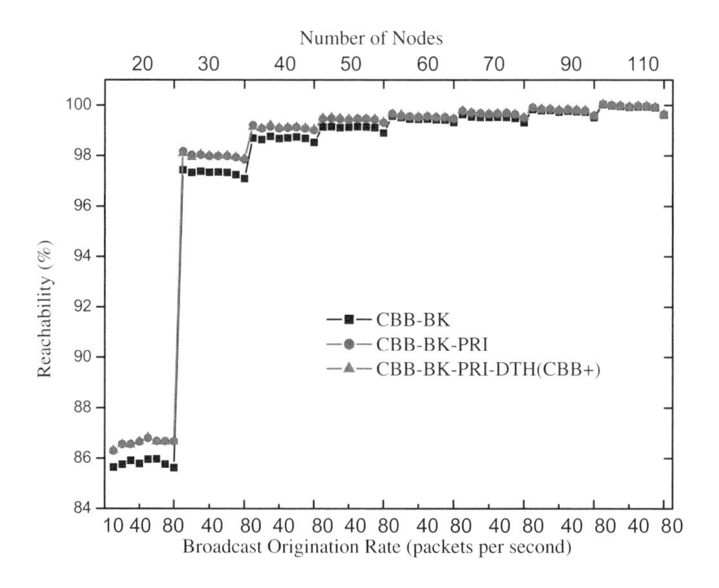


Figure 5: Reachability.

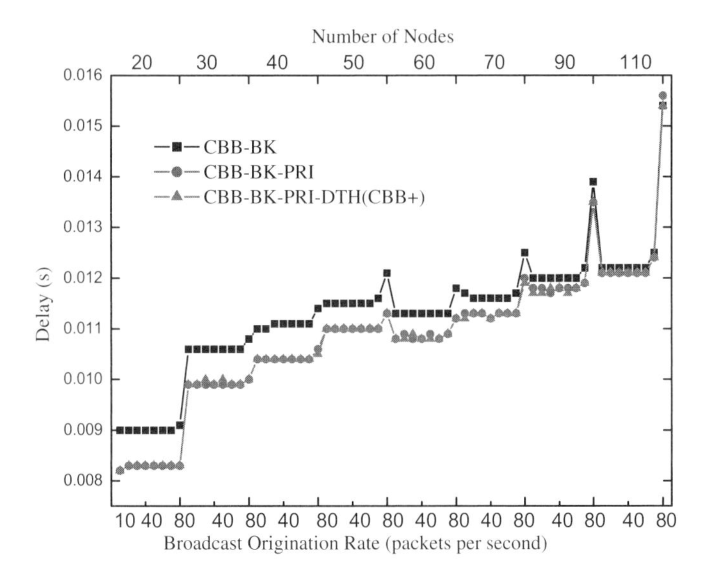


Figure 6: Delay.

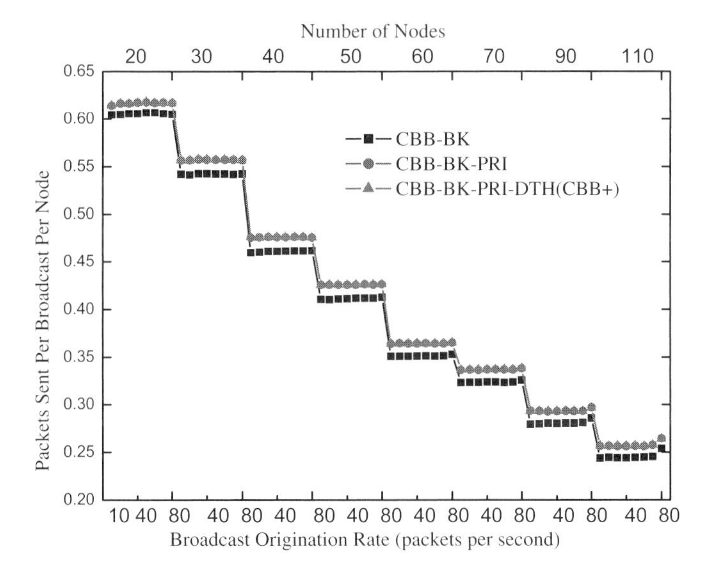


Figure 7: Packets sent per broadcast per host.

CBB-BK-PRI and CBB+. With higher reachability, more hosts receive the broadcast and rebroadcast the packet, which causes the PPB to increase. From Figures 5 and 7 we can see that the PPB of CBB-BK-PRI and CBB+ are 0.01 higher than that of CBB-BK on average, whereas their reachability is 0.7% higher than that of CBB-BK. It means, in the case of 100 hosts a networkwide broadcast with CBB+ or CBB-BK-PRI creates one more rebroadcast packet while covering 0.7 more host than that with CBB-BK. We think it is worthwhile.

CBB+ has the same reachability and delay with CBB-BK-PRI, whereas its PPB is smaller than CBB-BK-PRI. However, the improvement of CBB+ is not so obvious over CBB-BK-PRI. The reason is that in our simulations the host density of the network is very high. When the border host is privileged, the host located within distance D from the sender often receives more than C redundant broadcasts before it rebroadcasts. In such a case the dynamic threshold has no difference from the constant threshold.

Figures 8–10 show the performance comparison of CBB and CBB+ under various mobile environments where the moving speed of hosts varies from 2 m/s to 50 m/s. All the simulations are done with 60 hosts, and the broadcast origination rate is 10 packets per second. From these figures we can see that the mobility of hosts has little impact on the performance of

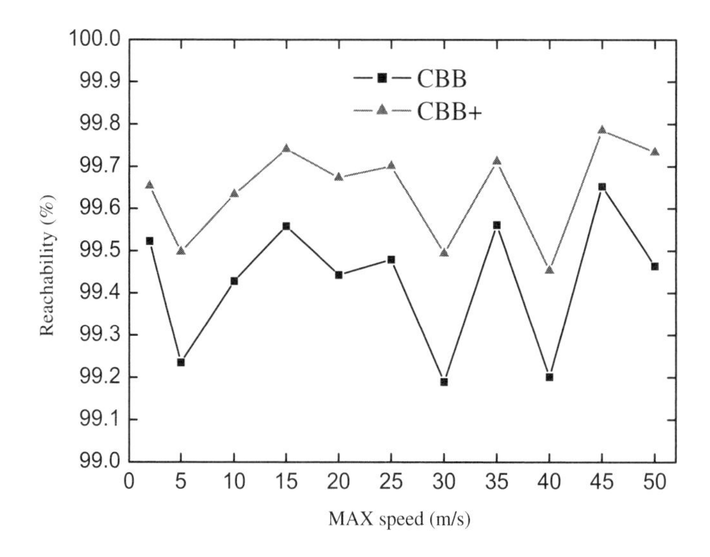


Figure 8: Reachability versus max speed.

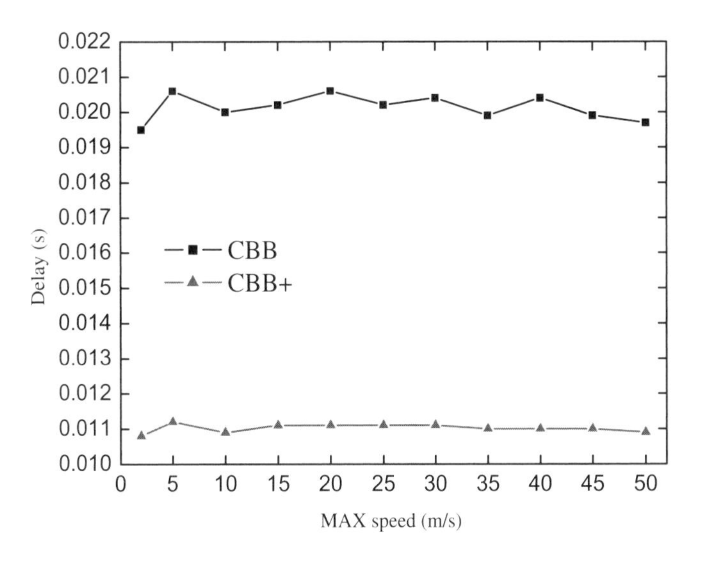


Figure 9: Delay versus max speed.

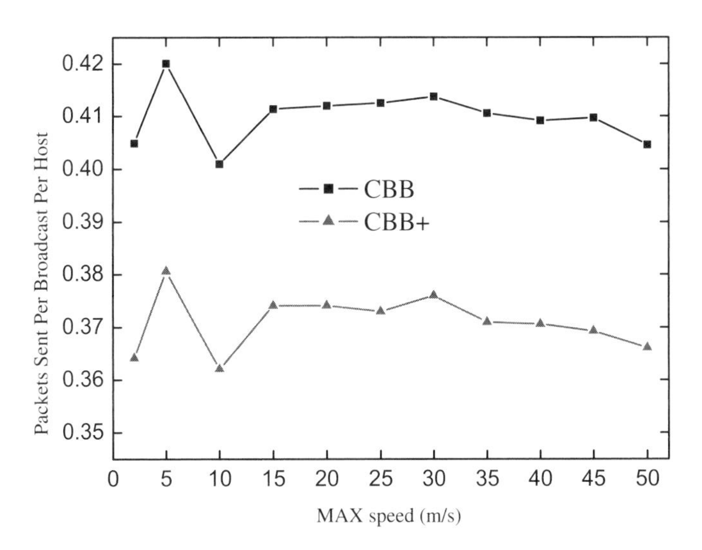


Figure 10: PPB versus max speed.

the counter-based schemes (including CBB and CBB+). The performances of CBB and CBB+ have only a slight change when the moving speed of the hosts increases. This also shows that the counter-based scheme is well suited for a highly mobile scenario.

## 5 Conclusions

The counter-based broadcasting scheme is simple. It does not need GPS or maintenance of neighbor information. It is well suited for highly mobile environments. The chapter presents several improvements on the counter-based broadcasting scheme and proposes the enhanced counter-based broadcasting scheme, CBB+. In order to reduce colliding probability and improve broadcast efficiency, CBB+ performs backoff only at the MAC layer, privileges the border hosts of a broadcast to rebroadcast in advance by assigning them shorter backoff time, and tunes the counter threshold dynamically according to the distance to the sender. Analyses and numerous simulations both show that the performance of CBB+ is highly superior to that of the original counter-based scheme proposed in [6].

## 6 Appendix

The proof of inequality:

$$\begin{aligned}(n-i)\left(n-i-\tfrac{1}{a+b}\right)\left(n-i-\tfrac{2}{a+b}\right)\cdots\left(n-i-\tfrac{a+b-1}{a+b}\right)\\ <\\ (n-i)\left(n-i-\tfrac{1}{a}\right)\left(n-i-\tfrac{2}{a}\right)\cdots\left(n-i-\tfrac{a-1}{a}\right)\\ \times\\ (n-i)\left(n-i-\tfrac{1}{b}\right)\left(n-i-\tfrac{2}{b}\right)\cdots\left(n-i-\tfrac{b-1}{b}\right).\end{aligned} \tag{3}$$

*Proof.* We focus on the following two sequences, which are extracted from the left and right expressions of the inequality, respectively.

Sequence $A(j): 0, \frac{1}{a+b}, \frac{2}{a+b}, \frac{3}{a+b}, \ldots, \frac{a+b-1}{a+b}$ $\qquad j = 0, \ldots, (a+b-1)$

Sequence $B(j): 0, \frac{1}{a}, \frac{a-1}{a}, \frac{2}{a} \ldots, 0, \frac{1}{b}, \frac{2}{b}, \frac{b-1}{b}, \ldots$ $\qquad j = 0, \ldots, (a+b-1)$.

Without loss of generality, assuming $a \geq b$ and sorting sequence $B(j)$ in ascending order, we obtain

Sequence $C(j): 0, 0, \frac{1}{a}, \frac{1}{b}, \ldots, \frac{b-1}{b}, \frac{a-1}{a}$ $\qquad j = 0, \ldots, (a+b-1)$

Obviously, $C(0) = A(0)$, $C(1) < A(1)$. If we can prove that

$$C(j) < A(j)(j = 2, \ldots, (a + b - 1)), \tag{4}$$

then the inequality (3) is proved. In the following, we prove the inequality (4).

It is clear that $A(j) = j/(a + b)$. The value of $C(j)$ is either $k/a$ or $k/b$. In the following we prove that when $C(j) = k/a$ the inequality holds. For the case that $C(j) = k/a$, we can prove the inequality in the same way.

In sequence C the number of the items that are less than $C(j)$ is

$$2 + k - 1 + \left[\frac{k}{a} \Big/ \frac{1}{b}\right],$$

where [ ] denotes the floor function.

So by the definition of the index we can have that

$$j = 2 + k - 1 + \left[\frac{k}{a} \Big/ \frac{1}{b}\right] = k + 1 + \left[\frac{kb}{a}\right]. \tag{5}$$

To prove that $C(j) < A(j)$ is to prove $k/a$ ¡ $j/(a + b)$, namely that

$$k(a + b) < ja. \tag{6}$$

Substituting (5) into the inequality (6), we obtain

$$k(a + b) < \left(k + 1 + \left[\frac{kb}{a}\right]\right) a.$$

It proves that $kb < (1 + [kb/a])\, a$.

It proves that $kb/a$ ¡ $1 + kb/a$. From the property of the floor function, we can see that this inequality holds.

So the inequality (3) is proved. □

## 7 Acknowledgments

This work was supported by the National Natural Science Foundation of China (Grant No.60603060, 60472067,60502037,90604019), the National Basic Research Pro-gram of China (Grant No. 2003CB314806).

## References


[1] IETF MANET work group, *http://www.ietf.org/html.charters/manet-charter.html*

[2] D. Johnson, D. Maltz, Y.-C. Hu, and J. Jetcheva, The dynamic source routing protocol for mobile ad hoc networks (DSR), *Internet Draft: draft-ietf-manet-dsr-07.txt*, Feb. (2002).

[3] C. Perkins, E. Royer, and S. Das, Ad hoc on demand distance vector (AODV) routing, *Internet Draft: draft-ietf-manet-aodv-12.txt*, November (2002).

[4] Z. Haas, M. Pearlman, and P. Samar, The zone routing protocol (ZRP) for ad hoc networks, *Internet Draft: draft-ietf-manet-zone-zrp-04.txt*, July 2002.

[5] Y. Ko and N.H. Vaidya. Location-aided routing (LAR) in mobile ad hoc networks. In *Proceedings of the ACM/IEEE International Conference on Mobile Computing and Networking (MOBICOM)*, pages 66–75, 1998.

[6] Yu-Chee Tseng, Sze-Yao Ni, et al., The broadcast storm problem in a mobile ad hoc networks, *Wireless Network* 2002.3.

[7] Y.-C. Tseng, S.-Y. Ni, and E.-Y. Shih, "Adaptive Approaches to Relieving Broadcast Storms in a Wireless Multihop Mobile Ad Hoc Network", IEEE Trans. on Computers, Vol. 52, No. 5, May 2003, pp. 545-557.

[8] B. Williams, and T. Camp, Comparison of broadcasting techniques for mobile ad hoc networks, *MOBIHOC*, 2002.9.

[9] J. Cartigny, and D. Simplot, Border host retransmission based probabilistic broadcast protocols in ad-hoc networks. In *In Proceedings of the 36th Annual Hawaii International Conference on System Sciences (HICSS'03)*, Hawaii, USA, 2003.1.

[10] LAN MAN Standards Committee of the IEEE Computer Society, IEEE Std 802.11-1999, Wireless LAN Medium Access Control (MAC) and Physical Layer (PHY) specifications, *IEEE Standard*, 1999.

[11] NS2, http://www.isi.edu/nsnam/ns/.

[12] T. Camp, J. Boleng, and V. Davies, A survey of mobility models for ad hoc network research, *Wireless Communications & Mobile Computing (WCMC)*, 2002.5.

Chapter 13

# Energy-Efficient Broadcast/Multicast Routing with Min-Max Transmission Power in Wireless Ad Hoc Networks

Deying Li
*School of Information*
*Renmin University of China, Beijing 100872, P.R. China*
Email: deyingli@ruc.edu.cn

Hongwei Du
*Department of Computer Science*
*City University of Hong Kong, Hong Kong*
Email: hongwei@cs.cityu.edu.hk

Xiaohua Jia
*Department of Computer Science*
*City University of Hongkong, Hong Kong*
Email: jia@cs.cityu.edu.hk

## 1 Introduction

Wireless ad hoc networks have received significant attention in many application areas, such as disaster relief operations, rescue operations, quick-setup conference meetings, and so on. Wireless ad hoc networks consist of a collection of mobile nodes dynamically forming a temporary network without

M.X. Cheng, D. Li (eds.) *Advances in Wireless Ad Hoc and Sensor Networks.*
Signals and Communication Technology, doi: 10.1007/978-0-387-68567-0_13.

the use of any existing network infrastructure. The lifetime of an ad hoc network depends on the lifetime of each individual node which is powered by a battery. If the energy of a node is drained out, the network could become disconnected. Thus, energy conservation becomes an important issue in wireless ad hoc networks.

There are two different objectives in energy-efficient broadcast/multicast routing in wireless ad hoc networks. One is to minimize the total energy consumption of all the nodes in the network [1–5]. The other focuses on balancing energy consumption by minimizing the maximum energy consumption of an individual node [6–8], such that the lifetime of the network could be maximized. In this chapter, we consider two fundamental energy-efficient problems related to minimizing the maximum energy consumption of nodes. One is the Min-Max broadcast routing problem; and the other is the Min-Max multicast routing problem. We assume each node has $k$ discrete levels of transmission power for adjustment. We propose two optimal algorithms for broadcast/multicast routing, respectively.

## 2 Related Work

Energy consumption has been an important issue in broadcast/multicast routing and extensive research has been done on this topic. Some of the works consider minimizing the total energy consumption [1–5]. In [3] and [4], several energy-efficient broadcast/multicast algorithms were proposed, namely, the Broadcast Incremental Power (BIP) and Multicast Incremental Power (MIP) algorithms, the Shortest Path Tree (SPT) algorithm, and the Minimums Spanning Tree (MST) algorithm. The proposed algorithms were evaluated through simulations. The authors in [5] got the quantitative characterization of performances of these three greedy heuristics.

The solutions developed in [1–5] are mainly based on geometry features of the nodes in the plane. Some other solutions are based on graph theory (i.e., based on the connectivity among the nodes in the network), such as in [2,9–10]. In [2], the minimum-energy broadcast problem was addressed and proved to be NP-hard in general, and an $O(n^{k+2})$ algorithm was proposed for the problem. The work in [9,10] assumes each node has a limited number of adjustable discrete levels of transmission power. In [9], the authors first gave a formal proof of the NP-hardness of the problem for both the geometry version and graph version. A heuristic based on the MST algorithm was proposed, but no performance ratio was given. In [10], another heuristic algorithm based on the directed Steiner tree method was proposed.

The drawback to minimizing the total consumed energy is that some nodes may be heavily used, and their energy could be drained out quickly, even though other nodes may still have abundant unused energy. Chang and Tassulas [6] first considered balancing the energy consumption in a network. Cheng et al.[7] used the idea of balancing energy consumption for broadcast routing. They proposed an algorithm to construct a broadcast tree with balanced energy consumption. Li et al.[8] proposed an optimal algorithm, the Minimum Weight Incremental Arborescence (MWIA), for constructing a broadcast tree where each node has different initial energy.

In this chapter, we consider balancing the energy consumption for broadcast/multicast in the network model so that each node has $k$ discrete levels of transmission power for adjustment. This network model is different from that in [6–8]. We propose an optimal algorithm for multicast and broadcast routing, respectively.

## 3 Network Model

A wireless ad hoc network is represented by a graph $G = (V, A)$, where $V$ is a set of nodes and $A$ is a set of arcs. We adopt the widely used transmission power model for radio networks. An arc $(v_1, v_2)$ exists if and only if $d^{\alpha}(v_1, v_2) \leq p(v_1)$, where $p(v_1)$ is the transmission power of node $v_1$, $d(v_1, v_2)$ is the distance between $v_1$ and $v_2$, and $\alpha$ is a constant typically taking a value between 2 and 4. A multicast request consists of a source and multiple destination nodes, and a broadcast has all nodes in $V$ except the source as its destination.

We assume that the transmission power at each node is finitely adjustable. Without loss of generality, we assume that there are $k$ adjustable discrete levels of transmission power at each node. Let $P_{\min}$ denote the minimum level of power and $P_{\max}$ the maximum operational power. Given a node $v_i \in V$, let $w_{i1}, w_{i2}, \ldots, w_{ik}$ be its $k$ discrete levels of transmission power. Assume that $w_{il_1} \leq w_{il_2}$ if $1 \leq l_1 < l_2 \leq k$. We assume that each node can choose its power level from level 1 (i.e, $P_{\min}$) to level $k$ (i.e, $P_{\max}$). The nodes in any particular broadcast/multicast tree do not necessarily use the same power level.

The Min-Max broadcast/multicast routing problem is defined as follows: *The Min-Max broadcast routing problem*: Given a wireless ad hoc network $G = (V, A)$, and a source $s$, decide the transmission power of each node to form a broadcast tree whose root is $s$ and spans all the nodes in $V$ such that the maximum transmission power of transmission nodes (nonleaf nodes) in the tree is minimized.

*The Min-Max multicast routing problem:* Given a wireless ad hoc network $G = (V, A)$, a source $s$, and a set of destinations $D$, decide the transmission power of each node to form a multicast tree, the root of which is $s$ and spans all nodes in $D$ such that the maximum transmission power of transmission nodes (nonleaf nodes) in the tree is minimized.

# 4 Our Solution

To decide the transmission power of each node to form a broadcast/multicast tree, we first construct an auxiliary arc-weight directed graph, then transform our problem to the problem of forming a broadcast/multicast tree on the auxiliary graph with the objective of minimizing the maximum arc weight in the tree.

## 4.1 Construct an Auxiliary Graph

Given $G = (V, A)$, and for each node $v_i \in V$, let $w_{i1}, w_{i2}, \ldots, w_{ik}$ be its $k$ power levels. We construct an auxiliary graph $G' = (V, A')$, where $V$ is the set of nodes which is the same as the original network. $\forall v_i, v_j \in V$, if $d^\alpha(v_i, v_j) \le w_{ik}$; then there is an arc $(v_i, v_j) \in A'$ and we assign its weight as the minimum power level of $v_i$ so that $v_i$ can use the level of power to reach $v_j$. That is, $w(v_i, v_j) = \min\{w_{il} | d^\alpha(v_i, v_j) \le w_{il}, \text{ for } 1 \le l \le k\}$ (see Figure 1). $G'$ is a directed graph with arc weight.

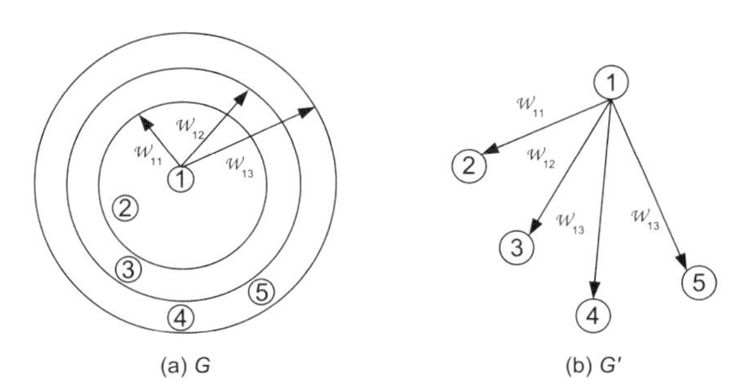


Figure 1: (a) A original graph $G$; (b) The auxiliary graph corresponding to $G$.

For example, in Figure 1, suppose node 1 has three discrete levels of transmission power: $w_{11}, w_{12}, w_{13}$. Power level $w_{11}$ can cover node 2; power level $w_{12}$ can cover nodes 2 and 3; and $w_{13}$ can cover nodes 2, 3, 4, and 5. In the auxiliary graph, $V$ is a set of nodes 1, 2, 3, 4, and 5. The weight of arc $(1, 2)$ is $w_{11}$; the weight of arc $(1, 3)$ is $w_{12}$; the weight of arcs (1, 4) and (1, 5) is $w_{13}$.

The Min-Max broadcast/multicast routing problem is transformed to the problem of forming a broadcast/multicast tree in an auxiliary graph rooted at the source with the objective of minimizing the maximum arc weight in the broadcast/multicast tree.

## 4.2 Min-Max Broadcast Routing Problem

Suppose $s$ is a source of broadcast. We propose an algorithm called the Min-Max broadcast algorithm. We first introduce some notation. Let $T = (V_1, A_1)$ be a tree selected, and $V_1$ is a set of the nodes in $T$ and $A_1$ is a set of arcs in $T$. $A'(V_1, V - V_1)$ is a set of arcs which is a subset of $A'$ and end nodes of the arcs are from $V_1$ to $V - V_1$. The algorithm grows a tree starting from the root $s$, and each time it adds an arc with the minimum weight from a node in $V_1$ to a node in $(V - V_1)$ and the nodes of the arc to $T$. This process continues until all the nodes in $V$ are included in $T$.

The Min-Max broadcast algorithm is formally presented as the following.

**Input:** $G' = (V, A')$ and a source $s$
  **Output** $T$: a broadcast tree
    $V_1 = s$;
    $A_1 = \emptyset$ ;
    **While** $(V_1 \neq V)$ do
      Choose $(a, b) \in A'(V_1, V - V_1)$ such that
        $w((a, b)) = \min\{w(a, b)|(a, b) \in A'(V_1, V - V_1)\}$;
      $V_1 = V_1 \bigcup \{b\}$;
      $A_1 = A_1 \bigcup \{(a, b)\}$;
    Construct the broadcast tree $T$ from $V_1$ and $A_1$.

**Theorem 1.** *The Min-Max broadcast algorithm produces an optimal solution for min-max broadcast routing problem.*

*Proof.* Suppose $T$ is a broadcast tree produced by the Min-Max algorithm (see Figure 1a). Let $(x, y)$ be an arc with the largest weight in the broadcast tree $T$. Let $T_{(x,y)}$ be the subtree at the time in the construction that arc

$(x, y)$ is added. Then $T_{(x,y)}$ is rooted at $s$ and $(x, y)$ is an arc with minimum weight from any node in $T_{(x,y)}$ to a node not in $T_{(x,y)}$ (see Figure 2b).

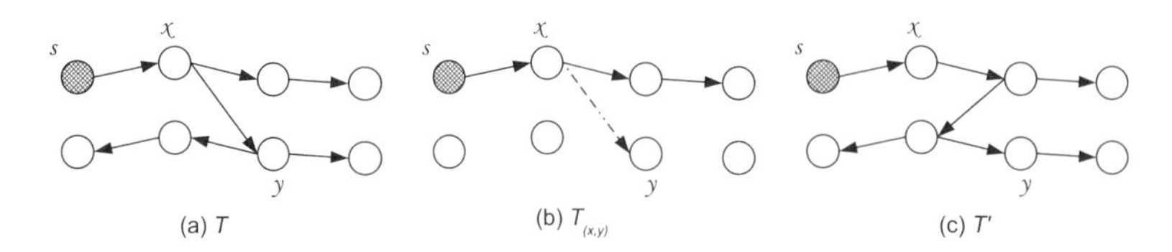


Figure 2: Two arborescences $T$ and $T'$.

Now suppose there exists another broadcast tree $T'$ rooted at $s$ (see Figure 2c) such that every arc of $T'$ has a weight less than $w(x, y)$ which is the weight of arc $(x, y)$. Because $T'$ contains a path from $s$ to $y$, it must contain an arc $(i, j)$ from a node $i$ in $T_{(x,y)}$ to a node $j$ not in $T_{(x,y)}$. However, $w(i, j) < w(x, y)$ contradicts the fact that $(x, y)$ is an arc with a minimum weight from any node in $T_{(x,y)}$ to a node not in $T_{(x,y)}$. As a conclusion, Theorem 1 is correct. □

### 4.3 Min-Max Multicast Routing Problem

In this section, we propose an algorithm called the Min-Max multicast algorithm to construct a multicast tree. Suppose that $(s, D)$ is a multicast request. The notations are same as in the above section. The algorithm also grows a tree starting from the root $s$, and each time it adds an arc with the minimum weight from a node in $V_1$ to a node in $(V - V_1)$ and the nodes of the arc to $T$. This process continues until all the nodes in $\{s\} \cup D$ are included in the tree.

The Min-Max multicast algorithm is formally presented as the following.

**Input:** $G' = (V, A')$ and $(s, D)$
  **Output** $T$: a multicast tree rooted from $s$
    $V_1 = s$;
    $A_1 = \emptyset$ ;
    **While** $(\{s\} \bigcup D$ is not subset of $V_1)$ do
      Choose $(a, b) \in A'(V_1, V - V_1)$ such that
        $w((a, b)) = \min\{w(a, b)|(a, b) \in (V_1, V - V_1)\}$;
      $V_1 = V_1 \bigcup \{b\}$;
      $A_1 = A_1 \bigcup \{(a, b)\}$;
    Construct the multicast tree $T$ from $V_1$ and $A_1$.

**Theorem 2.** *Every Min-Max multicast algorithm produces an optimal solution for min-max multicast routing problem.*

*Proof.* Suppose $T$ is a multicast tree produced by the Min-Max multicast algorithm. Let $(x, y)$ be an arc with the largest weight in a multicast tree $T$. Let $T_{(x,y)}$ be the subtree at the time in the construction that arc $(x, y)$ is added. Then $T_{(x,y)}$ is rooted at $s$ and $(x, y)$ is an arc with minimum weight from any node in $T_{(x,y)}$ to a node not in $T_{(x,y)}$.

Now suppose there exists another multicast tree $T'$ rooted at $s$ such that every arc of $T'$ has a weight less than $w(x, y)$ which is the weight of arc $(x, y)$. Because it contains a path from $s$ to $y$, it must contain an arc $(i, j)$ from a node $i$ in $T_{(x,y)}$ to a node $j$ not in $T_{(x,y)}$. However, $w(i, j) < w(x, y)$ contradicts the fact that $(x, y)$ is an arc with a minimum weight from any node in $T_{(x,y)}$ to a node not in $T_{(x,y)}$. As a conclusion, Theorem 2 is correct. □

# 5 Simulations

In simulations, four algorithms, namely, our algorithm Min-Max broadcast/multicast algorithm (MMB/MMM for short) and BIP/MIP ([4]) are simulated and compared. We study how the total energy cost and maximum power among all nodes are affected by varying two parameters over a wide range: the total number of nodes in the network ($N$), and the number of destination nodes in the multicast request ($M$).

The simulation is conducted in a $100 \times 100$ 2D free-space region by randomly distributing $N$ nodes. The value of $\alpha$ is set to 2 and the source node is randomly picked. For each node, there are $k$ discrete levels of transmission power for adjustment. Each level's power is denoted by $P_i$, which is defined by

$$P_i = i \times \left(\frac{R}{k}\right) \times r,$$

where $R$ is the size of the region and $i$ the number of levels. We use $r$ to adjust the power to make sure that we could construct a broadcast tree. The region has been partitioned into $k$ parts. The function represents transmission power of power levels of each node. We present averages of 100 separate runs for each result shown in the figures. In each run of the simulation, for given $N, M, r$, we randomly select $N$ nodes in the square, $M$ destination nodes, and a source. And we construct an auxiliary graph corresponding

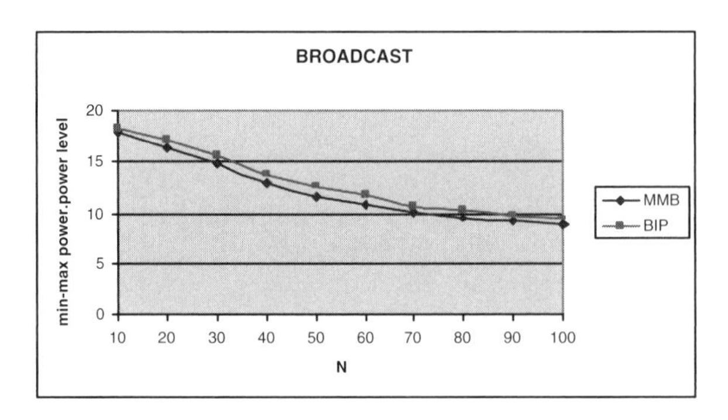


Figure 3: The min-max power. power level versus $N$: $\alpha = 2$, $k = 20$, $r = 0.5$.

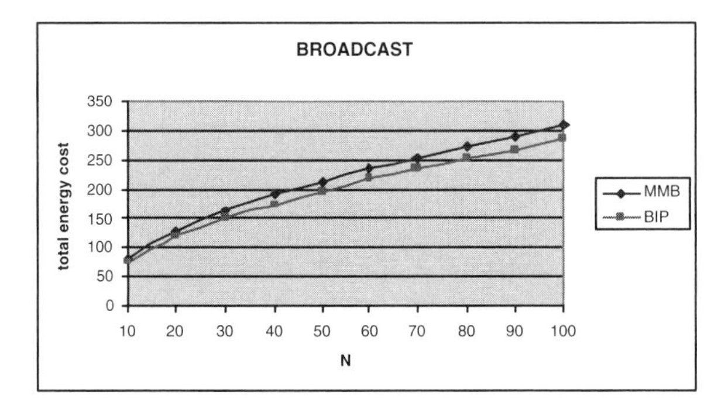


Figure 4: The total energy cost versus $N$: $\alpha = 2$, $k = 20$, $r = 0.5$.

to the original network. Then, we run the four algorithms on the auxiliary graph. Any topology where we cannot find any solution is discarded.

The simulation has two parts: the min-max broadcast routing, and the min-max multicast routing.

**PART I.**

In this part, MMB and BIP algorithms are simulated and compared. Figure 3 shows the min-max power versus the numbers of nodes. And Figure 4 shows the total energy cost versus the number of nodes.

From Figures 3 and 4, we can make following observations.

1. The min-max power decreases as the numbers of nodes increase in Figure 3. This is because when there are more nodes in a region, there are more choices to use less transmission power to reach neighbors

and nodes have to use more transmission power to reach neighbors in a sparse region. The total energy cost increases as the nodes increase in Figure 4.

2. MMB performs better than BIP in balancing energy consumption. This is because our algorithm is focused directly on increasing the lifetime of the energy consumption and BIP aims at reducing the total energy cost. Regarding the total energy consumption, BIP performs slightly better than ours.

**PART II.**

The min-max power of multicast routing is simulated in this part. In Figure 5, $M$, the destination of the multicast request, is set to $N \times 0.3$.

In Figure 5, we can see that MMM and MIP perform closely in min-max power and total energy cost.

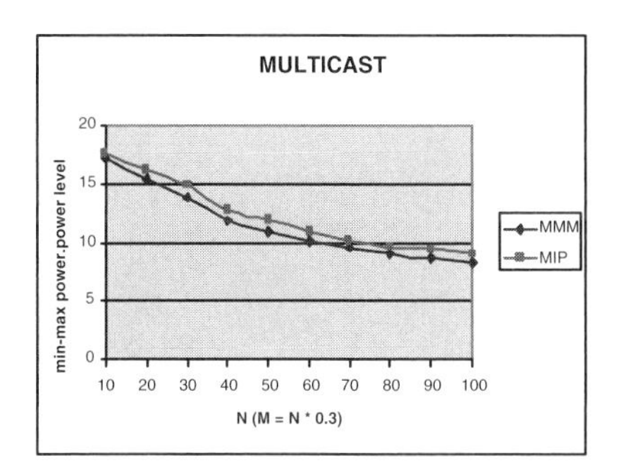


(a) The min-max power; power level versus $N$.

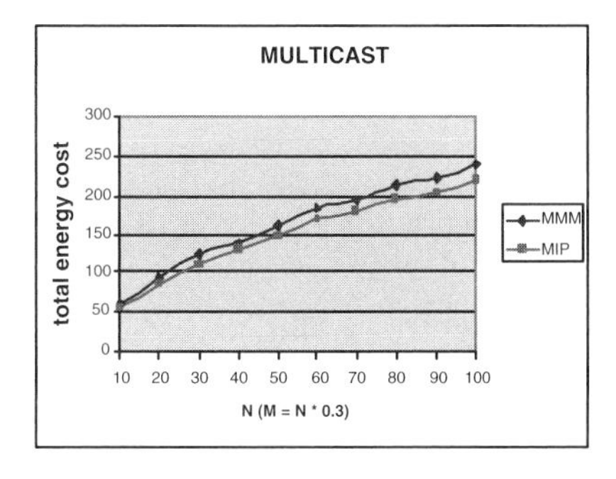


(b) The total energy cost versus $N$.

Figure 5: The min-max power and total energy cost in multicast: $\alpha = 2$, $k = 20$, $r = 0.5$, $M = N * 0.3$.

From Figure 6, we can make following observations.

1. In (a), the min-max power increases very little as the number of destination nodes increases. This is because the min-max power of a multicast tree comes from a node that has the largest transmission power among all the transmitting nodes in the tree, which is different from the total energy of all transmitting nodes in the tree. A multicast tree that has fewer destinations may not have a smaller value of

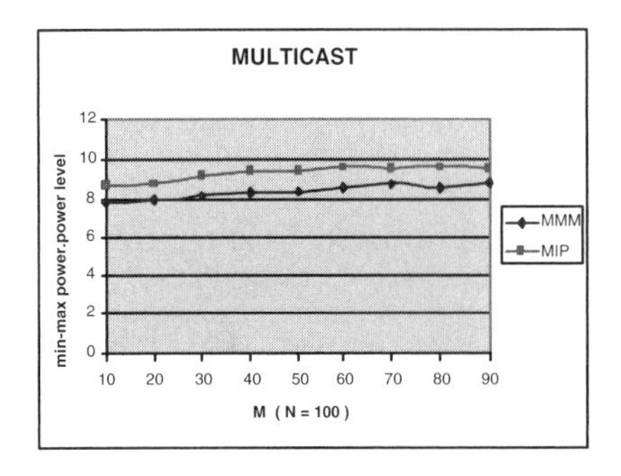


(a) The min-max power; power level versus $M$.

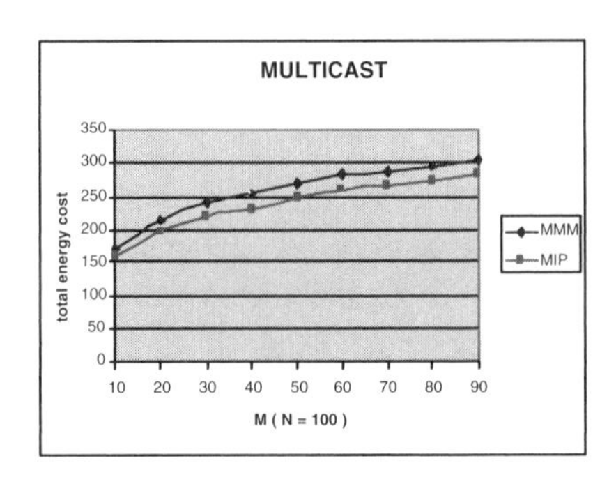


(b) The total energy cost versus $M$.

Figure 6: The min-max power and total energy cost in multicast: $\alpha = 2$, $k = 20$, $r = 0.5$, $N = 100$.

min-max power than a tree that has more destinations. Certainly, it has a higher chance to have a smaller value of min-max power.

2. In (b), while destination nodes increase, more nodes transmit powers. So the total energy cost increases.

Through the results in broadcast and multicast, our algorithms are good at balancing the energy consumption in increasing the lifetime of networks.

# 6 Conclusion

In this chapter, we have discussed the energy-efficient broadcast/multicast problem for balancing energy consumption. An optimal algorithm for Min-Max broadcast/multicast is proposed. From the simulation results, the MMB/MMM algorithms have good performance in balancing the energy consumption on networks.

# References

[1] J. Chang and L. Tassiulas, Energy conserving routing in wireless ad hoc networks, *IEEE INFOCOM(1)*, 2000, pp. 22–31.

[2] O. Egecioglu and T. F. Gonzalez, Minimum-energy broadcast in simple graphs with limited node power. In *Proceedings of IASTED International Conference on Parallel and Distributed Computing and Systems (PDCS 2001)*, pp. 334–338.

[3] J.E. Wieseltheir, G.D. Nguyen, and A. Ephremides, On the construction of energy-efficient broadcast and multicast trees in wireless networks. In *IEEE INFOCOM* 2002.

[4] J.E. Wieselthier, G. D. Nguyen, and A. Ephremides, Algorithm for energy-efficient multicasting in static ad hoc wireless networks, *Mobile Networks and Applications* 6, 2001, pp. 251–263.

[5] P. J. Wan, G. Calinescu, X. Y. Li and O. Frieder, Minimum-energy broadcast routing in static ad hoc wireless networks, *IEEE INFOCOM* 2001.

[6] J. Chang and L. Tassulas, Routing for maximum system lifetime in wireless ad hoc networks. In *Proceeding of 37th Annual Allerton Conference on Communication, Control, and Computing*, 1999.

[7] M. X. Cheng, J. Sun, M. Min, and D. Z. Du, Energy efficient broadcast and multicast routing in ad hoc wireless networks. In *22nd IEEE International Performance, Computing, and Communications Conference*, 2003.

[8] Y. Li, M. X. Cheng, and W. Wu, Optimal topology control for balanced energy consumption in wireless networks, *Journal of Parallel and Distributed Computing*, 65,(2), 2005, pp. 124–131.

[9] M. Cagalj, J.P. Hubaux, and C. Enz, Minimum-energy broadcast in all-wireless networks: NP-completeness and distribution issues. In *MOBICOM* 2002.

[10] W. Liang, Constructing minimum-energy broadcast trees in wireless ad hoc networks, In *MOBICOM* 2002.

## Signals and Communication Technology

(continued from page ii)

**Broadband Fixed Wireless Access**
A System Perspective
M. Engels and F. Petre
ISBN 0-387-33956-6

**Distributed Cooperative Laboratories**
Networking, Instrumentation, and Measurements
F. Davoli, S. Palazzo, and S. Zappatore (Eds.)
ISBN 978-0-387-29811-5

**The Variational Bayes Method in Signal Processing**
V. Šmídl and A. Quinn
ISBN 978-3-540-28819-0

**Topics in Acoustic Echo and Noise Control**
Selected Methods for the Cancellation of Acoustical Echoes, the Reduction of Background Noise, and Speech Processing
E. Hänsler and G. Schmidt (Eds.)
ISBN 978-3-540-33212-1

**EM Modeling of Antennas and RF Components for Wireless Communication Systems**
F. Gustrau and D. Manteuffel
ISBN 978-3-540-28614-1

**Interactive Video**
**Methods and Applications**
R.I. Hammoud (Ed.)
ISBN 978-3-540-33214-5

**ContinuousTime Signals**
Y. Shmaliy
ISBN 978-1-4020-4817-3

**Voice and Speech Quality Perception**
Assessment and Evaluation
U. Jekosch
ISBN 978-3-540-24095-2

**Advanced ManMachine Interaction**
Fundamentals and Implementation
K.-F. Kraiss
ISBN 978-3-540-30618-4

**Orthogonal Frequency Division Multiplexing for Wireless Communications**
Y. (Geoffrey) Li and G.L. Stüber (Eds.)
ISBN 978-0-387-29095-9

**Circuits and Systems Based on Delta Modulation**
**Linear, Nonlinear and Mixed Mode Processing**
D.G. Zrilic
ISBN 978-3-540-23751-8

**Functional Structures in Networks AMLn—A Language for Model Driven Development of Telecom Systems**
T. Muth
ISBN 978-3-540-22545-4

**RadioWave Propagation for Telecommunication Applications**
H. Sizun
ISBN 978-3-540-40758-4

**Electronic Noise and Interfering Signals**
**Principles and Applications**
G. Vasilescu
ISBN 978-3-540-40741-6

**DVB**
**The Family of International Standards for Digital Video Broadcasting, 2nd ed.**
U. Reimers
ISBN 978-3-540-43545-7

**Digital Interactive TV and Metadata Future Broadcast Multimedia**
A. Lugmayr, S. Niiranen, and S. Kalli
ISBN 978-3-387-20843-5

**Adaptive Antenna Arrays**
**Trends and Applications**
S. Chandran (Ed.)
ISBN 978-3-540-20199-1

**Digital Signal Processing with Field Programmable Gate Arrays**
U. Meyer-Baese
ISBN 978-3-540-21119-8

**Neuro-Fuzzy and Fuzzy Neural Applications in Telecommunications**
P. Stavroulakis (Ed.)
ISBN 978-3-540-40759-1

**SDMA for Multipath Wireless Channels**
**Limiting Characteristics and Stochastic Models**
I.P. Kovalyov
ISBN 978-3-540-40225-1

**Digital Television**
**A Practical Guide for Engineers**
W. Fischer
ISBN 978-3-540-01155-2

**Speech Enhancement**
J. Benesty (Ed.)
ISBN 978-3-540-24039-6

**Multimedia Communication Technology**
**Representation, Transmission and Identification of Multimedia Signals**
J.R. Ohm
ISBN 978-3-540-01249-8

**Information Measures**
**Information and its Description in Science and Engineering**
C. Arndt
ISBN 978-3-540-40855-0

## Signals and Communication Technology

**Processing of SAR Data**
**Fundamentals, Signal Processing, Interferometry**
A. Hein
ISBN 978-3-540-05043-8

**Chaos-Based Digital Communication Systems**
**Operating Principles, Analysis Methods, and Performance Evalutation**
F.C.M. Lau and C.K. Tse
ISBN 978-3-540-00602-2

**Adaptive Signal Processing**
**Application to Real-World Problems**
J. Benesty and Y. Huang (Eds.)
ISBN 978-3-540-00051-8

**Multimedia Information Retrieval and Management Technological Fundamentals and Applications**
D. Feng, W.C. Siu, and H.J. Zhang (Eds.)
ISBN 978-3-540-00244-4

**Structured Cable Systems**
A.B. Semenov, S.K. Strizhakov, and I.R. Suncheley
ISBN 978-3-540-43000-1

**UMTS**
**The Physical Layer of the Universal Mobile Telecommunications System**
A. Springer and R. Weigel
ISBN 978-3-540-42162-7

**Advanced Theory of Signal Detection**
**Weak Signal Detection in Generalized Obeservations**
I. Song, J. Bae, and S.Y. Kim
ISBN 978-3-540-43064-3

**Wireless Internet Access over GSM and UMTS**
M. Taferner and E. Bonek
ISBN 978-3-540-42551-9

Printed in the United States of America